# Light-Assisted Catalysis in Water and Indoor Air Cleaning: Challenges and Perspectives

# Light-Assisted Catalysis in Water and Indoor Air Cleaning: Challenges and Perspectives

Editors

**Ioan Balint**
**Monica Pavel**

MDPI • Basel • Beijing • Wuhan • Barcelona • Belgrade • Manchester • Tokyo • Cluj • Tianjin

*Editors*

Ioan Balint
Laboratory of Surface Chemistry and Catalysis
"Ilie Murgulescu" Institute of Physical-Chemistry of the Romanian Academy
Bucharest
Romania

Monica Pavel
Laboratory of Surface Chemistry and Catalysis
"Ilie Murgulescu" Institute of Physical-Chemistry of the Romanian Academy
Bucharest
Romania

*Editorial Office*

MDPI
St. Alban-Anlage 66
4052 Basel, Switzerland

This is a reprint of articles from the Special Issue published online in the open access journal *Catalysts* (ISSN 2073-4344) (available at: www.mdpi.com/journal/catalysts/special_issues/Light-Assisted_Water_Air-Cleaning).

For citation purposes, cite each article independently as indicated on the article page online and as indicated below:

LastName, A.A.; LastName, B.B.; LastName, C.C. Article Title. *Journal Name* **Year**, *Volume Number*, Page Range.

**ISBN 978-3-0365-8389-1 (Hbk)**
**ISBN 978-3-0365-8388-4 (PDF)**

# Contents

# About the Editors

**Ioan Balint**

Dr. Ioan Balint is a Senior Researcher and currently the Director of the "Ilie Murgulescu" Institute of Physical Chemistry of the Romanian Academy. Dr. Balint obtained his PhD in Chemistry in 1996 from the same institute. He obtained three postdoctoral stays, one of them at the Pierre et Marie Marie Curie University in France (1998–1999), and another two at the Department of Environmental Chemistry and Catalysis in the group of Prof. Ken-ichi Aika of the Tokyo Institute of Technology in Japan (2000–2005). He has been an invited Professor several times at Tokyo Institute of Technology and Tokyo Technical University in Japan, and he also gained an UNESCO Scholarship at the same institute. Dr. Balint has an extensive research background encompassing various topics within surface science, such as non-isothermal gas desorption kinetic from supported-metal catalysts; solid defect chemistry; and water–gas shift reaction on the surface of simple and doped ionic oxides, the field of material synthesis, such as mesoporous nano oxides; mono/bimetallic metal nanoparticles; catalysis: catalytic combustion of hydrocarbons; oxidative coupling of methane; hydrogenation and oxidative conversion of conversion hydrocarbons; structure-sensitive reactions; and nitrate and nitrite abatement, as well as the field of photocatalysis/light harvesting, such as water and air depollution; water splitting; light-induced reactive oxygen species generation; and solar cells. He has been the Project Director for five projects receiving national grants and an expert member of many national/international committees. Dr. Balint has published over 100 peer-reviewed papers in well-recognized scientific journals, and has an H-index of 21.

**Monica Pavel**

Dr. Monica Pavel is a senior researcher at the "Ilie Murgulescu" Institute of Physical Chemistry of the Romanian Academy. She obtained her BSc degree in Chemistry and Physics in 2005 from the University of Pitesti, Romania, and her MSc degree (2006) in "Sciences of Materials, Nanomaterials and Multimaterials" from the National Polytechnic Institute of Toulouse, France. During October 2007 to November 2010, she carried out her PhD studies in Materials Chemistry at the University of Lyon 1—the Institute of Researches for Catalysis and Environment of Lyon in France under the supervision of Dr. Pavel Afanasiev. She continued as a Postdoctoral Researcher (2011–2013) at the University of Bucharest, Romania, with a short internship at the Institute Charles Gerhardt Montpellier, France. Her research interests are on the synthesis of transition-mixed oxides and 2D materials; utilization of various methods (e.g., molten salts, organic solvents, co-precipitation, solid state route, ion-exchange reactions); photocatalytic removal of pollutants (chlorinated/organic compounds, antibiotics, etc.); combustion of light alkanes; and the total oxidation of volatile organic compounds (VOCs).

*catalysts*

MDPI

*Editorial*

# Light-Assisted Catalysis in Water and Indoor Air Cleaning: Challenges and Perspectives

**Ioan Balint * and Monica Pavel ***

"Ilie Murgulescu" Institute of Physical Chemistry, Romanian Academy, 202 Spl. Independentei, 060021 Bucharest, Romania
* Correspondence: ibalint@icf.ro (I.B.); mpavel@icf.ro (M.P.)

The detrimental effects of environmental pollution on human health, combined with global climate change, make it a critical contemporary problem. Despite the fact that water covers more than 71% of the Earth's surface, ensuring access to high-quality drinking water for everyone is a major concern that societies encounter in the 21st century [1]. Utilizing renewable solar light and a catalyst to mineralize various harmful chemicals present in indoor air and water sources into benign small molecules, such as $H_2O$ and $CO_2$, is an attractive approach. In this context, photocatalytic processes have consistently offered smart, green, and eco-friendly scale-up methods for environmental remediation. Numerous photocatalysts have proven successful in achieving the mineralization of chlorinated pollutants, organic contaminants, dyes, or antibiotics [2–6]. An analysis of the existing literature reveals the need for research studies to focus on developing efficient photocatalysts capable of mineralizing contaminants into non-toxic $CO_2$. Only such photocatalytic materials should be envisaged for environmental remediation.

This Special Issue devoted to "Light-Assisted Catalysis in Water and Indoor Air Cleaning: Challenges and Perspectives" is a collection of 10 papers, including 3 reviews and 7 research articles. The aim of this Special Issue is to present recent advancements in the photocatalytic removal of pollutants, elucidating the main factors contributing to their mineralization and the implication of reactive oxygen species (ROS) through dedicated experiments. Furthermore, it encompasses the design and physicochemical characterization of various photocatalytic systems employed in environmental remediation.

In their review paper, Pavel et al. [7] presented a detailed and comparative study of the photocatalytic removal of organic (e.g., alcohols, carboxylic acids, volatile organic compounds, and phenols) and inorganic (e.g., $NO_3^-$) contaminants. The efficiency of multiple UV–Vis-light active photocatalysts and their corresponding degradation pathways were described, emphasizing the key factors contributing to their mineralization. The reaction mechanisms, the identification and quantification of by-products, and the implication of reactive active species (ROS) were considered for each category of the model target pollutant. Applying BiOCl and $g\text{–}C_3N_4$–based photocatalysts for water purification was reviewed by Ren and co-workers [8]. The authors described the preparation methods of $g\text{–}C_3N_4$/BiOCl composites via hydrothermal, deposition–precipitation, solvothermal, and calcination techniques. Subsequently, the authors explained the potential application of $g\text{–}C_3N_4$/BiOCl heterojunctions (e.g., degradation of dyes, residual pharmaceutical agents, and plasticizers) and the distinct reaction mechanisms involved in improving their performance. A comprehensive review by Galloni et al. [9] addresses the latest contributions to olive mill wastewater treatments, highlighting the potentialities and drawbacks of each removal method discussed. The authors emphasized the necessity of developing sustainable, environmentally friendly photocatalysts that could serve as valid alternatives to conventional treatment methods when optimized. In this context, the authors proposed recoverable magnetic compounds, as well as floating- and membrane-based devices, as promising alternatives to conventional $TiO_2$–based systems.

**Citation:** Balint, I.; Pavel, M. Light-Assisted Catalysis in Water and Indoor Air Cleaning: Challenges and Perspectives. *Catalysts* **2023**, *13*, 1032. https://doi.org/10.3390/catal13071032

Received: 13 June 2023
Accepted: 21 June 2023
Published: 23 June 2023

Ali et al. [10] examined how the structural, morphological, and optical properties of pure $TiO_2$ and $TiO_2$-incorporated $Fe_2O_3$ influenced their respective photoelectrochemical performances. The study revealed that pure titania exhibited the highest activity in the photocatalytic degradation of Rose Bengal dye under UV light, along with a superior photocurrent response, compared to the incorporated hematite counterparts. Additionally, Yang et al. [11] proposed utilizing $TiO_2$/hectorite composites with varying molar ratios of lithium, magnesium, and silicon as photocatalysts for the photodegradation of methylene blue (MB) under UV light irradiation. Under optimized conditions (molar ratio of Li/Mg/Si = 1.32/5.34/8), the highest removal rate of MB dye (97.8%) was revealed, while reusability tests after five runs showed only a slight decrease in the photocatalytic activity. Patel and co-workers [12] provided further evidence that photocatalytic, sonocatalytic, and sonophotocatalytic approaches play an important role in wastewater treatment, using Mn-doped ZnS quantum dots (Qds) as catalysts to remove solochrome dark blue azo dye (SDB). The authors claim that the $Mn^{2+}$:ZnS Qds sample showed high activity for the sonophotocatalytic degradation of SDB (89%), surpassing the rates observed for sonocatalysis (69.7%) or photocatalysis (55.2%) alone. This behavior was attributed to the enhanced electron-holes separation, increased reactive radicals, and augmented active surface area. Another application of light-assisted processes is by removing pesticides from wastewater. In this regard, Kobkeatthawin et al. [13] provide insights into the photocatalytic degradation of imidacloprid (IMI) using g–$C_3N_4$/$TiO_2$ systems. The composites were prepared by subjecting g-$C_3N_4$ nanosheets and Ti precursor to a hydrothermal treatment, with various weight percentages of g–$C_3N_4$ in relation to $TiO_2$ (0.5, 1, 4, 10, and 15 wt.% of g–$C_3N_4$). In this study, the authors obtained enhanced photocatalytic performance for the composite materials compared to bulk materials due to a synergistic effect between $Ti^{3+}$–$TiO_2$ and g–$C_3N_4$. The 0.5$C_3N_4$/$TiO_2$ sample displayed the highest IMI removal efficiency, reaching up to 93% within 150 min, and good stability during multiple recycling tests. In addition, Sandulescu and co-workers [14] studied the photocatalytic oxidation of phenol under sunlight irradiation using both bare and noble metal-loaded $TiO_2$. The experiments revealed that the supported noble metals function as visible light absorbers, assisting the separation of photo-charges and the reduction of $O_2$ to $O_2^-$. The $O_2^-$ oxidizes mildly phenol to oxygenated products. In a parallel process, •OH radicals yielded by $TiO_2$ mineralized phenol to $CO_2$ via rapid reaction sequences. Ignat et al. [15] used active noble plasmonic metals/Ga-substituted MgAl–hydrotalcites for the photocatalytic degradation of p–dichlorobenzene and 4–nitrophenol under simulated solar irradiation. The results revealed the enhanced photocatalytic performances of the synthesized plasmonic heterostructures (Ag–MgGaAl and Au–MgGaAl) compared to both the calcined forms and hydrotalcite precursors. The paper also included a discussion on the kinetic models that governed the studied plasmonic catalysts. Zhou and co-workers [16] successfully coupled β-$NaYF_4$:Yb,Er,Gd fluorescent nanorods to a reduced $TiO_2$ nanocomposite and applied it to visible-light catalytic sterilization under 980 nm near-infrared (NIR) light illumination. The authors claimed that up to 98.1% of Escherichia coli were effectively eradicated within 12 min of NIR light irradiation at a minimum inhibitory concentration of 40 μg/mL. The high bacterial activity was attributed to the synergistic effect between the fluorescent nanorods and reduced $TiO_2$.

Collectively, these studies provide an overview of the latest advances in the development of photocatalytic materials for diverse chemical reactions utilized in the removal of pollutants from indoor air or wastewater. These research studies pave the way for improving catalytic systems and contributing to a cleaner environment.

Finally, we express our gratitude to the journal *Catalysts* for the opportunity to produce this Special Issue, and to the editorial team, especially Assistant Editors Mia Zhang and Janine Li, for their continuous support during the submission and publication process. We also extend our gratitude to the authors and referees for their diligence and contributions to the success of this Special Issue.

**Author Contributions:** Conceptualization, I.B. and M.P.; writing—original draft preparation, I.B. and M.P.; writing—review and editing, I.B. and M.P. All authors have read and agreed to the published version of the manuscript.

**Conflicts of Interest:** The authors declare no conflict of interest.

## References

1. Pratap, B.; Kumar, S.; Nand, S.; Azad, I.; Bharagava, R.N.; Ferreira, L.F.R.; Dutta, V. Wastewater generation and treatment by various eco-friendly technologies: Possible health hazards and further reuse for environmental safety. *Chemosphere* **2023**, *313*, 137547. [CrossRef] [PubMed]
2. Wetchakun, K.; Wetchakun, N.; Sakulsermsuk, S. An overview of solar/visible light-driven heterogeneous photocatalysis for water purification: $TiO_2$- and ZnO-based photocatalysts used in suspension photoreactors. *J. Ind. Eng. Chem.* **2019**, *71*, 19–49. [CrossRef]
3. Raciulete, M.; Papa, F.; Kawamoto, D.; Munteanu, C.; Culita, D.C.; Negrila, C.; Atkinson, I.; Bratan, V.; Pandele-Cusu, J.; Balint, I. Particularities of trichloroethylene photocatalytic degradation over crystalline $RbLaTa_2O_7$ nanowire bundles grown by solid-state synthesis route. *J. Environ. Chem. Eng.* **2019**, *7*, 102789. [CrossRef]
4. Zhang, X.; Li, J.; Fan, W.-Y.; Sheng, G.-P. Photomineralization of effluent organic phosphorus to orthophosphate under simulated light illumination. *Environ. Sci. Technol.* **2019**, *53*, 4997–5004. [CrossRef] [PubMed]
5. Sharma, P.; Kimar, N.; Chauhan, R.; Singh, V.; Srivastava, V.C.; Bhatnagar, R. Growth of hierarchical ZnO nano flower on large functionalized rGO sheet for superior photocatalytic mineralization of antibiotic. *Chem. Eng. J.* **2020**, *392*, 123746. [CrossRef]
6. Zheng, J.; Fan, C.; Li, X.; Yang, Q.; Wang, D.; Duan, A.; Pan, S.; Zhang, B.; Ding, J.; Rong, S.; et al. Effective mineralization and detoxification of tetracycline hydrochloride enabled by oxygen vacancies in g-$C_3N_4$/ LDH composites. *Sep. Purif. Techn.* **2023**, *305*, 122554. [CrossRef]
7. Pavel, M.; Anastasescu, C.; State, R.-N.; Vasile, A.; Papa, F.; Balint, I. Photocatalytic degradation of organic and inorganic pollutants to harmless end products: Assesment of practical application potential for water and air cleaning. *Catalysts* **2023**, *13*, 380. [CrossRef]
8. Ren, Q.; Liu, J.; Yang, Q.; Shen, W. A review: Photocatalysts based on BiOCl and g–$C_3N_4$ for water purification. *Catalysts* **2021**, *11*, 1084. [CrossRef]
9. Galloni, M.G.; Ferrara, E.; Falletta, E.; Bianchi, C.L. Olive mill wastewater remediation: From conventional approaches to photocatalytic processes by easily recoverable materials. *Catalysts* **2022**, *12*, 923. [CrossRef]
10. Ali, A.M.; Sayed, M.A.; Algarni, H.; Ganesh, V.; Aslam, M.; Ismail, A.A.; El-Bery, H.M. Synthesis, characterization and photoelectric properties of $Fe_2O_3$ incorporated $TiO_2$ photocatalyst nanocomposites. *Catalysts* **2021**, *11*, 1062. [CrossRef]
11. Yang, D.; Chen, J.; Hong, X.; Cui, J.; Li, L. One-pot synthesis of $TiO_2$/hectorite composite and its photocatalytic degradation of methylene blue. *Catalysts* **2022**, *12*, 297. [CrossRef]
12. Patel, J.; Singh, A.K.; Jain, B.; Yadav, S.; Carabineiro, S.A.C.; Susan, M.A.B.H. Solochrome dark blue azo dye removal by sonophotocatalysis using $Mn^{2+}$ doped ZnS quantum dots. *Catalysts* **2022**, *11*, 1025. [CrossRef]
13. Kobkeatthawin, T.; Trakulmututa, J.; Amornsakchai, T.; Kajitvichyanukul, P.; Smith, S.M. Identification of active species in photodegradation of aqueous imidacloprid over g–$C_3N_4$/$TiO_2$ nanocomposites. *Catalysts* **2022**, *12*, 120. [CrossRef]
14. Sandulescu, A.; Anastasescu, C.; Papa, F.; Raciulete, M.; Vasile, A.; Spataru, T.; Scarisoreanu, M.; Fleaca, C.; Mihailescu, C.N.; Teodorescu, V.S.; et al. Advancements on basic working principles of photo-driven oxidative degradation of organic substrates over pristine and noble metal-modified $TiO_2$. Model case of phenol photo oxidation. *Catalysts* **2021**, *11*, 487. [CrossRef]
15. Ignat, E.C.; Lutic, D.; Ababei, G.; Carja, G. Novel heterostructures of noble plasmonic metals/Ga-substituted hydrotalcite for solar light driven photocatalysis toward water purification. *Catalysts* **2022**, *12*, 1351. [CrossRef]
16. Zhou, H.; He, F. Using Gd-enhanced β-$NaYF_4$:Yb,Er fluorescent nanorods coupled to reduced $TiO_2$ for the NIR-triggered photocatalytic inactivation of Escherichia coli. *Catalysts* **2021**, *11*, 184. [CrossRef]

## Short Biography of Author

**Ioan Balint.** Dr. Ioan Balint is a Senior Researcher and currently, the Director of the "Ilie Murgulescu" Institute of Physical-Chemistry of the Romanian Academy. Dr. Balint obtained his PhD in Chemistry in 1996 at the same institute. He obtained three postdoctoral stays, one of them at the Pierre et Marie Marie Curie University in France (1998–1999), and another two at the Department of Environmental Chemistry and Catalysis in the group of Prof. Ken-ichi Aika, of the Tokyo Institute of Technology in Japan (2000–2005). He was several times, an invited professor at Tokyo Institute of Technology and Tokyo Technical University, in Japan and also gained an UNESCO Scholarship at the same institute. Dr. Balint has an extensive research background encompassing various topics such as surface science: non-isothermal gas desorption kinetic from supported-metal catalysts; solid defect chemistry; water–gas shift reaction on surface of simple and doped ionic oxides; material synthesis: mesoporous nano oxides; mono/bimetallic metal nanoparticles; catalysis: catalytic combustion of hydrocarbons; oxidative coupling of methane; hydrogenation and oxidative conversion of conversion hydrocarbons; structure-sensitive reactions; nitrate and nitrite abatement; photocatalysis/light harvesting: water and air depollution; water splitting; light-induced reactive oxygen species generation; solar cell. He was project director for 5 national grants, and has been as expert member in many national/international committees. Dr. Balint has published over 100 peer-reviewed papers in well-recognized scientific journals, and has an H-index of 21.

**Monica Pavel.** Dr. Monica Pavel is a senior researcher at the "Ilie Murgulescu" Institute of Physical-Chemistry of the Romanian Academy. She obtained a BSc degree in chemistry and physics in 2005 at the University of Pitesti, Romania, and a MSc degree (2006) in materials science, nanomaterials and multimaterials from the National Polytechnic Institute of Toulouse, France. From October 2007 to November 2010, she carried out her PhD studies in materials chemistry at the University of Lyon 1—Institute of Researches for Catalysis and Environment of Lyon in France under the supervision of Dr. Pavel Afanasiev. She continued as a postdoctoral researcher (2011–2013) at the University of Bucharest, Romania, with a short internship at the Institute Charles Gerhardt Montpellier, France. Her research focuses on the controllable synthesis of transition-mixed oxides and 2D materials by various methods (e.g., molten salts, organic solvents, co-precipitation, solid state reactions, ion-exchange reactions); the photocatalytic removal of pollutants (chlorinated/organic compounds, antibiotics, etc.); combustion of light alkanes, and the total oxidation of volatile organic compounds (VOCs).

catalysts

Review

# Photocatalytic Degradation of Organic and Inorganic Pollutants to Harmless End Products: Assessment of Practical Application Potential for Water and Air Cleaning

Monica Pavel, Crina Anastasescu, Razvan-Nicolae State, Anca Vasile, Florica Papa and Ioan Balint *

"Ilie Murgulescu" Institute of Physical Chemistry of the Romanian Academy, 202 Spl. Independentei, 060021 Bucharest, Romania
* Correspondence: ibalint@icf.ro

**Abstract:** It is well-documented that large-scale pollution generated by human activity has a dramatic impact on ecosystems. In this context, removing harmful chemicals via photocatalysis has tremendous potential as a depollution method, utilizing freely available solar light and catalytic materials with low or negligible ecotoxicity. The main drawbacks, which aren't often addressed in the available literature, are the formation of harmful intermediate products, low reaction rates, limited catalyst stability, and difficult catalyst recovery. In most cases, published works assess the efficiency of tested photocatalysts from pollutant degradation studies, whereas identifying and quantifying by-products is not often conducted. This review summarizes the recent advances reported for the photocatalytic removal of some organic (e.g., alcohols, carboxylic acids, volatile organic compounds, phenol) and inorganic (e.g., $NO_3^-$) contaminants. The efficiency of various UV- and visible-light active photocatalysts and the reaction degradation pathways were explained, emphasizing the main factors contributing to their mineralization. The reaction mechanisms, the identification and quantification of degradation intermediates, and the implication of reactive active species (ROS) were discussed and analyzed for each category of model target pollutant. Additionally, the actual challenges and future photocatalytic approaches to improve environmental remediation were presented.

**Keywords:** photocatalysis; air cleaning; wastewater treatment; reactive oxygen species; degradation pathways

**Citation:** Pavel, M.; Anastasescu, C.; State, R.-N.; Vasile, A.; Papa, F.; Balint, I. Photocatalytic Degradation of Organic and Inorganic Pollutants to Harmless End Products: Assessment of Practical Application Potential for Water and Air Cleaning. *Catalysts* **2023**, *13*, 380. https://doi.org/10.3390/catal13020380

Academic Editor: Ioannis Konstantinou

Received: 5 January 2023
Revised: 1 February 2023
Accepted: 7 February 2023
Published: 9 February 2023



## 1. Introduction

The economic prosperity brought by industrialization is associated with the dramatic degradation of the environment (i.e., water and air pollution, loss of natural resources, climate change, etc.). The uncontrolled release of numerous hazardous contaminants, such as dyes, chemicals, heavy metals, organic solvents, petroleum products, and solid wastes, is strongly contaminating the environment.

Advanced oxidation processes (AOPs) have attracted considerable interest due to their significant potential for environmental remediation [1]. Among them, heterogeneous photocatalysis employing semiconductor materials and various light sources is a promising route for the removal of persistent pollutants to produce harmless end products. During the photocatalytic process in the presence of the light of suitable energy (with higher energy than the respective band gap of the material), an electron ($e^-$) is excited from the valence band (VB) of a semiconductor to the conduction band (CB), generating a positive hole ($h^+$) in the valence band. Photogeneration of charge carriers ($e^-/h^+$) initiates the photocatalytic degradation process. The valence band hole oxidize surface absorbed water molecules or $OH^-$ to produce hydroxyl radicals ($\bullet OH$). The photoexcited electrons reduce oxygen molecules and produce hydroperoxyl radicals ($HO_2\bullet$) or superoxide radicals ($\bullet O_2^-$). During the photocatalytic process (Equations (1)–(4)), these reactive oxygen species

(ROS) and free electrons/holes react with the surface adsorbed molecules (e.g., organic, inorganic compounds) and convert the pollutants to unharmful products.

$$\text{Semiconductor} + h\nu \rightarrow h^+ + e^- \quad (1)$$

$$e^- + h^+ \rightarrow \text{energy} \quad (2)$$

$$h^+ + H_2O \rightarrow \bullet OH + H^+ \quad (3)$$

$$e^- + O_2 \rightarrow \bullet {O_2}^- \quad (4)$$

The efficiency of a photocatalytic reaction is mediated by the capability of the photocatalyst to generate longer-lived electrons and holes, leading to the formation of reactive free radicals. It also depends on the type of photocatalytic material and the operational parameters such as solution pH, irradiation time, and the presence of holes scavengers (sacrificial electron donors). Another factor affecting photocatalysis is the state of the material, e.g., powder form or its immobilization on support. Organic pollutants may contain a great diversity of elements, such as C, N, O, S, Cl, etc. During the successive degradation steps, each carbon atom requires at least four photo electrons ($e^-$) and four photo holes ($h^+$) in order to be mineralized into $CO_2$. Consequently, it is clear that the high number of electrons and holes required for the mineralization of large organic molecules impede the efficiency of the depollution process. During the successive degradation steps, the photo charges break the large molecule down into smaller units via intricate redox processes. Consequently, photocatalytic technologies are easier for pollutants with one carbon atom in their molecule, such as formic acid, methanol, or formaldehyde. For larger organic molecules, the photocatalytic technologies should be associated with other methods, such as treatment with strong oxidants (e.g., ozone, persulfate ion, hydrogen peroxide) to raise the mineralization extent of the pollutant [2–5]. In depollution technologies, the fundamental problem is the extent to which the pollutant is photomineralized. Most papers analyze only the degradation of the target compounds into intermediates, which may be as harmful as the pollutant itself, without assessing the amount of carbon dioxide.

This work surveys the basic mechanisms involved in photocatalytic mineralization of the most common gaseous/liquid pollutants and a large area of efficient engineered materials used for their removal under light exposure. The novelty comes from the discrimination between the photocatalytic mechanisms/materials inducing the unknown or scarcely quantifiable intermediates and those leading to harmless end products such as $CO_2$, $Cl^-$, or $N_2$. The manuscript is divided into four main subsections (Figure 1). The first subsection describes the photocatalytic removal of alcohols and carboxylic acids in gaseous and liquid phases. Methanol, ethanol, and oxalic acid were described as model molecules. The second subsection is related to the removal of volatile organic compounds (VOCs) from indoor air and wastewater. The class of the chlorinated VOCs is exemplified by trichloroethylene (TCE), perchloroethylene (PCE), and dichloroacetic anion (DCA) compounds, while the group of the aromatic VOCs is illustrated by benzene, toluene, xylene, and ethylene pollutants. The third subsection presents the photocatalytic mineralization of harmful aromatic compounds from wastewater, focusing on phenol model molecules. Finally, the fourth subsection refers to the photocatalytic removal of aqueous inorganic nitrogen-based compounds from wastewater concentrating on nitrate (${NO_3}^-$) reduction. All these target pollutants were chosen due to their widespread usage, toxicity, and environmental pollution.

Challenges and future directions of photocatalytic environmental remediation are discussed. The potential of novel emerging photocatalytic technologies transferable to industrial applications is also analyzed.

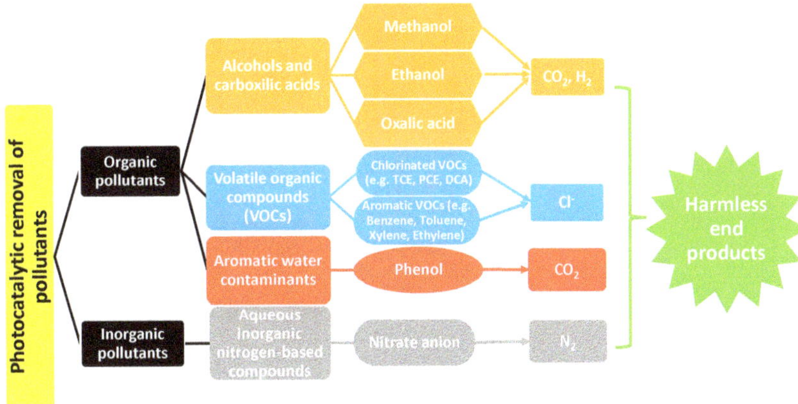


**Figure 1.** Schematic illustration of the manuscript structure and its objectives.

## 2. Photocatalytic Removal of Organic and Inorganic Pollutants

Environmental pollution is a pervasive problem with consequences for human health, living organisms, and ecosystems. To solve this issue, the total transformation of the parent harmful compounds to eliminate their toxicity and persistence is necessary. In the following, recent advancements reported for the removal of recalcitrant organic and inorganic molecules are presented, taking a look at their reaction degradation mechanisms.

### *2.1. Photooxidation of Primary Alcohols and Carboxylic Acids in Gaseous and Liquid Media*

The photocatalytic oxidation (PCO) of organic compounds to $CO_2$ should be the ideal degradative process in depollution technologies, especially when solar light, which is a cheap, regenerable energy source, is used [6]. The photooxidation of alcohols is investigated as a model reaction for the abatement of organic pollutants [7]. The degradation of the intermediary carboxylic acids, which resulted in alcohol's photodegradation pathway, is also of great interest. In the meantime, alcohols can be an efficient and convenient hydrogen source via dehydrogenation or sacrificial reagents in photocatalytic water splitting.

Methanol is often chosen for fundamental studies concerning primary alcohols due to its structural simplicity but also as a prototype for organic pollutants in applicative research devoted to the environmental field. The oxidative sequence of methanol photomineralization to $CO_2$ involving only a few organic intermediates is an appropriate choice for the investigation of oxidative degradation mechanism [8,9] but also for establishing the activity ranking of various photocatalysts. The numerous studies on methanol photooxidation, both in gaseous phase and liquid media, centered especially on titania [10], support the idea that the driving force over semiconductor-type materials is the capacity of the catalyst to photo generate appropriate density of charges ($e^-/h^+$) combined with prevention of charge recombination.

#### 2.1.1. Basic Data on Methanol Photodegradation on $TiO_2$

In what comes, we should list the key factors controlling the photomineralization of methanol on different engineered materials. The methanol degradation pathway is related to methanol surface coverage on photocatalysts as well as to the density of surface hydroxyl groups and adsorbed oxygen. For example, by exposing the adsorbed methanol on anatase (101) to UV irradiation, Setvin et al. [11] observed that the main reaction products are formaldehyde ($CH_2O$) and methyl formate ($CH_3{-}O{-}CHO$), formed by distinct reaction pathways. The first results from methanol interaction with the co-adsorbed oxygen or terminal $OH^-$ groups leading to methoxy anion formation:

$$CH_3OH + OH^- \rightarrow CH_3O^- + H_2O \quad (5)$$

By accepting a hole, methoxy radical is generated:

$$CH_3O^- + h^+ \rightarrow CH_3O\bullet \quad (6)$$

This is further converted to formaldehyde. On the other hand, methyl formate ($CH_3-O-CHO$) is obtained at high methanol surface coverages. Shen et al. [12] reported an increased photoactivation rate of methanol on annealed rutile (110) surface, facilitating both oxygen and methanol adsorption. The authors identified the thermally activated cleavage of the O−H bond as being the first step, followed by the photo-catalytically driven C−H bond cleavage. Also, the surface defects of $TiO_2$ photocatalyst proved to increase the photo dissociation rate of methanol [13]. Shen et al. [8] showed that the methoxy group, formed on the catalyst surface, acts as a more efficient hole scavenger compared to the molecular methanol. On the methanol-saturated surface of rutile nanoparticles exposed to UV irradiation [14], the conversion of the methoxy group to formate, requiring two photo-electrons, is enhanced in the presence of oxygen. In this case, oxygen also acts as the main electron scavenger, competing with methanol oxidation.

A great number of semiconductor materials have been developed for the removal of organic pollutants from air and water by photomineralization. Noble (Au, Pt, Ag, Pd) and d-(Cu) metals deposition on $TiO_2$ is typically used to extend the light absorption range of $TiO_2$ and ZnO [15–17] but to bust the photocatalytic performances via better charge separation at Schottky domains created at metal/semiconductor interfaces [18–20]. The great majority of works concerning the photooxidation of alcohols (methanol) advance the following mechanistic pathways: (a) direct hole oxidation route and/or (b) oxidation intermediated by hydroxyl radicals (•OH) resulting from the following reactions:

$$h^+ + \text{-OH (surface hydroxyl)} \rightarrow \bullet OH \quad (7)$$

$$O_2 + e^- \rightarrow (O_2^-)_{ads} \text{ (on catalyst surface)} \quad (8)$$

$$(O_2^-)_{ads} + 2H^+ + e^- \rightarrow H_2O_2 \rightarrow 2\bullet OH \quad (9)$$

When $O_2$ is not able to reach the surface because of high organic coverage, the direct hole route is favored [21].

#### 2.1.2. Methanol and Ethanol Oxidation in Gaseous Phase on Bare and Modified Catalysts

Despite the great number of papers focusing on the photooxidation of methanol, only a few of them discriminate between degradation (oxidative degradation to intermediates) and mineralization (oxidation to $CO_2$). El-Roz et al. [22] carried out a mechanistic study using an operando-FTIR system coupled with gas phase analysis techniques (gas-IR and MS) on $TiO_2$ P25 photocatalyst. The authors found that methanol concentration is a key factor in tailoring oxidation selectivity. Under light irradiation of 365 nm, the maximum methanol conversion to $CO_2$ and $H_2O$ was observed for 500 ppm $CH_3OH$ in the gas phase (20% $O_2$/Ar), according to the reaction:

$$CH_3OH + 3/2O_2 \rightarrow CO_2 + 3H_2O \quad (10)$$

For 1200 ppm $CH_3OH$ in the gas phase, the favored product was methyl formate:

$$2CH_3OH + O_2 \rightarrow CH_3OCHO + 2H_2O \quad (11)$$

According to the authors, the oxidation sequence starts with the dissociative chemisorption of methanol, leading to the formation of surface methoxy groups which are oxidized to formate and finally to $CO_2$, the last step being considered as rate limiting for the methanol photooxidation. Additionally, methyl formate secondary product can result from a reaction between adsorbed formate and methoxy group.

Photocatalytic oxidation of methanol in visible light over AuNPs modified $WO_3$ was studied by DePuccio et al. [23] using a continuous flow gas-phase reactor. Distinct photocatalytic tests performed comparatively on $AuNPs/SiO_2$ and bare $WO_3$ indicated that the surface plasmon resonance (SPR) phenomenon (induced by AuNPs) and the band gap excitation of $WO_3$ are responsible for two distinct mechanisms involved in methanol oxidation. One mechanism is responsible for methyl formate generation, whereas the other leads to the formation of formaldehyde. When AuNPs are deposited on $WO_3$, the SPR triggered by light absorption enhances the separation of the photogenerated charges by $WO_3$, increasing thus the photocatalytic activity. Over bare $WO_3$, the formation of $CO_2$ was not observed, contrasting with $Au/WO_3$, where $CO_2$ was the major reaction product (39%).

Ethanol photocatalytic oxidation is also highly important for depollution applications since it largely emerges from industrial activity. In addition, it is noteworthy to investigate the differences in photoreactivity brought by the carbon–carbon bond in comparison to methanol.

Muggli et al. [24] studied the photocatalytic oxidation of ethanol using transient reaction techniques and isotope labeling. The Carbon-13 labeled ethanol ($CH_3\,^{13}CH_2OH$) adsorbed on Degussa P-25 was exposed to light (maximum intensity ≈ 390 nm) and 0.2% $O_2$ in He. The $CO_2$, $H_2O$, and acetaldehyde formation have been observed for ethanol saturation coverage, whereas for lower coverage, only $CO_2$ and $H_2O$ were obtained. The α-carbon of ethanol proved to be preferentially oxidized. This finding is also certified by the resulting intermediates (acetaldehyde and acetic acid) [25]. The reactions below depict the whole process.

$$\text{Ethanol-derived Acetaldehyde}_{\text{(ads)}} \rightarrow \text{(a) \& (b)} \tag{12}$$

$$\text{(a)} \rightarrow \text{acetic acid} \rightarrow CO_2 + \text{formaldehyde} \rightarrow \text{formic acid} \rightarrow CO_2 \text{ (fast)} \tag{13}$$

$$\text{(b)} \rightarrow \text{formic acid} + \text{formaldehyde} \rightarrow \text{formic acid} \rightarrow CO_2 \text{ (slow)} \tag{14}$$

According to the authors, the presence of acetaldehyde on the catalyst surface, together with other intermediates, decreases the reactivity of (b). The enhancement of the acetaldehyde oxidation rate is required to increase the overall $CO_2$ formation.

Yu et al. [26] studied the photocatalytic conversion of ethanol to $CO_2$ in the presence of $O_2$ by using in situ infrared (IR) spectroscopy. The surface coverage of adsorbed $H_2O$ and ethanol on P25 Degussa was identified as a key parameter in deciding the favored reaction pathway. The involved oxidizing species involved in the reaction are formed according to the following equations:

$$TiO_2 + h\nu \rightarrow h^+ + e^- \tag{15}$$

$$H_2O_{ads} \rightarrow OH^- + H^+ \tag{16}$$

$$OH^- + h^+ \rightarrow \bullet OH \tag{17}$$

$$O_{2ads} + e^- \rightarrow O_2^- \tag{18}$$

$$2O_2^- + 2H_2O_{ads} \rightarrow 2\bullet OH + 2OH^- + O_2 \tag{19}$$

The authors summarized the ethanol photo-oxidation by using $h^+$ and •OH as it comes:

$$(CH_3CH_2OH_{ad}/CH_3CH_2O_{ad} \rightarrow H_2O + CO_2) \tag{20}$$

I. For low ethanol coverage and adsorbed $H_2O$ on the $TiO_2$ surface, an •OH-initiating oxidation mechanism leading to $HCOO^-{}_{ads}$ as a major intermediate was proposed by the authors. The envisaged parallel/series reactions were:

$$CH_3CH_2OH_{ads}/CH_3CH_2O_{ads} \rightarrow, \tag{21}$$

$$\text{(a)} \rightarrow \text{adsorbed C}_1\text{-oxygenated species } (HCHO_{ads}, HCOOH_{ads}, \text{ and } HCOO^-{}_{ads}) \tag{22}$$

$$\text{(b)} \rightarrow \text{adsorbed C}_2\text{-oxygenated species } (CH_3COOH_{ads} \text{ and } CH_3COO^-{}_{ads}) + \text{C}_1\text{-oxygenated species} \tag{23}$$

$$\text{(c)} \rightarrow \text{adsorbed } CH_3CHO_{ads} \tag{24}$$

II. For high ethanol coverage, decreased in the amount of adsorbed water on the $TiO_2$ surface was registered due to the fact that the direct interaction between the photogenerated holes and adsorbed ethanol is favored. This leads to the hydrogen abstraction from α-carbon ($CH_3CH_2OHads/CH_3CH_2O_{ads}$) and the formation of $CH_3COO^-{}_{ads}$ as primary, intermediate species, according to the reactions:

$$CH_3CH_2O_{ads} + h^+ + O_{(lattice)} \rightarrow CH_3COO^- + 2H^+ \quad (25)$$

$$CH_3CH_2OH_{ads} + 2h^+ + O_{(lattice)} \rightarrow CH_3COOH + 2H^+ \quad (26)$$

In this system, the oxidation of ethanol to $CO_2$ is carried out preferentially by highly oxidizing •OH radicals to the detriment of holes.

Modifying P25 with Ag nanoparticles, Fukuhara et al. [27] obtained an active photocatalyst for the degradation of ethanol in the UV and visible range. They monitored the heat released by ethanol partial and total oxidation (Equations (27)–(29)):

$$C_2H_5OH + 1/2O_2 \rightarrow CH_3CHO + H_2O,\ \Delta_rH^0 = -172.91\ \text{kJ mol}^{-1} \quad (27)$$

$$C_2H_5OH + 3O_2 \rightarrow 2CO_2 + 3H_2O,\ \Delta_rH^0 = -1277.38\ \text{kJ mol}^{-1} \quad (28)$$

$$CH_3CHO + 5/2O_2 \rightarrow 2CO_2 + 2H_2O,\ \Delta_rH^0 = -1104.47\ \text{kJ mol}^{-1} \quad (29)$$

The results revealed different light-induced acting mechanisms for Ag nanoparticles and $TiO_2$. Under UV irradiation, $TiO_2$ promotes the partial oxidation of ethanol to acetaldehyde, whereas $O_2$ activation accounts for $CO_2$ and $H_2O$ yield. Under visible light, the complete oxidation of ethanol to $CO_2$ and generation of $H_2O$ occur over $Ag^0$ nanoparticles.

The presence of oxygen was highlighted as a key factor for $CO_2$ production on Ag-$TiO_2$ catalyst under UV-Vis irradiation (Figure 2). Figure 2b shows that $O_2$ is activated under UV light on $TiO_2$, whereas the surface of the Ag nanoparticle promotes the partial oxidation of $C_2H_5OH$ to $CH_3CHO$ and the cleavage of the C−C bond leading to $CO_2$ and $CH_4$ formation.

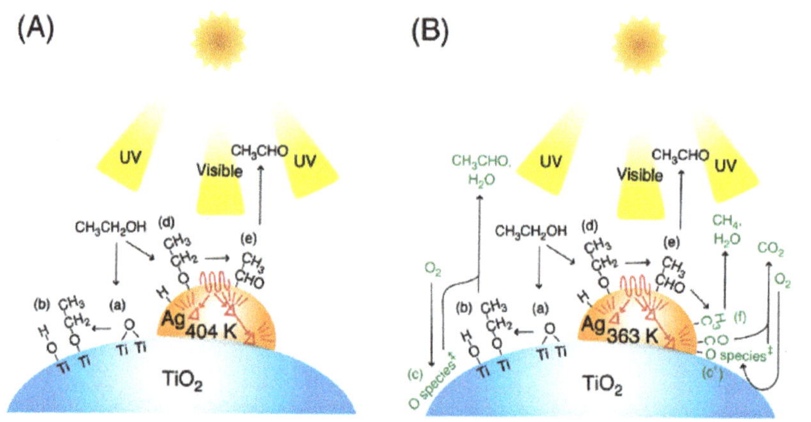


**Figure 2.** Photocatalytic reaction steps involving: (**A**) Ag-$TiO_2$, ethanol, UV-Vis light irradiation and (**B**) Ag-$TiO_2$, ethanol, $O_2$, UV-Vis light irradiation. Reproduced with permission from ref. [25]. Copyright 2023 American Chemical Society.

#### 2.1.3. Methanol, Ethanol, and Oxalic Acid Oxidation in Liquid Phase on Pristine and Modified Catalysts

Kawai et al. [28] proposed the following mechanism for methanol degradation on $TiO_2$-suspended powder in deaerated aqueous media under 500 W Xe-lamp light irradiation:

$$MeOH \rightarrow HCHO + H_2, \quad (30)$$

$$HCHO + H_2O \rightarrow HCO_2H + H_2, \quad (31)$$

$$HCO_2H \rightarrow CO_2 + H_2O, \quad (32)$$

At the beginning of the process, $H_2$ was the main gaseous product obtained along with small amounts of $CO_2$. The photogenerated holes are used for the oxidation of MeOH, HCHO, and $HCO_2H$ intermediates, whereas the photogenerated electrons are responsible for $H_2$ production via proton reduction.

Villareal et al. [29] explored the photoelectrochemical oxidation of methanol and formic acid dissolved in water on polycrystalline anatase electrodes. The authors found that methanol oxidation to $CO_2$ is carried out by surface-bound hydroxyl radicals. This mechanism suggests weak interaction between methanol and oxide surface. In contrast, formic acid appears to be oxidized directly to $CO_2$ by the valence band-free holes, indicating strong adsorption on the catalyst surface, despite the competitive adsorption of water.

Methanol photooxidation in aqueous media has also been studied by Haselman et al. [30] on platinum-modified $TiO_2$. An ATR-FTIR setup was used for time-resolved investigations of both Pt particle growths during in situ photo deposition on $TiO_2$ thin films and the photooxidation of methanol under UV irradiation in aqueous environments. The authors certified that methanol is oxidized in aqueous solutions to $H_2$ and $CO_2$ by the photogenerated holes. Also, for 2.7% Pt loading on $TiO_2$, the methanol photooxidation to $CO_2$ takes place via the formation of formaldehyde intermediate, while for lower Pt loading (1.4%), methyl formate is obtained. The reaction mechanism depends on the availability of specific active sites (on platinum and defects).

The solar light-driven generation of $CO_2$ and $H_2$ from a water/methanol mixture has been carried out on anatase and sodium titanates with tubular morphology [31]. The tubular morphology proved to have a beneficial effect on charge separation at the interface of semiconducting crystalline domains. The radical trapping experiments certified the formation of reactive oxygen species (•OH and $O_2^-$) by reaction between photogenerated electrons and holes with adsorbed $O_2$ and hydroxyl groups. The essential contribution ROS to the overall photocatalytic activity showed by tubular titania-based materials was clearly emphasized by experimental results. Papa et al. [32] synthesized PVP-protected bimetallic nanoparticles (Pt-Cu and Pt-Ag) by a modified protocol of the alkaline polyol method. Active photocatalysts for aqueous methanol mineralization have been obtained by supporting them on $TiO_2$ (2.16 mmol $CO_2$ $h^{-1}$ $g_{cat}^{-1}$ for Pt-Ag/$TiO_2$ and 1.68 mmol $CO_2$ $h^{-1}$ $g_{cat}^{-1}$ for Pt-Cu/$TiO_2$) under 125 W medium pressure Hg lamp and 10% $O_2$/Ar.

Photocatalytic experiments confirmed that not only semiconductor materials show photocatalytic activity but also large band-gap insulators such as tubular $SiO_2$ may work as extremely active photocatalysts in the methanol/water mixture when exposed to solar light [33]. $SiO_2$ nanotubes with a high surface density of light-absorbing defects were obtained by a modified sol-gel method, using DL tartaric acid as an organic template.

The formation of $Si^{3+}$ defects by calcination enhanced light absorption characteristics, the importance of thermal treatment being illustrated in Figure 3a.

The $CO_2$ generation rate in Ar flow was 2.4 μmol $h^{-1}$ and increased at 12 μmol $h^{-1}$ in the presence of $O_2$. The evolvement of $CO_2$ and $H_2$ from aqueous methanol exposed to solar (AM 1.5) and visible light irradiation ($\lambda > 420$ nm) demonstrated that the light absorbing defects, having the energy levels located within the forbidden gap of $SiO_2$, are able to work as photocatalytic sites (Figure 3b).

ROS photogeneration over the $SiO_2$, $SiO_2$-$TiO_2$ nanotubes and P25, together with their impact on the aqueous methanol photodegradation, were also investigated [34]. The $TiO_2$

proved to work as a photocatalyst by intermediation of •OH radicals, while the $SiO_2$-$TiO_2$ generated $O_2^-$. In contrast, the organic substrate was activated and degraded on the surface of $SiO_2$ by the intra-band gap, isolated surface quantum defects.

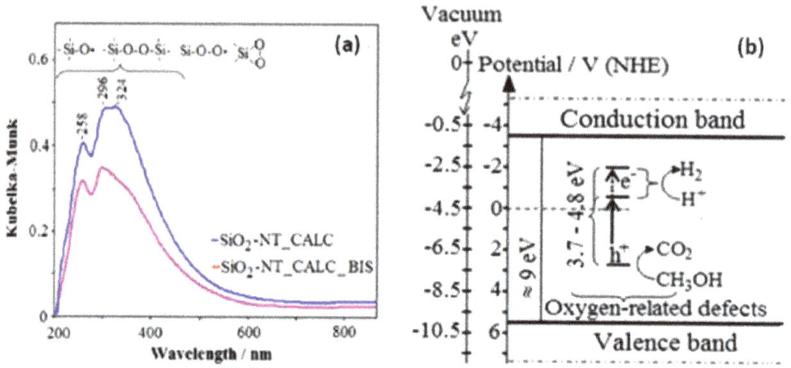

**Figure 3.** (**a**) Comparative UV-Vis spectra of silica nanotubes subjected to calcination in air for 3 h ($SiO_2$-NT CALC) and 1 h ($SiO_2$-NT CALC–BIS) (**b**) Representation of intraband gap defects in tubular $SiO_2$ Reproduced with permission from ref. [33]. Copyright 2023 Elsevier.

Oxalic acid is found in biological systems but also as an emerging residue from industrial activities (textile industry, etc.). Oxalic acid is often chosen as a model pollutant for testing the photoreactivity of dicarboxylic acids because its photodegradation process conducts mainly to $CO_2$. Other presumable intermediates, such as formic acid and carbonate, have not been clearly revealed [35].

The earlier mentioned $SiO_2$ material with different morphologies and modified with platinum proved to be efficient for the photomineralization of oxalic acid in the 200–800 nm range [36]. For this system, the degradation mechanism proposed the following sequences:

- Involving of the photogenerated holes in the oxalate oxidation at the catalyst surface (Equation (33))

$$2h^+ + C_2O_4{}^{2-} \rightarrow 2CO_2 \tag{33}$$

- Enhancing of the photoelectron transfer to the oxygen and proton by the platinum presence (Equation (34))

$$2H^+ + 0.5O_2 + 2e^- \rightarrow 2H_2O \tag{34}$$

The spherical $SiO_2$ particles were inactive, whereas the rate of $CO_2$ evolvement over tubular-shaped $SiO_2$ particles was 45 μmol $g^{-1}{}_{cat}$ $h^{-1}$. Platinum deposition on tubular $SiO_2$ further increased the $CO_2$ formation rate to 428 μmol $g^{-1}{}_{cat}$ $h^{-1}$, the efficiency of the modified catalyst being comparable with that of semiconductor-based materials.

Kosanić et al. [37] conducted the same photocatalytic process in oxygenated aqueous solution over $TiO_2$ powder (Degussa P25) under UV irradiation (300–400 nm) and correlated the $CO_2$ production with the illumination time. The authors advanced the following mechanism initiated by the •OH radicals:

$$HC_2O_4{}^- + \bullet OH \rightarrow OH^- + \bullet COOH + CO_2 \tag{35}$$

$$\bullet COOH + O_2 \rightarrow \bullet HO_2 + CO_2 \tag{36}$$

$$\bullet HO_2 + \bullet HO_2 \rightarrow H_2O_2 + O_2 \tag{37}$$

Doping of $TiO_2$ nanopowders with iron allowed the use of simulated solar light irradiation (AM 1.5) for the mineralization of oxalic acid in an aqueous solution. The redox sites created by introducing $Fe^{3+}$ into the $TiO_2$ lattice were able to absorb the visible component of solar light [38]. The photocatalytic activity was found to increase due to the enlargement of catalyst surface area and to the decrease in optical band-gap. The highest recorded $CO_2$ formation rate was 245 μmol $h^{-1}$. The proposed mechanism considered that the $C_2O_4^{2-}$ mineralization takes place either by •OH attack or by direct reaction with holes.

Kiatkittipong et al. [39] compared the photocatalytic generation of $CO_2$ in an aqueous solution under UV irradiation (λmax = 254 nm) over various titanate nanoribbons. It was found the following sequence of the photocatalytic activity, in accordance with the increasing of the crystallinity (decreasing of bulk defects) due to the calcination step: $TiO_2 > Na_2Ti_6O_{13} > Na_{1.48}\ H_{0.52}\ Ti_3O_7$~$H_2Ti_3O_7$.

Cauxa et al. [40] used oxalic acid as an electron donor for water splitting performed on g-$C_3N_4$ and loaded with platinum (0.37 wt.%) under UV-Vis irradiation (using a cut-off filter for obtaining λ > 380 nm). Based on registered $CO_2$: $H_2$ ratios, the authors assumed that oxidation of water also occurs in addition to the photooxidation of oxalic acid since more hydrogen was produced. Also, the presence of hydrogen peroxide was identified in the oxalic solution after photocatalytic tests suggesting the free radical formation before the complete degradation of oxalic acid to $CO_2$. This is in accordance with the mechanism proposed by Kosanić et al. [37].

Karunakaran et al. [41] performed a systematic study on oxalic acid photomineralization under natural sunshine triggered by various particulate semiconductors ($TiO_2$, CuO, ZnO, $Pb_2O_3$, $PbO_2$, $Bi_2O_3$). The authors identified the operational parameters influencing $CO_2$ production as being the following: oxygen presence, the concentration of the oxalic acid solution, and the area of the catalyst bed. The photocatalytic efficiency relative to the oxalic acid mineralization was the following: ZnO > CuO = $TiO_2$ = $Bi_2O_3$ = $Pb_2O_3$ > $PbO_2$ for the next reaction conditions: 50 mL of 0.25 M acid, $[O_2]$ dissolved = 24.7 mg $L^{-1}$, 1.0 g—catalyst loading, 15.68 $cm^2$—catalyst bed, 10 min sunshine. Additionally, for each catalyst and the same reaction conditions, the degradation of the formic, acetic and citric acid was also performed. Table 1 presents the registered reactivity sequence:

**Table 1.** The registered reactivity over the studied photocatalysts.

| Photocatalyst | Reactivity Sequence |
|---|---|
| ZnO | oxalic acid > acetic acid > citric acid<br>The catalyst is not stable in formic acid solution |
| CuO | formic acid > oxalic acid > acetic acid > citric acid |
| $TiO_2$ | formic acid > oxalic acid > acetic acid > citric acid |
| $Bi_2O_3$ | formic acid > oxalic acid = acetic acid > citric acid |
| $Pb_2O_3$ | formic acid > oxalic acid = acetic acid > citric acid |
| $PbO_2$ | formic acid > oxalic acid = acetic acid > citric acid |

### 2.2. *Volatile Organic Compounds (VOCs) from Air and Wastewater*

This section deals with the latest discoveries on VOC abatement from air and water using various photocatalysts and photocatalytic techniques.

#### 2.2.1. Overview of Volatile Organic Compounds (VOCs)

Volatile organic compounds (VOCs), such as alcohols, aromatics (e.g., benzene, toluene, xylene), aldehydes, and halocarbons are a major group of air pollutants [42] produced mostly from outdoor (industrial processes and transportation) or indoor sources (domestic products, building materials, and food industry). They can cause major health issues like respiratory diseases, heart issues or even cancer [42–44]. VOCs also represent a precursor of PM 2.5 (particulate matter are fine inhalable particles with diameters of 2.5 μm or smaller), which determines important environmental challenges all over the world [43]. On the other hand, chlorinated VOCs cause severe contamination in aquatic environments, TCE (trichloroethylene) and PCE (perchloroethylene), used at a large scale as solvents or in dry cleaning, are the most encountered types of VOC contaminants [45,46]. In the specific case of chlorinated compounds, the monitoring of $Cl^-$ release in time in the water phase or TOC (total organic compound) evolution can be a measure of pollutant removal efficiency [47].

With the exception of a few studies [48–50], most investigations ignore this aspect focusing only on the quantification of the chlorinated compound by GC measurement.

Many techniques have been used to remove those volatile organic compounds from the air or from water. These include adsorption, thermal catalysis, ozone oxidation or photocatalytic oxidation, the latter attracting more and more attention lately. The heterogeneous catalysis can offer various advantages if compared to other AOPs (advanced oxidation processes), such as moderate operation temperatures or pressure and low cost if sunlight is used as an irradiation source. On the other hand, even if photocatalysts offer good results, at least in lab-scale experiments, other AOPs proved to be as efficient, especially ozonation. For example, ozonation and activated carbon adsorption are accepted technologies for water remediation [51], but they do not meet the aim of our study.

Generally, three mechanisms regulate the photocatalytic mineralization of VOCs: light harvesting, photogenerated charge carrier (electrons and holes) separation, and charge injection into VOCs and their intermediates from the surface of the catalyst. In order to improve the effectiveness of charge carrier separation, various methods have been applied, such as metal doping, nonmetallic element doping, and the production of heterojunction from several materials with matching energy bands [43]. This can also be explained as follows: when a semiconductor photocatalyst is irradiated by light, an electron-hole pair is generated. Thus, an electron is excited from the valence band to the conduction band resulting in a hole ($h^+$) remaining in the valence band. This oxidizing hole can interact directly with an adsorbed molecule or can oxidize water to produce hydroxyl radicals, which are themselves oxidizing agents [51].

Therefore, in this review we focused on chlorinated and aromatic VOCs removal from the environment (water and air) using various photocatalysts working under different irradiation sources.

#### 2.2.2. Chlorinated VOCs Abatement from Air and Water

Monteiro et al. [52] studied the photocatalytic oxidation of perchloroethylene using a photoreactor having a compound parabolic collector. The $TiO_2$-based paint was used as a coating for two configurations of monolithic structured cellulose acetate. The photocatalytic experiments were performed under solar irradiation (spectral range of $300 < \lambda < 800$ nm). They obtained in the best experimental conditions (initial concentration = 1100 ppm, feed flow rate = 75 $cm^3\ min^{-1}$, relative humidity $\approx$ 40% and incident UV irradiance $\approx$ 38.4 W $m^{-2}$ in the presence of oxygen) a PCE conversion of around 60%. Also, they observed that if the relative humidity is low, Cl radical chain propagation reactions could be included in the photocatalytic oxidation mechanism of PCE and that the photoreaction can still occur in the absence of oxygen.

The authors proposed a reaction mechanism according to the schemes below (Figure 4). It relies on the supposition that the reaction of PCE degradation starts with the addition of •OH radicals, thus leading to a dechlorination reaction where •Cl radicals are formed. In detail, Figure 4a shows the attack of hydroxyl radical to PCE pursued by chlorine radical release yielding trichloroethenol. Figure 4b represents the addition of •Cl to PCE, thus forming chloroalkyl radical, which is then oxidized by superoxide radicals creating a peroxy radical. The peroxy radical can be transformed into chloroethoxy radical by reacting with a second peroxy radical which suffers a CC bond scission forming $CCl_2O$ and $CCl_3$ radicals. $CCl_3$ converts into chloroform or carbon tetrachloride by reacting with H+ or •Cl, while $CCl_2O$ produces phosgene that may be hydrolyzed into $CO_2$ and HCl. Figure 4c shows the chlorination of PCE resulting in chloroalkanes. Here, PCE suffers a reaction with •Cl forming chloroalkyl radical that can be hydrogenated or chlorinated, resulting in pentachloroethane or perchloroethylene. Finally, Figure 4d represents the formation mechanism of trichloroethyl acetate and trichloromethyl acetate. If •OH is added to chloroalkyl radical and dichloroacetyl chloride, pentachloroethanol and dichloroacetic acid could be obtained. Dichloroacetic acid could be further chlorinated to obtain trichloroacetic acid. At the surface of the catalyst, dichloromethanol can be formed. In the end, ethyl, trichloroacetate and methyl trichloroacetate may be produced after reactions between trichloroacetic acid and pentachloroethanol or dichloromethanol.

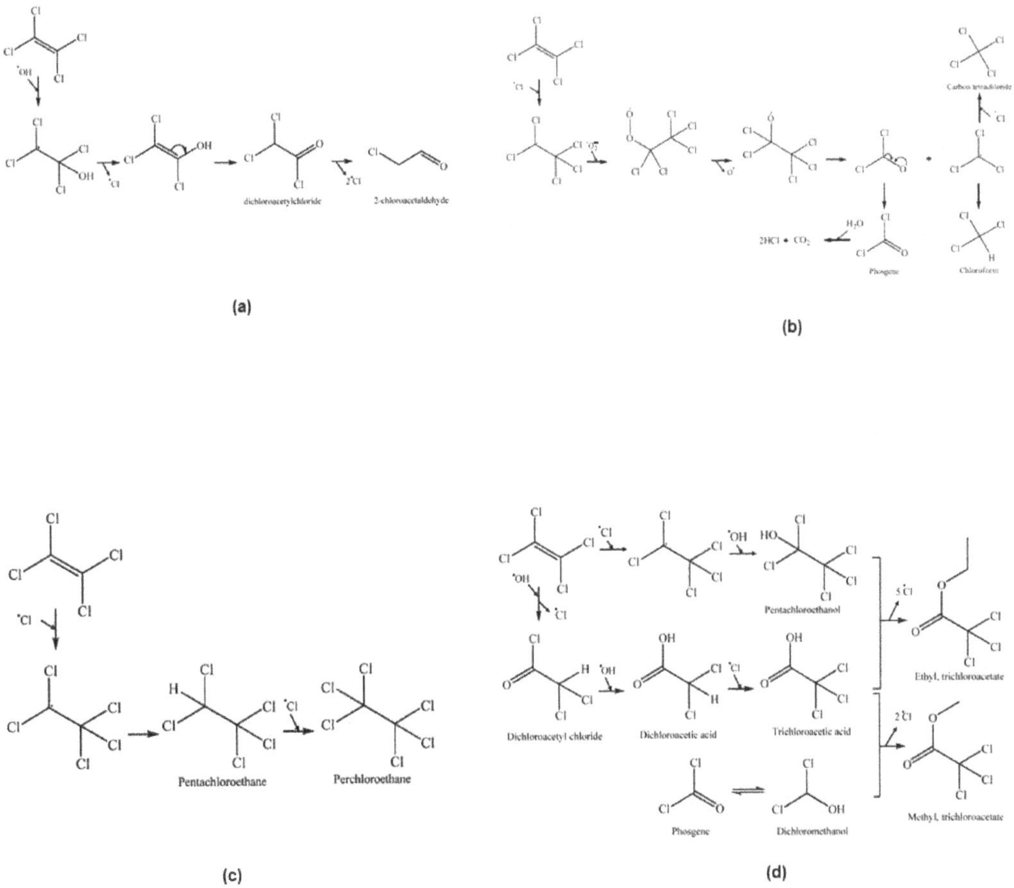


**Figure 4.** (**a**) •OH radical addition to PCE followed by •Cl radical generation; (**b**) •Cl addition to PCE followed by $O_2^-$ radical addition forming chloroalkanes and phosgene; (**c**) chlorination of PCE by addition of •Cl producing chloroalkanes; (**d**) esterification of ethyl, trichloroacetate and methyl, trichloroacetate. Reproduced with permission from ref. [52]. Copyright 2015 Elsevier.

Egerton et al. [53] presented their work regarding the influence of platinum on the UV (irradiation by a Philips PL-L 36W 09 actinic lamp, UV peak intensity at ~360 nm) photocatalytic degradation of dichloroacetate anion (DCA) by rutile and anatase forms of titania. They did the Pt deposition photochemically, with Pt(II) being more active than Pt(0). The activity of undoped rutile was lower than that of anatase, while after doping, the activity of rutile increased. In the end, both titania catalysts performed similarly. The authors observed that undoped Pt/anatase catalysts did not oxidize DCA when under visible light irradiation. On the other hand, Pt/rutile managed to oxidize DCA, being three times faster than experiments where Pt-free catalysts were used under UV irradiation.

Grzechulska-Damszel et al. [54] worked on a study regarding the photocatalytic decomposition of very low amounts (15 μg $dm^{-3}$, these concentration levels corresponding to those available in groundwater) of trichloroethylene (TCE) and tetrachloroethylene (PCE) in water using $TiO_2$. The authors performed various tests, including blank tests and tests without light irradiation and observed that low (when catalysts were used in dark mode) or no degradation occurred. For the photocatalytic tests, the authors used commercial titania and a commercial reaction system (Trojan UVMax, Trojan Technologies, London, ON, Canada, UV/Vis radiation in the range of 250–800 nm, with high maximum at 254, 436 and 546 nm) commonly used for water disinfection. The photocatalyst was used either as suspended particles or as an immobilized refill (titania was coated onto a glass fabric). When subjected to light irradiation (UV-Vis lamp), either TCE or PCE alone was removed at about 95% in under 90 min, while when a PCE/TCE mixture was tested, around 95% degradation was still obtained after 150 min. Therefore, the authors concluded that low amounts of TCE and PCE can be removed from the water when using a titania-based photocatalytic process and recommend the use of the immobilized system, being more beneficial from an economic point of view.

Suarez et al. [55] studied the photocatalytic activity of $TiO_2$/zeolite hybrids for VOC oxidation. Two UV-A lamps (8 W Philips and 6.5 mW $cm^{-2}$ irradiance) were used as the irradiation source. The authors synthesized titania with various morphologies (nanoparticles of almost 5 nm and decahedral anatase particles (DAP) having around 100 nm) and used commercial $TiO_2$ for comparison reasons. They obtained the catalytic hybrid by incorporating $TiO_2$ on ZSM5 using either freeze drying, incipient wetness impregnation or mechanical mixing. After characterization, they evaluated the photocatalytic performance of the studied catalysts on trichloroethylene photooxidation (and also formaldehyde) under UV-A irradiation and continuous airflow. The authors observed that the incorporation of titania, regardless of its type, into ZSM5 improved the photocatalytic performance up to 10 times compared to single titania particles. Even though DAPs had the highest VOCs reaction rate, it was concluded that $TiO_2$ nanoparticles homogeneously distributed on the zeolite material showed the highest VOCs photooxidation and $CO_2$ formation rates of the series (titania NP/ZSM5 > DAP/ZSM5 > commercial titania/ZSM5).

A study regarding the effect of photocatalysis on the degradation of trichloroethylene (TCE) in aqueous solutions using a photocatalyst-coated plastic optical fiber (POF) was done by Chen et al. [56]. They used $TiO_2$ and ZnO as photocatalysts and two diodes (LEDs) with low light intensity as the light source (40 mW $cm^{-2}$, 395 nm and 20 mW $cm^{-2}$, 365 nm). Parachlorobenzoic acid (pCBA) was utilized as a hydroxyl radical for the calculation of hydroxyl radical conversion rate (ROH, UV). After performing experimental tests, the authors observed that titania-coated POF was more efficient in degrading TCE in basic solutions, while ZnO-coated POF had better results in acid solutions. This was expected due to the fact that the mechanism of TCE removal by titania is photocatalysis, while by zinc oxide is adsorption. Another noticeable fact was that if the coating time increases, thus increasing the coating thickness, the degradation efficiency decrease. On the other hand, the enhancement of light intensity improved the photocatalytic efficiency. pH played an important role in the photocatalytic tests and was observed by the authors that if it is increased from 4 up to 10, the (ROH, UV) increases from $2 \times 10^3$ to $8 \times 10^3$ $M{\cdot}s{\cdot}cm^2{\cdot}mJ^{-1}$ for titania and from $8 \times 10^2$ to $2 \times 10^3$ $M{\cdot}s{\cdot}cm^2{\cdot}mJ^{-1}$ when zinc oxide was used.

State et al. [45] investigated the photocatalytic removal of TCE (trichloroethylene) under simulated solar irradiation (AM 1.5) using Au/$TiO_2$ and Pd-Au/$TiO_2$. The authors obtained the mentioned photocatalyst using incipient wetness and deposition precipitation methods. After performing the photocatalytic tests, it was observed that TCE was converted to more than 80% of $Cl^-$ and $CO_2$ no matter the catalyst used. Also, in order to obtain TCE conversion directly to $Cl^-$ and $C_2$ (ethane, ethylene) using the $H_2$ generated photo catalytically in situ, the authors performed the photocatalytic degradation of TCE over the same catalytic materials in the presence of methanol traces. Thus, it was evidenced that when Pd-Au/$TiO_2$ was used, hydro dechlorination (HDC) and photomineralization reactions of TCE took place simultaneously (Figure 5), while Au/$TiO_2$ was inactive. A reaction mechanism was proposed: the organic substrate was the source of protons (and thus of $H_2$) and carbon of $CO_2$. The •OH radicals supply the $O_2$ for the development of oxidized organic intermediates and, finally, of $CO_2$.

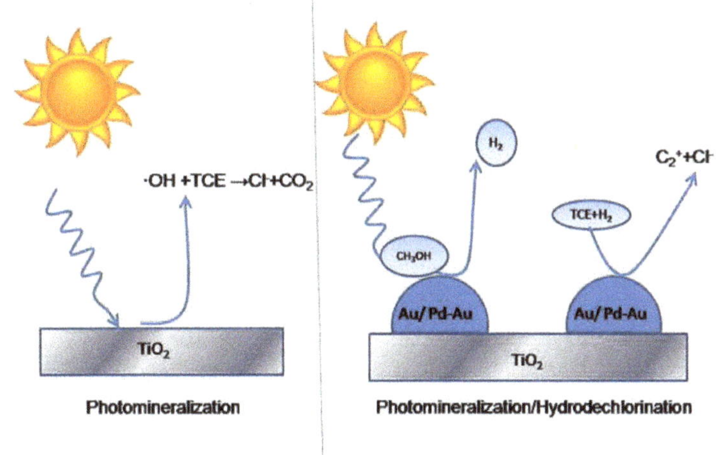


**Figure 5.** Photomineralization and photomineralization/hydro dechlorination mechanisms. Reproduced with permission from ref. [45]. Copyright 2017 Elsevier.

Hsu et al. [46] evaluated the possibility of using $LaFeO_3$ as a photocatalyst for trichloroethylene (TCE) degradation from the water via heterogeneous oxidation. The authors synthesized $LaFeO_3$ using the sol-gel method. TCE was chosen as an organic pollutant to be removed because it is one of the most encountered water pollutants in Taiwan. The authors' results indicated that up to 95% removal efficiency from water could be obtained when using 2 g per liter photocatalyst ($LaFeO_3$) and 1 h illumination from a Xe lamp up to 400 W. Other parameters were also studied, such as light intensity, catalyst loading, the influence of TCE concentration or pH contribution on TCE removal. They reported the highest energy efficiency to be 10.8 mg TCE/kWh. The TCE removal efficiency decreases if the initial TCE concentration increases, while the removal efficiency increases if the catalyst loading increases. Also, by increasing the pH value, the removal efficiency rises due to the addition of $OH^-$ groups. In the end, the authors revealed that the processes of adsorption, photodegradation and photocatalysis take place simultaneously and reach equilibrium after 1 h, and a removal mechanism had also been identified, suggesting that the overall removal efficiency reached 82% (20% accounting for adsorption and 39% for photolysis).

In the study of Raciulete et al. [48], the photocatalytic activity for TCE removal under simulated sunlight irradiation (using a 150 W short-arc Xe lamp (1000 $W \cdot m^{-2}$, Peccell-L01) over $RbLaTa_2O_7$ perovskites with mostly nanowire and platelet morphologies was investigated. Two RbLaTa-based layered perovskite samples were prepared via a solid-state synthesis route. The authors observed that the sample synthesized in mild conditions (e.g., 1200 °C for 18 h) favored the photo-mineralization of TCE to $Cl^-$ and $CO_2$ due to the presence of a high density of hydroxyl groups (Figure 6). Contrariwise, the activity of the sample annealed in harsh conditions (e.g., 950 and 1200 °C, for 36 h) remained modest for TCE removal, whereas its surface carbonate was beneficial for the formation of intermediate products. With the purpose of enhancing the overall photocatalytic performances of RbLaTa-based layered perovskites, the samples were subjected to protonation [47]. The strategy involved the slow replacement of the interlayer $Rb^+$ of $RbLaTa_2O_7$ hosts by $H^+$ via the cation exchange route. The authors showed that the obtained $HLaTa_2O_7$ protonated perovskites were able to photomineralize TCE under simulated solar irradiation. The enhanced activity of protonate perovskites was ascribed to favorable roles played by their increased specific surface area and high density of hydroxyl groups.

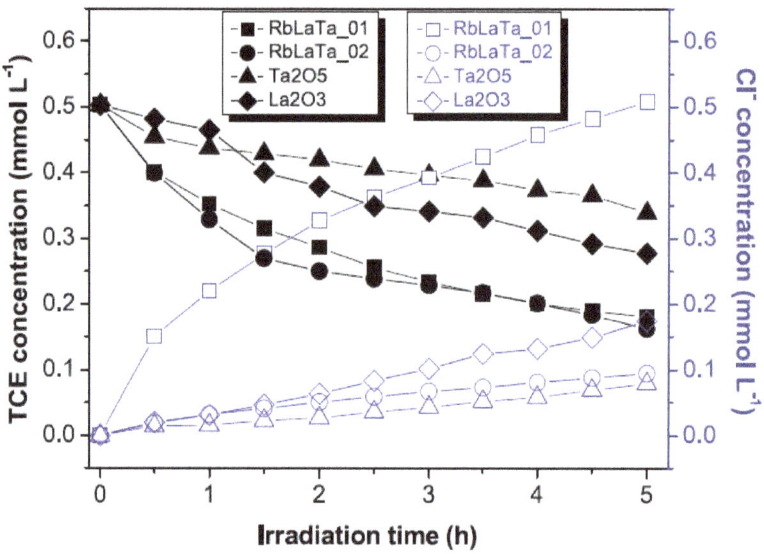


**Figure 6.** Time-evolution of TCE (closed symbols) and $Cl^-$ (open symbols) concentrations over RbLaTa_01(02) photocatalysts and precursors in the absence of methanol. Reaction conditions: initial concentration of TCE = 5 mg·$L^{-1}$; simulated solar light, T = 18 °C, mass of catalyst = 0.05 g, volume of water = 110 mL. Reproduced with permission from ref. [48]. Copyright 2019 Elsevier.

Photocatalytic degradation of various chlorinated environmental pollutants (VOCs), such as various chlorinated ethene and methane derivatives, in real groundwater samples, was studied by Dutschke and coworkers [57]. They developed an appropriate experimental setup in order to apply advanced oxidation processes (AOP) to real groundwater samples. Their setup used an $O_3$-bubble column reactor with a carrier-bound $TiO_2$/UV system (365 nm LEDs). The authors did a comprehensive study and discussed the influence of flow rate, $O_3$ concentration and radiation dose on the process performance. After parameter optimization (shown in Table 2) [57] using Box–Behnken experimental design, they obtained almost complete degradation rates for DCE: 99%, TCE: 99%, and PCE: 98%. A degradation rate of 85% was obtained for TCM (trichloromethane) without the formation of transformation products. The formation of tetrachloromethane (PCM) due to induced chlorination represented a problem during their photocatalytic experiments, but this was overcome by using suitable $O_3$ doses and irradiation in order to produce enough hydroxyl radicals, thus PCM will not appear as a transformation product during the degradation of other chlorinated organic pollutants.

**Table 2.** Optimized parameters for chlorinated VOCs degradation. Reproduced with permission from ref. [57]. Copyright 2022 Elsevier.

| Parameter | | Value | |
|---|---|---|---|
| Flow rate | | 30 | L $h^{-1}$ |
| Ozone concentration | | 1289 | mg $m^{-3}$ |
| Irradiance | | 33.6 | W $m^{-2}$ |
| VOC | | Degradation (SE) in % | |
| DCE | $C_2H_2Cl_2$ | 99.4 | (±0.3) |
| TCE | $C_2HCl_3$ | 99.3 | (±0.2) |
| PCE | $C_2Cl_4$ | 98.0 | (±0.2) |
| TCM | $CHCl_3$ | 84.8 | (±7.8) |
| PCM | $CCl_4$ | 14.0 | (±11.4) |

#### 2.2.3. Aromatic VOCs Abatement

Lyu et al. [43] tried to improve photocatalytic mineralization under UV irradiation (using four UV lamps at 8 W, 254 nm) of volatile organic compounds by developing a homojunction-adsorption layer on anatase $TiO_2$. Injection and separation of photogenerated charge carriers can improve the mineralization efficiency of VOCs. Therefore, they used toluene as model VOC, and they grew microporous $TiO_2$ onto the surface of anatase $TiO_2$ to obtain a homojunction-adsorption layer, thus optimizing the adsorption ability and photoactivity of the catalyst for photocatalytic mineralization of the model VOC (toluene). Various techniques were used to analyze the physical properties and mineralization of toluene. Results showed that the growth of microporous $TiO_2$ increased the surface area of the catalyst by 160% compared with anatase $TiO_2$, as well as induced microporous structure. This enhanced the adsorption of toluene at a low concentration of 5.5 mg $m^{-3}$ and limited the release of the adsorbed organic compounds. The authors also reported that a homojunction occurred at the interface of microporous $TiO_2$ and anatase $TiO_2$, leading to improved separation of photogenerated charge carriers. They concluded that due to this enhanced adsorption ability and photoactivity, the mineralization efficiency of toluene over micropore-anatase $TiO_2$ composite was 1.78 and 2.12 times higher than that over M-$TiO_2$ and anatase $TiO_2$, respectively.

Wang et al. [42] prepared the rGO-$TiO_2$ composites through a modified refluxing-solvothermal method and used them as catalysts for photocatalytic degradation of single VOC (p-xylene and ethylene) and VOC mixtures (benzene, toluene, p-xylene) under simulated solar irradiation (using 250 W and 500 W xenon lamps). The authors reported that their developed catalyst had higher photocatalytic degradation activity for a single VOC and even 2.6 times higher activity for a VOCs mixture compared to commercial P25 $TiO_2$. Also, the photodegradation efficiency of P25 decreased from 80% to 63.8%, while the synthesized catalyst kept its efficiency unchanged (around 93%). The improved performance of rGO-$TiO_2$ was attributed to an enhanced separation efficiency of electron-hole, better light harvesting ability and increased VOC adsorption capacity.

Ji et al. [58] studied the photocatalytic oxidation of gaseous benzene using as photocatalyst mesoporous $TiO_2$ prepared by one-step hydrolysis method and varying the calcination temperature. They observed that the calcination temperature interferes with the catalytic activity of synthesized titania, also affecting its structural properties. The authors also used commercial P25 titania for comparison reasons, but this had poor efficiency and deactivated quickly, while their obtained mesoporous titania had a much better stability and photocatalytic activity for benzene abatement. They reported that samples calcined at 400 °C had the best photocatalytic activity. Additionally, the authors used VUV (vacuum UV) irradiation and observed that it enhances benzene degradation, obtaining a removal efficiency of around 80%, while under UV irradiation (2 × 4 W, 254 nm, Cnlight) was just 10%. They concluded that this procedure, where VUV photolysis together with ozone generated from the VUV irradiation (2 × 4 W, Cnlight), is suitable for benzene degradation at room temperature if mesoporous titania is used.

Fiorenza et al. [59] developed ZnO-based photocatalysts for various VOCs mineralization. For the solar photocatalytic tests, they used an irradiation source, a solar lamp (Osram Ultra Vitalux 300 W, irradiance of 10.7 mW/$cm^2$). They synthesized core-shell ZnO@Au NPs and used them for gas-phase oxidation of toluene ($C_7H_8$) (but also of formaldehyde ($CH_2O$) and ethanol ($C_2H_5OH$)), obtaining 95% conversion for toluene and 85% conversion for formaldehyde resulting only in water and $CO_2$ as by-products. The same catalyst used in the photooxidation of ethanol also performed well, leading to a conversion of almost 60% having as by-products acetaldehyde with its subsequent oxidation to $CO_2$ (attaining up to 72% selectivity to $CO_2$). When performing the stability of the catalysts, the authors reported very good stability even after five consecutive runs. In the end, they concluded that the performance of the developed catalysts was due to the interaction between the zinc shell and gold core that prevented Au agglomeration, thus improving the photo-stability of the material and total VOC oxidation.

Nanodiamond (ND)-decorated zinc oxide photocatalysts were developed by Liu et al. [60] and used for the photooxidation of toluene utilizing UV-365 irradiation. They prepared various types of ZnO with different exposed crystal faces in order to evaluate the performance of the catalysts and the differences in activity. They observed that bare nanodiamond decorated ZnO having a higher amount of active (0001) exposed crystal faces obtained a total toluene removal in 2 h, the active species being superoxide radicals and photogenerated holes. After performing DRIFT analysis, the authors also explained a decomposition pathway for toluene; namely, toluene reacts with $\bullet O_2^-$ generating benzoic acid and benzaldehyde, and the benzoic acid is oxidized to oxalic acid and decomposed to carbon dioxide and water. In the end, the authors proposed a mechanism for toluene photodegradation by nanodiamond decorated zinc oxide catalyst (Figure 7). Thus, zinc oxide and nanodiamonds can generate $e^-$ and $h^+$ if excited under UV light. After obtaining the ND-ZnO, the electrons can pass from the ND to ZnO and can form an electric field. Afterward, the $h^+$ transfers from ZnO to ND and reacts to toluene. At the same time, $e^-$ from ND transfers to zinc oxide and interacts with $O_2$ forming $\bullet O_2^-$, thus mineralizing gaseous toluene and intermediates. As the valence band edge potential of the mentioned photocatalyst is more negative than the redox potential of $\bullet OH/H_2O$, there were no $\bullet OH$ radicals generated during the process. Therefore, the photo-generated $h^+$ could have contributed directly to the reaction of toluene oxidation.

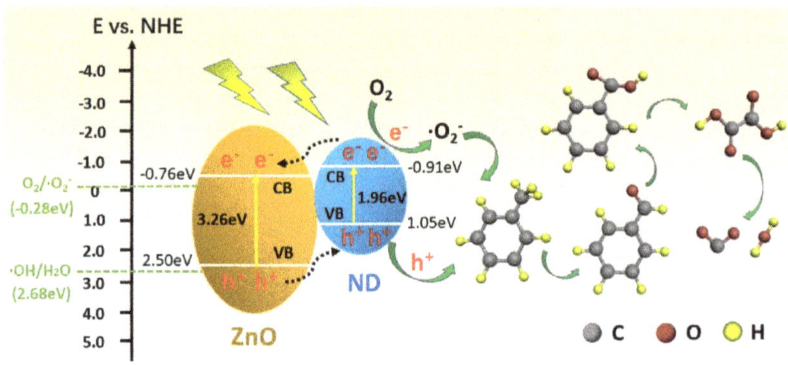


**Figure 7.** Proposed mechanism for photocatalytic degradation of toluene upon ND-decorated ZnO photocatalysts. Reproduced with permission from ref. [60]. Copyright 2019 Elsevier.

Zhang et al. [61] reported the preparation of Ag/ZnO/nBC photocatalyst consisting of zinc oxide, cellulose nanocrystal-derived nano biochar (nBC) and silver nanoparticles, which was used in the photodegradation of various volatile organic compounds using a 300 W xenon lamp. They observed that VOC molecules are adsorbed on the surface of the mentioned catalyst via oxygen-containing groups (–OH, –C=O, and –CO), thus obtaining high photocatalytic efficiencies for the degradation of methyl alcohol (92%), acetone (81%), formaldehyde (89%) and phenol (90%).

### *2.3. Aromatic Water Contaminants*

Nowadays, the removal of low-concentration organics recalcitrant for minimizing environmental and human health risks is of growing concern [62,63]. Among different aromatic substrates, phenol is a representative pollutant due to its presence in petroleum refinery wastewater affecting the ecosystems. It is often used as a model chemical for investigating the degradation of more complex and harmful compounds. Therefore, we will consider this pollutant focusing on its photocatalytic mineralization. This section of the paper outlines the general physicochemical properties of the phenol and its usages, describing the mechanism of the photocatalytic reaction. It also discusses and evaluates various types of photocatalysts and their performance in the photomineralization of phenol.

2.3.1. Phenol: General Properties and Uses

Phenol is a well-known contaminant introduced continuously into the environment through its usage in industry and human activities (Figure 8) [64]. Phenol is an aromatic compound, colorless to light-pink, with a burning taste and a distinctive odor. It is moderately soluble in water, soluble in most organic solvents, and less soluble in aliphatic hydrocarbons.

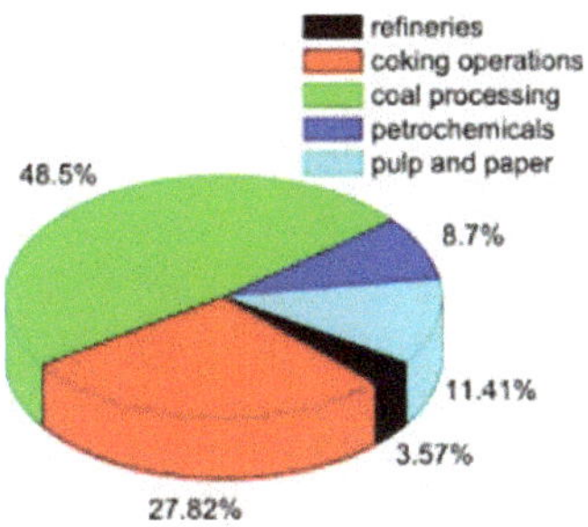


**Figure 8.** Phenol concentration in the effluent of major industries. Reproduced with permission from ref. [64]. Copyright 2019 Elsevier.

Phenols are produced naturally during the decomposition of plants and microorganisms, being released slowly in soils compared to other soil organic matter [65]. The oxidation of cumene (Hock process) is by far the dominant technology for the synthesis of phenol, with a production rate of more than $6.7 \times 10^6$ tons per year [66]. Other industrial processes include the toluene oxidation with benzoic acid as an intermediate product and the recovery of phenol from coal tar. Phenol is commonly used in different branches of industry comprising plastics and related materials, production of polycarbonates, epoxies, Bakelite, detergents, nylon, pesticides, explosives, dyes, textiles production, and pharmaceuticals [67]. Because of its impact on human health, the United States Environmental Protection Agency (USEPA) enlisted the phenol as a pollutant of priority concern. It is highly irritating to the skin and genotoxic, while repeated exposure causes weight reduction and infertility; it has endocrine-disrupting effects and can damage the liver, kidneys, and nervous system [68]. The World Health Organization (WHO) recommended a maximum permissible concentration of phenol in drinking water of 2 mg $L^{-1}$ [69]. However, even at very low concentrations, contamination of drinking water supplies represents a human threat, directly affecting the health of ecosystems.

2.3.2. Mechanism of Photocatalytic Removal of Organic Pollutants

Most authors agreed that the photocatalytic degradation of organic substrates (in this case, the phenol) follows Langmuir–Hinshelwood kinetics (Equation (38)) [70]:

$$r = -\frac{dC}{dt} = \frac{kKC}{1 + KC} \quad (38)$$

where $r$ represents the initial rate of reaction, $C$ is the concentration of the reactant, $t$ is the irradiation time, $k$ is the rate constant, and $K$ is the adsorption coefficient of the reactant. When the concentration of the pollutant is in the millimolar range, Equation (38) can be simplified to the apparent rate order equation (Equation (39)) [71]:

$$\ln \frac{C_0}{C} = kKt = K_{app}t \quad (39)$$

where C and $C_0$ are the concentrations of pollutants at time $t$ and $t_0$, respectively, and $K_{app}$ is the apparent-first order rate constant (in $\min^{-1}$). It is assumed that the reaction occurs on the surface of the photocatalyst. The mechanism of the photocatalytic reaction implies two steps: (i) the degradation of phenol to intermediates (e.g., hydroxylated and short-chain compounds) and (ii) the mineralization of the by-products to carbon dioxide and water [72].

Sobczynski et al. [73] admitted that during the UV irradiation of $TiO_2$, the phenol reacted with photogenerated holes ($h^+$) or photoinduced hydroxyl radicals, resulting in a variety of hydroxylated reaction by-products. The photocatalytic reaction proceeded via multiple steps, in which the principal intermediates were hydroquinone, catechol, and p-benzoquinone. Also, another variety of aliphatic intermediates formed in the reaction mixture, which finally converted to non-toxic end products (Figure 9).

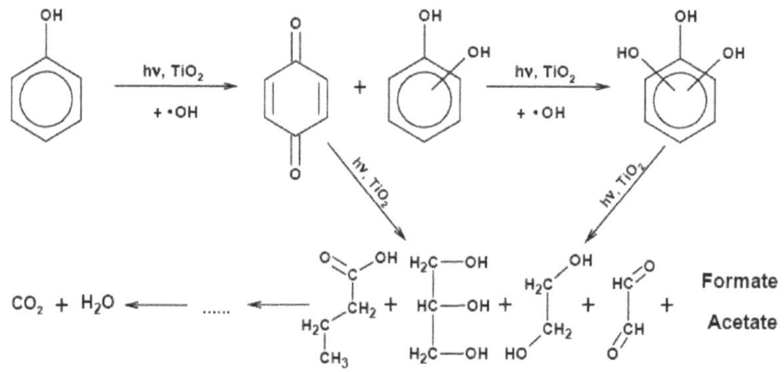


**Figure 9.** Possible mechanism of the phenol mineralization in the presence of $TiO_2$ under UV-light irradiation. Reproduced with permission from ref. [73]. Copyright 2004 Elsevier.

Guo and co-workers [74] focused on the identification of the intermediate produced by phenol photodegradation on $TiO_2$ using GC-MS and HPLC tools. They agreed that the main reaction intermediates were hydroquinone, resorcinol, catechol, 1,2,3-benzenetriol, (E)-2-butenedioic acid, 2-hydroxy-propaldehyde hydroxy-acetic acid, 3-hydroxy-propyl acid, and glycerol. The •OH radicals are highly reactive species that attack the phenyl ring of the phenol yielding to catechol, resorcinol, and hydroquinone. Further, the phenyl rings will break up to give malonic acid, followed by short-chain organic acids (e.g., maleic, oxalic, acetic, formic acids, glycolic acid), and finally, $CO_2$. Based on the detected by-products, the authors established that besides hydroxyl radicals, the •H was also a significant active free radical in the degradation pathways. During the photocatalytic process, $H^+$ or •H is scavenged by oxygen to form $HO_2$• radicals, which finally convert to hydroxyl radicals. However, these authors concluded that the principal reactive species responsible for organics degradation was •OH radical.

Wysocka et al. [75] carried out the photocatalytic degradation of phenol over Me-modified $TiO_2/SiO_2@Fe_3O_4$ nanocomposites (Me = Pd, Au, Pt, Cu) obtained by ultrasonic-assisted sol-gel method. In their study, the Pd- and Cu-$TiO_2/SiO_2@Fe_3O_4$ photocatalysts displayed the highest photo-oxidation rate of phenol and mineralization. •$O_2^-$ and •OH were the active species involved in the photodegradation process. These radicals attack the phenyl ring yielding catechol, hydroquinone, and benzoquinone generation, followed by oxalic acid and $CO_2$ formation (Figure 10). Conversely, different pathways occurred for the hydroquinone and catechol oxidation. Catechol was directly oxidized to oxalic acid and further mineralized to unharmful products. The hydroxylated by-products were oxidized to aliphatic carboxylic acids and finally to $CO_2$. The authors found that for Pt-$TiO_2/SiO_2@Fe_3O_4$ nanocomposite, a lack of catechol after 60 min of irradiation resulted in a low mineralization rate. It has been postulated that the enhanced photocatalytic activity of Pd- and Cu-modified photocatalysts were due to increases in the number of adsorption sites and efficient charge carrier separation. In the case of the Au-$TiO_2/SiO_2@Fe_3O_4$ sample, the keto-enol tautomeric equilibrium retarded the rate of the phenol mineralization.

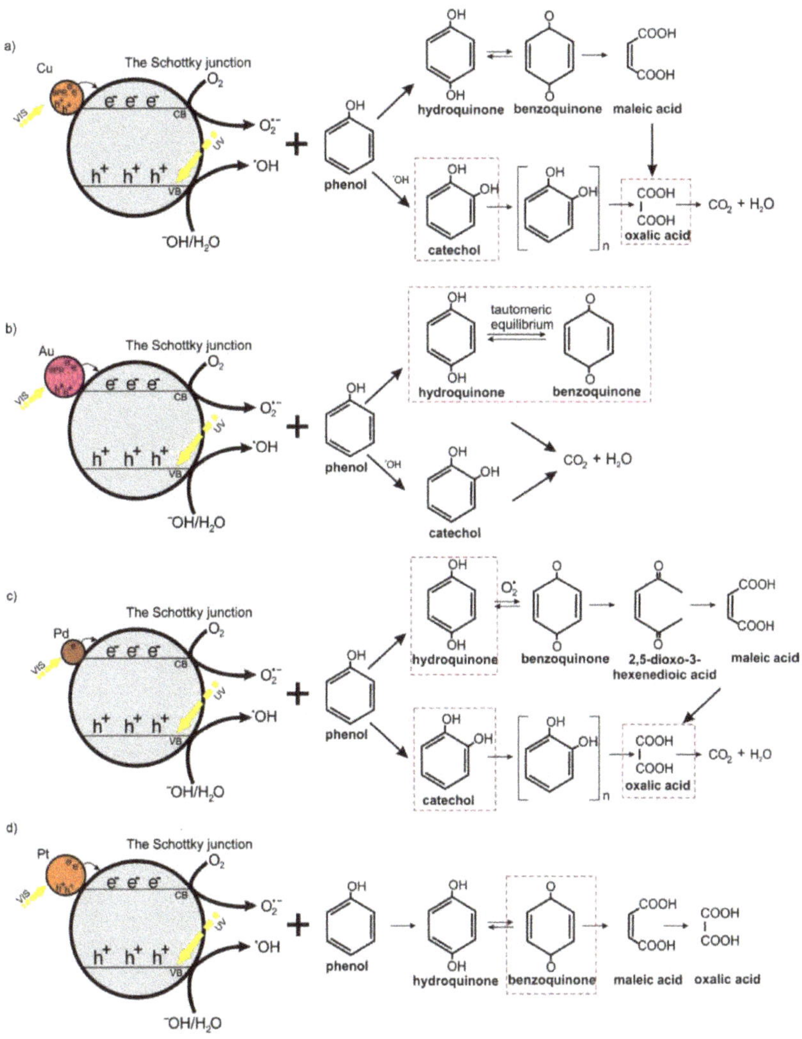


**Figure 10.** Schematic illustration of the phenol photocatalytic degradation mechanism for $TiO_2/SiO_2@Fe_3O_4$ magnetic catalysts modified with: (**a**) copper; (**b**) gold; (**c**) palladium, and (**d**) platinum nanoparticles. Reproduced with permission from ref. [75]. Copyright 2018 MDPI.

The photomineralization reaction depends on several factors, such as (i) the synthesis conditions (e.g., synthesis route, thermal treatment), (ii) the physicochemical properties of the catalyst (e.g., phase composition, morphology, particle size, surface area, porosity, band-gap energy), and (iii) the operational parameters (e.g., solution pH, initial concentration of the organic substrate, the mass of catalyst, wavelength, reaction temperature, radiant flux, and design of the reactor) [76].

### 2.3.3. Phenol Removal over Various Catalytic Materials

Titanium dioxide is by far a benchmark for numerous photocatalytic applications due to its high chemical stability, low cost, possibility to tune the band gap with other semiconductors, and biological inertness. However, photocatalysis on $TiO_2$ is still limited by its UV band wavelengths, weak adsorption of hydrophilic pollutants, and agglomeration of nano-sized particles. Therefore, it is a permanent rush to develop new light harvesters which aim to outperform titanium dioxide. The most common cleaning technologies are (i) biological (aerobic/anaerobic), (ii) physical (sedimentation, screening, filtration, floating, flocculation, and aeration), and (iii) chemical (neutralization, ozonation, precipitation, adsorption, and oxidation) processes.

A novel Z-scheme-based photocatalyst composed of $Bi_2O_3/Bi_2MoO_6$ heterojunction was proposed by Fu et al. [77] for efficient photodegradation of phenol with 96.4% degradation efficacy. In this system, a TOC removal efficiency of 75.5% was achieved. The spatially separated redox charge carriers, the excellent light harvesting capability, and the fast-charge transportation features of the catalysts were the factors determining their enhanced performances.

Zhang and co-workers [78] prepared a $Bi_2S_3/Bi_2WO_6$ composite by a hydrothermal method which efficiently weakens the recombination rate of photoinduced charge carriers and narrows the photoresponse range. The composite photocatalyst exhibited enhanced photocatalytic activity in the degradation of phenol under visible light irradiation, being 6.2 times higher compared to bare $Bi_2WO_6$. The photodegradation rate of phenol (Figure 11a) reached 51.6% in the presence of the $Bi_2S_3/Bi_2WO_6$ composite after 2 h of visible light irradiation, whereas only 12% of phenol was degraded by $Bi_2WO_6$. The photogenerated electrons (Figure 11b) were introduced from the conduction band (CB) of $Bi_2S_3$ to the CB of $Bi_2WO_6$ due to the intimate contact between the two semiconductors. Simultaneously, holes on the valence band (VB) of $Bi_2WO_6$ were transferred to that of $Bi_2S_3$ under the band energy potential difference. Therefore, efficient separation of the photoinduced $e^-/h^+$ pairs took place while their recombination was hindered.

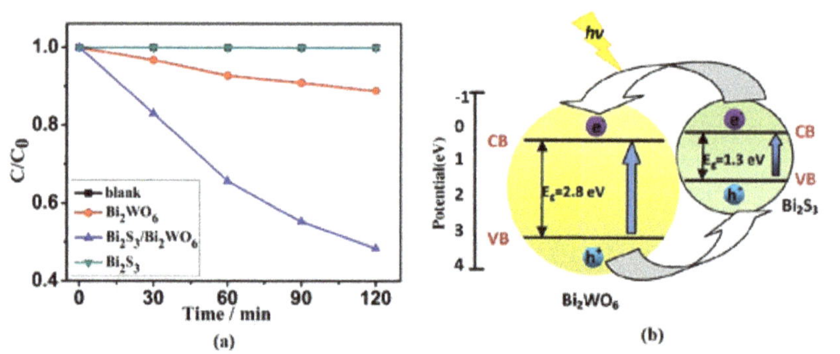


**Figure 11.** (**a**) Photocatalytic degradation of phenol under visible light irradiation as a function of time; and (**b**) Diagram for energy band levels of $Bi_2S_3/Bi_2WO_6$ composites and the possible charge separation process. Reproduced and adapted with permission from ref. [78]. Copyright 2012 American Chemical Society.

Different $BiFeO_3$ morphologies synthesized by three synthesis routes (e.g., co-precipitation CP, hydrothermal HT, sol-gel SG) were investigated by Chien et al. [79] for the photodegradation of phenol as a model organic pollutant. The SG-$BiFeO_3$ sample exhibited remarkable direct sunlight photocatalytic degradation of phenol (98.95%), superior to those of the HT-$BiFeO_3$ (77.4%) and CP-$BiFeO_3$ (66.9%) in 120 min. The radical scavenger studies implied that the photogenerated hole ($h^+$), hydrogen peroxide ($H_2O_2$) and hydroxyl (•OH) radicals were the dominant reactive species. Under direct solar irradiation, the photogenerated electron on the $BiFeO_3$ surface migrated from the filled VB to the CB band and left an equal number of holes in VB (Figure 12). The promoted CB electrons of $BiFeO_3$ (+0.53 eV) reacted with the oxygen molecules to generate $H_2O_2$ ($E^0(O_2/H_2O_2)$ = +0.695 eV). Simultaneously, the holes situated in the VB band of $BiFeO_3$ (+2.59 eV) would react with the $OH^-$ ($E^0$(•OH/$OH^-$) = +1.99 eV) to generate the •OH radicals. Subsequently, these radicals ($h^+$, $H_2O_2$ and •OH) react with the surface-adsorbed phenol, converting it into mineralized products.

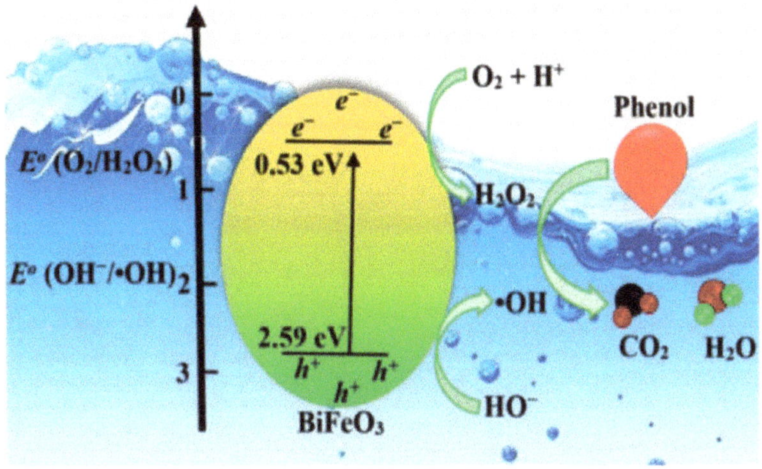


**Figure 12.** Mechanism of the phenol photocatalytic removal over $BiFeO_3$ under direct sunlight irradiation. Reproduced with permission from ref. [79]. Copyright 2022 Elsevier.

Jiang et al. [80] fabricated a series of BiOI-ZnO nanocomposites with various BiOI contents and tested their photoreactivity for phenol degradation under simulated solar irradiation. The phenol photodegradation rate reached 99.9% within 2 h, whereas only 40% of phenol removal took place over pristine ZnO. In the BiOI-ZnO system, the internal electric field formed between n-p heterojunctions of two oxide phases forced the electron and hole charge carriers to move in the opposite direction. Thus, the internal electric field between the component oxides facilitates the separation and transfer of the photocarriers. Since the CB of BiOI is much more negative than that of ZnO, the generated electron in BiOI favors the diffusion through into the CB of ZnO. Concomitantly, the photogenerated $h^+$ in VB of ZnO moves to p-type BiOI. As a result, more photogenerated carriers migrated to the catalyst surface, contributing to the reaction. Figure 13 illustrates the proposed mechanism for the enhanced removal of phenol by the BiOI/ZnO photocatalyst. In this research study, the superoxide species (•$O_2^-$) and holes were established as the main reactive species in the photocatalytic reaction.

Zhang et al. [81] take a look at the photocatalytic mineralization of phenol over a single $BiPO_4$ under UV-C irradiation. After studying the influence of several operating parameters, it was established that the mineralization of phenol was favorable in acidic conditions; the catalytic process decreased with increasing initial phenol concentration, and the chloride ions promoted the rate of mineralization. The $BiPO_4$ photocatalyst mineralized more than 95% of phenol (10 mg $L^{-1}$) after 5 h of illumination. Further after, Wang et al. [82] investigated the photodegradation of phenol under simulated solar irradiation of $CeO_2$, $Bi_4O_7$ and 10% $CeO_2/Bi_4O_7$ photocatalysts. The authors indicated that for the individual $CeO_2$ and $Bi_4O_7$, the phenol removal rates were only 12% and 40%, respectively. The 10% $CeO_2/Bi_4O_7$ photocatalyst degrades 92% phenol within 120 min, corresponding to the TOC value of 53%. Since the Fermi energy level of $CeO_2$ is higher than that of $Bi_4O_7$, the electrons in $CeO_2$ will be transferred to $Bi_4O_7$ until the Fermi levels of $Bi_4O_7$ and $CeO_2$ are equalized. The energy band of $Bi_4O_7$ bends downward, and the energy band of $CeO_2$ bends upward. Hence, a built-in electric field is formed at $Bi_4O_7/CeO_2$ interface. The electrons flow from $Bi_4O_7$ to $CeO_2$ and holes from $CeO_2$ to $Bi_4O_7$, forming thus a typical type II heterojunction. The trapping experiments of active species evidenced that $h^+$ and •$O_2^-$ played significant roles in phenol removal. However, the single $Bi_4O_7$ was almost deactivated, while the 10% $CeO_2/Bi_4O_7$ demonstrated improved stability after three cycling experiments.

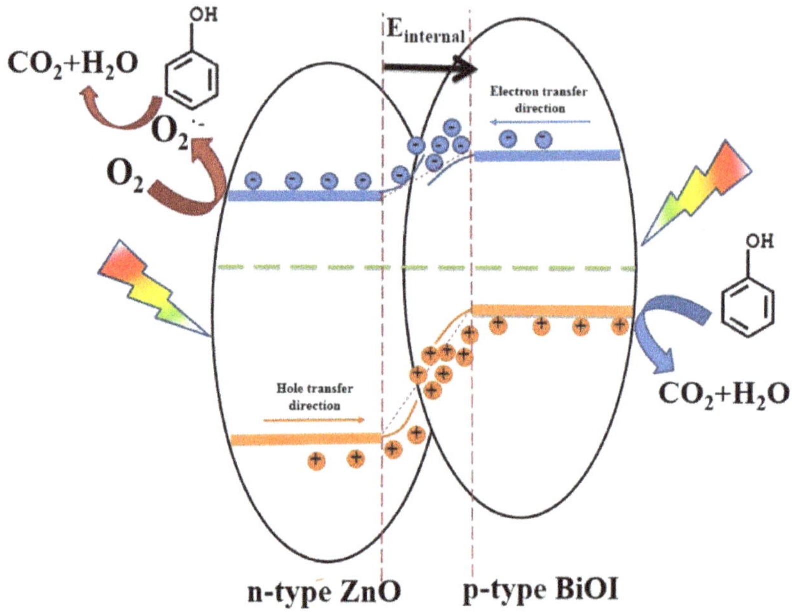


**Figure 13.** Photocatalytic pathway for the separation and transfer of the photogenerated carriers under simulated solar irradiation over BiOI/ZnO photocatalyst. Reproduced with permission from ref. [80]. Copyright 2017 Elsevier.

A series of ternary $Bi_7O_9I_3/g\text{-}C_3N_4/Bi_3O_4Cl$ photocatalysts were synthesized via the oil bath method by Yuan et al. [83] and tested for phenol photocatalytic removal. The optimal TOC removal rate reached up to 93.57% under visible irradiation within 160 min. After performing the trapping-species experiments and EPR characterization, the authors indicated that •OH and $•O_2^-$ were the oxidizing species responsible for the pollutant removal. The same study indicated that a dual S-scheme charge migration was generated at the interface of $Bi_7O_9I_3$, $g\text{-}C_3N_4$, and $Bi_3O_4Cl$, which favors efficient charge separation.

Table 3 summarizes various studies regarding the experimental conditions and the main reactive species participating in phenol photomineralization over some photocatalysts.

**Table 3.** Photocatalytic degradation of phenol over various materials.

| Type of Photocatalyst | Experimental Conditions | Light Source | Reaction Time | Degradation Efficiency | TOC Removal Efficiency | Main Active Species | Ref. |
|---|---|---|---|---|---|---|---|
| $TiO_2/MnO_2$ follow sphere | [phenol] = 5 mg $L^{-1}$ [photocatalyst] = 20 mg | Simulated solar light (XHA 300 W Xe lamp AM 1.5 G filter | 180 min | 100% | 91% | superoxide radical | [84] |
| 2% $Au/TiO_2$ | [phenol] = 30 mg $L^{-1}$ | Solar light | 3.5 h | 100% | - | - | [85] |
| Pr(0.072%)-$TiO_2$ | [phenol] = 50 ppm, [photocatalyst] = 1 g $L^{-1}$, pH = 6.58 | UV light | 2 h | 94.4% | - | - | [86] |
| EY-$TiO_2$/Pt(0.5%) | [phenol] = 40 ppm, [photocatalyst] = 0.8 g $L^{-1}$, 0.2 M TEOA, neutral pH | Visible solar light | 2 h | 100% | - | superoxyde radical | [87] |
| $TiO_2$-BiOBr-Bentonite | [phenol] = 20 ppm, [photocatalyst] = 150 mg/100 mL solution | Visible light | 70 min | 100% | 83% | superoxide radical and $e^-$ (major role) $h^+$, •OH (appreciable role) | [88] |
| $Fe_3O_4$@rGO@AgI | [phenol] = 50 ppm [photocatalyst] = 0.2 g/350 mL solution | UV-C light, λ = 254 nm | 9 h | 99% | - | mainly •OH | [89] |
| mont-La(6%)-$Cu_{0.6}Cd_{0.4}S$ nanocomposite | [phenol] = 20 mg $L^{-1}$ [photocatalyst] = 60 mg/L | Medium-pressure Hg-vapor lamp, near UV-Vis irradiation | 4 h | 86% | 77.8% | •OH and $h^+$ radicals | [90] |
| $TiO_2$/rGO | [phenol] = 20 ppm | UV light | 180 min | 97.87% | - | •OH, superoxide radical, and $h^+$ | [91] |
| $CuFe_2O_4$/rGO | [phenol] = 200 ppm [photocatalyst] = 60 mg/L | Visible light | 15 min | 100% | - | - | [92] |
| PAN-CNT/$TiO_2$-$NH_2$ | [phenol] = 10 ppm, pH = 5 [photocatalyst] =20 mg | UV light | 7 min | 99.2% | - | •OH and superoxide radical | [93] |
| CdS/rGO/$Fe^{2+}$ | [phenol] = 10 ppm [photocatalyst] = 20 mg/20 mL | 300 W Xe lamp irradiation (λ > 420 nm) | 60 min | 100% | 43.66% | •OH | [94] |
| N-$TiO_2$/HPCF (Hierarchical Porous Carbon Foam) | [phenol] = 30 mg $L^{-1}$ [photocatalyst] = 50 mg/50 mL aq. Ph | 500 W Xe lamp, simulated sunlight irradiation | 2 h | 97% | 78 % after 6 h of illumination | - | [95] |
| (2:1) Bi-$Bi_7O_9I_3$/Ag-AgI | [phenol] = 10 ppm [photocatalyst] = 50 mg/100 mL aq. Ph | 300 Xe lamp, λ > 420 nm | 120 min | 100% | 95.38% | •OH (more significant role than $h^+$) | [96] |
| Pt/$TiO_2$-ZnO@ZIF-8 | [phenol] = 5 ppm | UV light | 24 min | 99.7% | - | - | [97] |
| $CeO_2/TiO_2$ | [phenol] = 38 ppm [photocatalyst] = 20 mg/40 mL aq. Ph | 300 W high-pressure UV mercury lamp, UV irradiation | 120 min | 99.1% | - | •OH, $h^+$ | [98] |

Hydrothermal synthesis of BiOCl-activated carbon (AC) was reported by Sharma et al. [99] as an efficient photocatalyst for phenol removal. After 120 min of UV-light irradiation, more than half of the phenol photodegraded by the sample with 1 wt.% AC/BiOCl. The results showed a TOC value of 11.92 mg $L^{-1}$ for phenol mineralization. $SnO_2$:Sb nanoparticles (with 0.2%, 0.4%, and 0.6% concentration of Sb) turned out to be competent catalysts for phenol removal under UV and solar light irradiation [100]. The authors claimed that the change in the phenol concentration influenced the solution pH due to the formation of by-products during the reaction. The degree of mineralization reached 97% over 0.6% $SnO_2$:Sb nanoparticles within 120 min reaction time, while 71% and 45% values were achieved for 0.4% and 0.2% $SnO_2$:Sb samples, respectively.

Sandulescu et al. [101] showed a comprehensive view of the photocatalytic oxidation of phenol under sunlight irradiation over bare and noble metal-loaded $TiO_2$. Experiments indicated that the supported noble metals act as a visible light absorber, assisting the separation of photo-charges and reduction of $O_2$ to $O_2^-$. The $O_2^-$ oxidizes mildly phenol to oxygenated products. In a parallel process, •OH radicals yielded by $TiO_2$, mineralized phenol to $CO_2$ by fast reaction sequences.

Photocatalytic removal of phenol under UV light was investigated by Mendoza−Damian et al. [102] by studying the effect of $Sn^{4+}$ content on the $SnO_2$-ZnAl LDH photocatalytic properties. The 0.3 mol% of the $Sn^{4+}$−containing ZnAl LDH displayed the highest photocatalytic activity, with a phenol mineralization efficiency of 90.98%. The improved efficiency was due to a higher light absorption capacity and synergistic effect between the $SnO_2$ and ZnAl LDH heterostructure.

Raciulete et al. [103] developed a multi-step ion-exchange methodology by exchanging $Rb^+$ with $Cu^{2+}$ spacer in the layered $RbLaTa_2O_7$ host to achieve photocatalysts capable of wastewater depollution. The photocatalytic degradation of phenol under simulated solar irradiation, employed as a model reaction, showed that Cu−modified layered perovskites displayed an increased photocatalytic activity compared to the $RbLaTa_2O_7$ host. Experiments demonstrated that the product intermediates over Cu-modified perovskites were hydroquinone (HQ), 1,2-di-hydroxy-benzene (1,2-DhBZ), and benzoquinone (BQ). Among the Cu−modified layered perovskites, the sample reduced at 800 °C was the most effective photocatalyst regarding the efficiency of phenol mineralization, yielding 2.82 μmoles $h^{-1}$ of $CO_2$ and 1.78 μmoles $h^{-1}$ of $H_2$.

#### 2.3.4. Photocatalysis Coupled with Ozonation Process

Purification of wastewater, including phenolic compounds, has been realized by several treatment methods, such as photocatalytic degradation, electrochemical methods, adsorption, Fenton's reaction, and ozonation processes [104–106]. Ozone is a powerful oxidizing agent that destroys organic pollutants in wastewater by producing reactive oxygen species (ROS). Photocatalytic ozonation proved to be an efficient and promising advanced oxidation process available to remove widely spread organic contaminants in wastewater. The difference between photocatalytic ozonation and catalytic ozonation in an aqueous solution lies in the chain reactions initiation. The photochemical reaction is triggered by an electron transfer from a semiconductor to oxygen or ozone. The catalytic ozonation mechanism usually begins with the reaction between the hydroxyl anion with ozone [107]. In both processes, superoxide anion ($O_2^-$) is primarily formed and subsequently reacts with ozone to give ozonide ion ($O_3$•), consequently resulting in the formation of hydroxyl radical [108]. To a lesser extent, superoxide anion ($O_2^-$) can also act as an oxidant, ultimately leading to the mineralization of the organic compounds. Photocatalytic ozonation shows a synergetic effect since it can decrease the electro-hole recombination due to the great electron trapping of ozone, together with the interaction of $O_3$ with the superoxide radical. Both mechanisms are responsible for forming the ozonide radical, which further transforms into a hydroxyl radical, showing more powerful oxidant behavior [109]. Therefore, this technology can enhance the hydroxyl radical's generation, even at low pH, increasing the mineralization rate. The application of photocatalytic ozonation on wastewater treatment is expected to be more effective than photocatalysis and/or ozonation technologies alone. During the ozonation process, the resulting hydroxyl radical in the water phase reacts with organic contaminants leading to their mineralization [110].

Very recently, Yu et al. [111] synthesized single-crystal $WO_3$ nanosheets (NSs) by a hydrothermal method and checked their photocatalytic activity for phenol photomineralization under visible light. The excellent performances of $WO_3$ NSs were attributed to their lamellar morphology with single-crystal microstructure and good dispersion, providing continuous interior channels for the charge carrier transportation from the bulk to the surface of $WO_3$ nanosheets. The authors investigated the degradation efficiency ($R_D$) and mineralization ratio ($R_m$) of phenol under different systems (Figure 14a), including ozonation alone ($O_3$), ozonation combined with visible light exposure in the absence of photocatalyst (Vis/$O_3$), catalytic ozonation in the presence of photocatalyst ($WO_3$ NSs/$O_3$ and $WO_3$ NPs/$O_3$), photocatalysis ($WO_3$ NSs/Vis) and photocatalytic reaction conducted in the presence of ozone ($WO_3$ NSs/Vis/$O_3$ and $WO_3$ NPs/Vis/$O_3$). They found that the mineralization ratio for $WO_3$ NSs/Vis/$O_3$ reached 96% after 150 min, and it rose continuously to 98% at 240 min, while the $R_m$ was 83% for $WO_3$ NPs/Vis/$O_3$ at 240 min. After $WO_3$ nanoparticles (NPs) and nanosheets exposure to visible light, the electrons and holes were generated (Figure 14b). These photogenerated carriers were separated and transferred from the bulk to the surface of $WO_3$ and reacted with $O_3$ and $HO^-/H_2O$ to produce hydroxyl radicals. The produced hydroxyl radicals attack the intermediates of phenol degradation. Subsequently, the complete mineralization of phenol occurred.

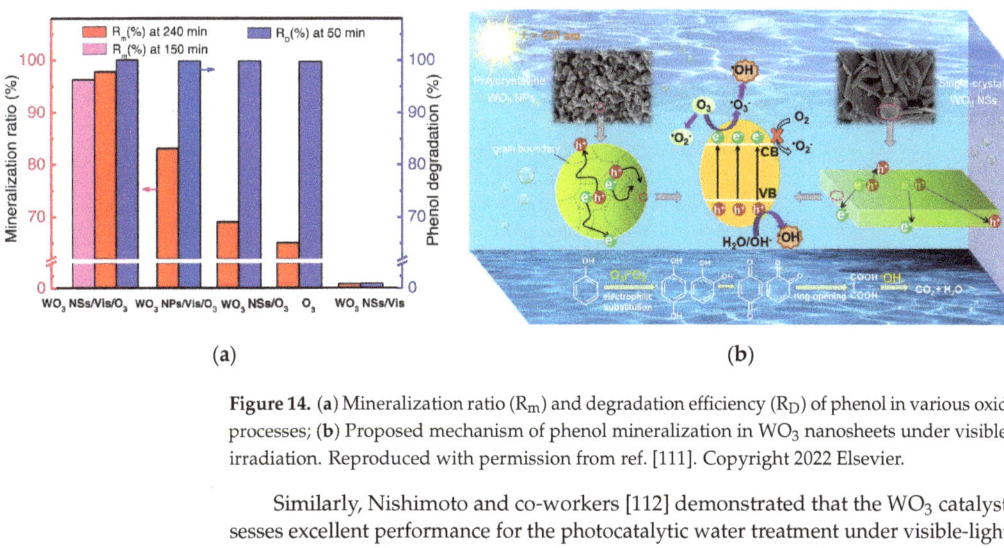


(a) (b)

**Figure 14.** (**a**) Mineralization ratio ($R_m$) and degradation efficiency ($R_D$) of phenol in various oxidation processes; (**b**) Proposed mechanism of phenol mineralization in $WO_3$ nanosheets under visible light irradiation. Reproduced with permission from ref. [111]. Copyright 2022 Elsevier.

Similarly, Nishimoto and co-workers [112] demonstrated that the $WO_3$ catalyst possesses excellent performance for the photocatalytic water treatment under visible-light irradiation combined with ozonation. The authors employed two different catalysts (e.g., $WO_3$ and N-doped $TiO_2$), comparing their capability for TOC removal. Bare $WO_3$ exhibited a superior response for the photomineralization of phenol in the presence of ozone, which readily reacted with its photogenerated electrons in the conduction band. Tawabini and Zubair [113] presented a combined UV and ozone process for phenol removal while inhibiting the formation of bromate in water. Photolysis by UV partially degrades the pollutant. Although combining the UV/$O_3$ techniques, total removal of 50 ppm of phenol in less than 5 min occurs. The authors observed that after the optimization of the operational parameters (e.g., continuous ozonation rate of 1 L/min, addition of 1.5 ppm ammonia for adjusting the pH), the bromate formation was diminished drastically to non-detected levels. In the same way, by coupling catalytic ozonation with photocatalysis, nearly 100% degradation performance for phenol removal within 2 min was obtained over MgO/g-$C_3N_4$ catalysts by An et al. [114]. The operational conditions were a visible-light source equipped with a

300 W xenon vertical irradiation, the concentration of the pollutant was 30 mg $L^{-1}$, and a reactor volume of 250 mL. For the developed photocatalyst, the MgO played a dual role: (i) accelerating the photogenerated charges separation of g-$C_3N_4$ and (ii) facilitating the conversion of ozone into •OH, thus enhancing the catalytic ozonation process.

### 2.4. *Aqueous Inorganic Nitrogen-Based Pollutants*

This section has as its main objective the presentation of the mechanisms underlying the photocatalytic degradation reaction of nitrates ions ($NO_3^-$) and the factors that lead to obtaining final products that do not affect the environment and life in general. In addition to the traditional physical treatment (adsorption, membrane filtration, ion exchange, reverse osmosis) and (electro)chemical treatment (conventional chemical denitrification, zero-valent metal nanoparticles, hydrogen driven catalytic denitrification, electrochemical reduction, electrocoagulation, electrodialysis) processes for nitrate removal [115,116], investigations related to nitrate controlling have been carried out in recent years on the possibility of using photocatalytic processes [117].

#### 2.4.1. Overview of Nitrate Anion and Its Reaction Intermediates (Nitrite and Ammonium Ions)

Photocatalytic reduction of nitrate to molecular nitrogen presents a promising approach to removing nitrate from drinking water sources. However, unwanted products such as nitrite ($NO_2^-$) and ammonium ($NH_4^+$) result from the nitrate reduction reaction. The values of these unwanted products have been regulated over the years. The standard recommended levels of nitrates, nitrite and ammonium concentration in drinking water, 50 ppm $NO_3^-$, 3 ppm $NO_2^-$ and 0.5 ppm $NH_4^+$, respectively, are introduced by the World Health Organization (WHO) [118]. To reduce the recombination of photo-generated electrons and holes, metal loading or holes scavengers are generally needed, causing, in this way, secondary pollution of drinking water [119].

Continuous efforts have been devoted to the achievement of heterogeneous photocatalytic nitrate reduction, which is potentially a green and low-cost operation; however, achievements are not as much as expected. In general, photocatalytic nitrate reduction includes two successive reactions, from nitrate to nitrite and then from nitrite to nitrogen radical. The nitrogen radicals will further combine to form $N_2$. Sometimes ammonia ($NH_3$) may be generated as the over-reduced product. The unwanted $NH_3$ is more toxic than $NO_3^-$, and the problem is that the use of the majority of traditional photocatalysts cannot avoid the formation of $NH_3$, leading to a low selectivity to $N_2$ [120–130]. If the formation of undesired by-products (e.g., $NO_2^-$ and $NH_4^+$) is prevented, photocatalytic technology could be one of the most promising options to solve the problems of environmental pollution and energy shortage due to its advantages such as adaptability, low cost, and no secondary pollutants [131]. Dark catalytic hydrogenation of nitrate has been considered a promising alternative to overcome the economic and environmental disadvantages of separation technologies without generating waste streams [132]. $N_2$ selectivity and water matrix effects are important factors that should be assessed for any technology based on catalytic and/or photocatalytic processes proposed for nitrate removal [133,134].

#### 2.4.2. Efficient Nitrate Removal from Wastewater over Different Materials

Photocatalytic nitrate reduction is one of the emerging transformative technologies capable of yielding harmless gaseous products. The $e_{CB}^-/h_{VB}^+$ pair recombination is the main drawback of photocatalytic processes and affects their overall efficiency. The strategies for improving the charge carrier separation have already been reported and discussed thoroughly in several reviews [117,135–138]. It is widely accepted that $NO_2^-$ is the first stable intermediate product obtained from nitrate reduction, and it can remain in solution as $NO_2^-$ or undergo further reduction to $N_2$ or to $NH_4^+$. Some authors have been unable to detect quantifiable amounts of $NO_2^-$ at the end of the photocatalytic reduction of $NO_3^-$ because it can be easily reduced after its formation [139–143]. Because of its faster reduction, several works study the direct reduction of $NO_2^-$. The reduction of nitrite is the divergent point that defines the selectivity towards harmless $N_2$ or undesired $NH_4^+$.

Few studies quantify the direct yield towards $N_2$. The quantification of $N_2$ by gas chromatography was reported by Kominami et al. [141] and by Zhang et al. [144]. Experimental work by Zhang et al. [144] concluded that $N_2$ was the only gas product released using an Ag/$TiO_2$ photocatalyst. $N_2O$ has been identified as an intermediate released in other reductive treatments, such as hydrogenation [145–147]. Even though $N_2$ is an inert species, other nitrogen-containing gas species such as $N_2O$, NO, and other $N_xO_y$ are hazardous species with high environmental implications in atmospheric chemistry [148,149]. The last major product obtained during $NO_3^-$ and $NO_2^-$ reduction is the ammonium cation, released from an alternative pathway to the one of the HNO intermediate. The selectivity of nitrate depends on the ratio of surface coverage of N species to reductant species. A low coverage or high concentration of reducing mediators could deteriorate the selectivity for the formation of $NH_4^+$. An appreciable pseudo-concentration of adsorbed nitrogen intermediates, mainly HNO and NO•, would favor the pathway leading to N−gas species. Many reactions are highly pH−dependent; therefore, acidic pH is necessary to ensure sufficient $H^+$ to allow fast kinetic rates. Therefore, the pH dependence is not only related to the surface charge that modulates the adsorption of species on the photocatalyst surface but also to the $H^+$ source to ensure the complete reduction.

Doped semiconductor photocatalysts can provide higher conversion of nitrate and selectivity to nitrogen gases than pristine $TiO_2$, but results are comparable to composite catalysts, raising the question of whether interstitial/substitutional doping is necessary compared to photo deposition methodologies [150]. An excessive metal loading in the composite can be detrimental, becoming a recombination center instead of the desired electron sink [151]. Optimizing metal loads to about 1.0% *w/w* typically minimizes this detrimental effect [152]. Monometallic composites are the most prevalent types reported in the literature. The metals' performance was also related to the intrinsic capability of each platinoid to stabilize $H_{ads}$ because materials with higher overpotential for $H_2$ evolution presented a predominant yield of $NH_4^+$. The most influential factors affecting the efficiency of proton reduction on the metal surface are (i) the hydrogen overpotential during water splitting and (ii) the $H_{ads}$ stabilization [153]. In support of this hypothesis, Hamanoi et al. [154] proved experimentally that a decrease in $NO_3^-$ conversion is observed when hydrogen evolution is increased. Furthermore, bubbling $H_2$ enhances the reduction of $NO_3^-$ to $NH_4^+$, demonstrating that the adsorption of $H_2$ on platinoids surface catalytic sites as $H_{ads}$ contributes to the reduction process [117,127].

Photocatalysts based on pristine $TiO_2$ and related mono/bimetallic composites or bimetallic systems are intensively studied [155–158], but conventional metal-modified photocatalysts usually suffer from metal leaching, aggregation, and gradual deactivation and need to be significantly improved in terms of $N_2$ selectivity. New materials were developed or used for the first time for photocatalytic denitration (perovskite-based photocatalysts, layered double hydroxides (LDHs) with hydrotalcite-like structures, nonlinear optical material $LiNbO_3$) [159].

It is well known that photocatalytic oxidation has been investigated extensively for its capability of producing highly oxidative •OH, but little attention has been paid to the photocatalytic reduction of oxidative pollutants such as nitrate in water. Photocatalytic denitrification appeared as a feasible approach to accomplish this aim since it was first reported by Schlögl and co−workers in 1999 [160]. During the photocatalytic denitrification process activated by light irradiation, the photocatalyst generates electrons ($e_{CB}^-$) in the conduction band (CB) and holes ($h_{VB}^+$) in the valence band (VB) of the semiconductor. Then, the nitrate is reduced through direct interaction with $e_{CB}^-$ or reaction with reductive $CO_2$•$^-$ radicals produced from the reaction between $h_{VB}^+$ and hole scavengers (e.g., formic acid) [161–163]. According to literature, the latter mechanism generally rules the photocatalytic denitrification for several materials like conventional $TiO_2$, ZnO, ZnS, CdS, and $SrTiO_3$ [128,164–168]. It is difficult to control the formation of $CO_2$•$^-$ radicals due to the dependence on the used hole scavengers. Liu et al. [169] reported the photocatalytic denitrification by nonlinear optical (NLO) material, i. e. lithium niobate ($LiNbO_3$) in the presence of formic acid (FA) serving as a hole scavenger. A 110 W high−pressure Hg lamp was employed as a 365 nm UV light source. $LiNbO_3$ achieved 98.4% total nitrate removal and 95.8% $N_2$ selectivity under neutral pH conditions. During the process, the nitrate may be reduced by (i) reductive $CO_2\bullet^-$ radicals produced from the reaction between $h_{VB}^+/\bullet OH$ and hole scavengers, (ii) electrons generated at CB, as well as (iii) hydrogen produced from water splitting at CB. They concluded that photocatalytic denitrification should be dominated by reactions involving the conduction band (CB) either through interaction with electrons or hydrogen produced from water splitting. They showed that the role of $H_2$ is very limited, and more than 98% of $NO_3^-$ is reduced directly by electrons at the conduction band of $LiNbO_3$.

Photocatalytic denitration using various non-toxic hole scavengers is the most common technique reported in the literature. The photocatalytic degradation of nitrates in an aqueous solution has been examined by Anderson and co-workers [118], using different $Au/TiO_2$ photocatalysts and oxalic acid as a hole scavenger. It has been shown that oxalic acid and nitrate can be simultaneously degraded over $Au/TiO_2$ to produce predominantly $CO_2$ and nitrogen, but complete nitrate removal was not achieved.

Luiz et al. [139] studied $TiO_2$ and $TiO_2$ doped with $Zn^{2+}$, $Cu^{2+}$ and $Cr^{3+}$ (metal doped-$TiO_2$ (Cu-$TiO_2$, Cr-$TiO_2$ and Zn-$TiO_2$).) The prepared materials were used as photocatalysts to reduce nitrate and oxidize formic acid under the irradiation of a low-pressure mercury lamp (UV radiation at 254 nm, output power of 17 W). The results obtained from the nitrate photoreduction experiments indicated that the metal-doped $TiO_2$ activity decreases in the order: 4.4% Zn-$TiO_2$ > 4.4% Cu-$TiO_2$ > 4.4% Cr-$TiO_2$. Zn-$TiO_2$ exhibits the greatest selectivity towards $N_2$ (95.5%), a nitrate conversion up to 92.7%, and a high reaction rate (14.2 μmol $NO_3^-$ (min $g_{catalyst})^{-1}$).

Doudrick et al. [128] examined the photocatalytic reduction of nitrate in water using titanium dioxide (Evonik P90) loaded with silver nanoparticles and formate as a hole scavenger (electron donor). Photocatalytic experiments were performed using a UV light source (450-W medium-pressure mercury–vapor lamp). Under acidic conditions (pH = 2.5), nitrogen gases (~85%) and ammonium (~15%) were the final by-products. The authors evidenced that radicals are unlikely to be responsible for nitrate reduction, so a photocatalyst with the proper Fermi level must be selected to meet the thermodynamic requirements. Because the pH was a factor in their experiments, proton localization at the reaction sites was important for treatment at ambient pH and for achieving harmless by-products, which can be accomplished by selecting the proper co-catalysts (e.g., Ag, Cu). Although photocatalysis is not fully suitable for drinking water applications yet, P90/Ag removes nitrate efficiently and with high selectivity.

There are some reports presenting the photocatalytic reduction of $NO_3^-$ in the absence of sacrificial agents, but their activity is not sufficiently satisfactory [130,157,170–180]. Wei et al. [181] aimed to develop photocatalysts for the chemical reduction of $NO_3^-$ in visible light (fluorescent lamps irradiated with a power intensity of 2.64 mW/cm$^2$, λ = 419 nm) and in the absence of sacrificial agents. The use of $Ni_2P$ as a potential base-metal alternative to precious metals as a catalyst for the hydrogenation of $NO_3^-$ under mild, near-ambient conditions (1 atm, 60 °C) has been demonstrated [182]. This potential catalyst exhibited complete $NO_3^-$ reduction with very high selectivity for ammonia ($NH_3$) [182]. Considering that the light source and the photocatalyst are two key factors in the photocatalytic reduction of $NO_3^-$, Wei et al. [181] synthesized $Ni_2P$/semiconductors ($Ni_2P/Ta_3N_5$, $Ni_2P$/TaON, and $Ni_2P/TiO_2$) and used these heterostructures as photocatalysts for the reduction of $NO_3^-$ in water (Figure 15a,b).

Starting with a 2 mM (28 g/mL $NO_3^-$-N) solution at pH 2, $Ni_2P/Ta_3N_5$ and $Ni_2P$/TaON achieve 79% and 61% $NO_3^-$ conversion, respectively, and conversion rates of 196 μmol $g^{-1}$ $h^{-1}$ and 153 μmol $g^{-1}$ $h^{-1}$, respectively, after 12 h under 419 nm irradiation. Control experiments confirmed that $Ni_2P$/semiconductor heterostructures and light illumination are requisites for the photocatalytic reduction of $NO_3^-$. Based on these findings, Wei et al. proposed two possible electron migration pathways and assumed that the dominant pathway in these heterostructures is light absorption by the semiconductor followed by electron injection into $Ni_2P$ (Figure 15c,d) [181].

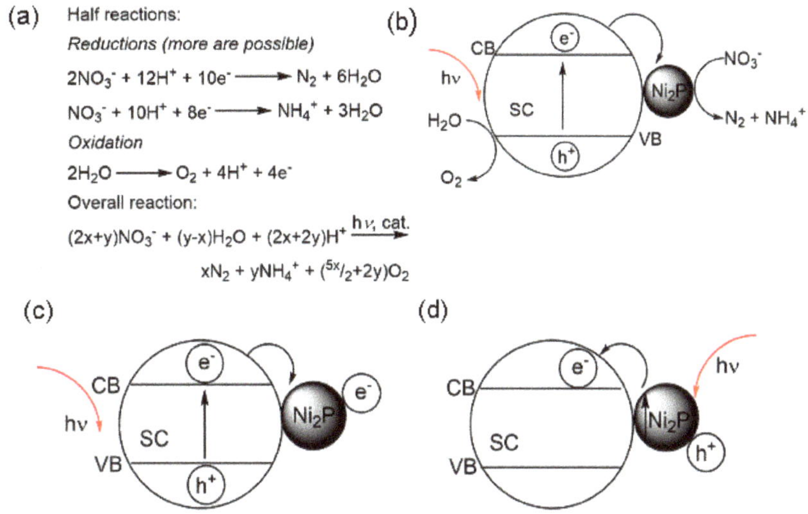


**Figure 15.** (**a**) Half and overall reactions for photocatalytic reduction of $NO_3^-$ to the most desirable product(s) $N_2$ and/or $NH_3$. (**b**) Mechanistic scheme showing energy flow during the photocatalytic reduction of $NO_3^-$ over a $Ni_2P$-modified semiconductor. (**c**,**d**) Schematic illustrations of two possible charge separation pathways during photocatalysis. In mechanism (**a**), light absorption by the semiconductor results in photo-generated electrons and holes, with the electrons getting trapped by $Ni_2P$. The resulting $Ni_2P$ Fermi level upshift leads to a higher driving force for $NO_3^-$ reduction. In mechanism (**b**), photogenerated "hot" electrons from $Ni_2P$ are injected into the semiconductor. Reproduced and adapted with permission from ref. [181]. Copyright 2020 John Wiley and Sons.

Silveira and co-workers [134] presented the promising use of $FeTiO_3$ and oxalic acid as reducing agents for the selective photo-reduction of nitrate to $N_2$. They studied the feasibility of using natural ilmenite as a catalyst for $NO_3^-$ photo-reduction with oxalic acid as a reducing agent. The generation of $NO_{x(g)}$ via $NO_3^-$ and $NO_2^-$ reduction is also observed. The complete $NO_3^-$ and $C_2O_4^{2-}$ removal and a selectivity towards $N_2 > 93\%$ was achieved by using the stoichiometric $C_2O_4^{2-}$ amount after 210 min, without the generation of undesirable $NH_4^+$.

Formic acid is known as one of the most efficient hole scavengers for nitrate reduction. The deep reduction to $N_2$ is quite difficult because the process requires a significantly high density of electrons at the catalytic sites. Yue et al. [120] proposed that the reactions occur on the surface of the catalyst particles, as shown in Figure 16.

The authors [120] systematically investigated the performance of $CuInS_2$ in photocatalytic nitrate reduction under visible light irradiation by loading co-catalysts. A 300 W Xe lamp was used to provide visible light irradiation. Band-pass or cut-off filters were applied to obtain monochromatic beam light ($\lambda$ = 400, 450, 500, 550, 600, or 650 nm) or pure visible light ($\lambda > 400$ nm), respectively. In particular, with the assistance of the LSPR effect of Au, the high record of the nitrate conversion rate of 8.32 mg N $h^{-1}$ was achieved under pure visible light. Overall, $CuInS_2$ holds high potential in the application of photocatalytic nitrate removal under solar irradiation. Yue et al. [120] advanced the idea that the reaction mechanism takes place via adsorption–reduction reactions where nitrate ions are reduced directly by photo-generated $e^-$. This mechanism is supported by the fact that the introduction of additional halide anions in an aqueous solution reduces the photocatalytic efficiency due to the competition between the adsorption of ion species. Yue and co-workers [120] The efficiency of solid formate as a hole scavenger was evaluated. It was shown that both glucose and sucrose appear to be effective, with only a slight decrease in photocatalytic efficiency. On the contrary, harmful molecules (including benzene, phenol, and benzoic acid) and other typical $h^+$ scavenger agents (such as methanol and ethylene glycol) were also applicable with the rationale of simultaneously decomposing two pollutants.

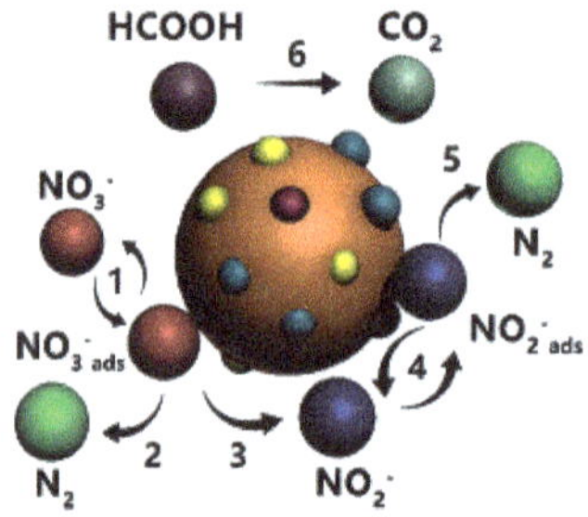


$$NO_3^-(aq) \leftrightarrow NO_3^-(ads) \quad (1)$$

$$2NO_3^-(ads) + 10e^- + 12H^+(aq) \rightarrow N_2(g) + 6H_2O(l) \quad (2)$$

$$NO_3^-(ads) + 2e^- + 2H^+(aq) \rightarrow NO_2^-(aq) + H_2O(l) \quad (3)$$

$$NO_2^-(aq) \leftrightarrow NO_2^-(ads) \quad (4)$$

$$2NO_2^-(ads) + 6e^- + 8H^+(aq) \rightarrow N_2(g) + 4H_2O(l) \quad (5)$$

$$HCOOH + 2h^+ \rightarrow CO_2(g) + 2H^+(aq) \quad (6)$$

**Figure 16.** Photocatalytic nitrate reduction on $CuInS_2$ loaded with co-catalysts in the presence of sacrificial agents. The abbreviations (aq), (ads), and (g) mean ions in an aqueous solution, adsorbed on the surface and in the gas form, respectively. Reproduced with permission from ref. [120]. Copyright 2016 Royal Society of Chemistry.

Titania ($TiO_2$) and metal-loaded titania using Pt [123,151,165], Pd [141,151,165], Rh [151], Ru [151], Au [118,121], and Ag [127,128,144] are widely used and effective in the reduction of nitrate with high selectivity toward $N_2$ [125,127,128,144,165].

Zhang et al. [144] obtained high conversion (98%) and almost 100% selectivity for nitrogen for nitrate photocatalytic reduction by using as catalyst nontoxic fine Ag clusters obtained by photo-deposition of silver precursors on nano-sized titanium dioxide particles (denoted as $Ag/TiO_2$), formic acid as hole scavenger, and a 125-W high-pressure Hg lamp, main wavelength around 365 nm, as a light source. The formation of more detrimental products, nitrite and ammonium, was thereby avoided, and residual formic acid can be completely decomposed into a harmless $CO_2$ by further irradiation. Hou et al. [183] presented the novel core-shell structured $Ag/SiO_2@cTiO_2$ composites for photocatalytic reduction of high-concentration nitrate (2000 mg $L^{-1}$). Photocatalytic denitrification experiments were performed with the light source of a 500 W high-pressure mercury lamp (main wavelength around 365 nm). Due to the electron sink effect, Ag NPs in the $TiO_2$ shell could trap the photogenerated electrons and prolong the lifetime of charge carriers. The photogenerated electrons could be transferred from the CB of the $TiO_2$ shell to Ag NPs for prevention of its oxidation to $Ag^+$. Therefore, $Ag/SiO_2@cTiO_2$ could reduce high-concentration nitrate to $N_2$ effectively.

Lin et al. [119] developed a bio electro-photocatalytic system under UV irradiation (Figures 11 and 17a) which exhibits a high selectivity for photocatalytic reduction of nitrate to $N_2$.

A 30-W low-pressure mercury lamp was used as the light source. The proposed nonconventional bio electro-photocatalytic system has the advantage of a greater denitrification rate, higher selectivity to $N_2$, absence of harmful by-products formation (nitrite or ammonium), the introduction of hole scavenger in nitrate solution is avoided, and cost-effectiveness (Figure 17b–d). Compared with the conventional denitrification mechanism shown in their work [119], this type of a bio electro-photocatalytic reaction pathway has a lower energy barrier (Ea) (Figure 17b), suggesting that the complete photocatalytic reduction of nitrate to $N_2$ without cumulation of harmful byproducts is energetically possible.

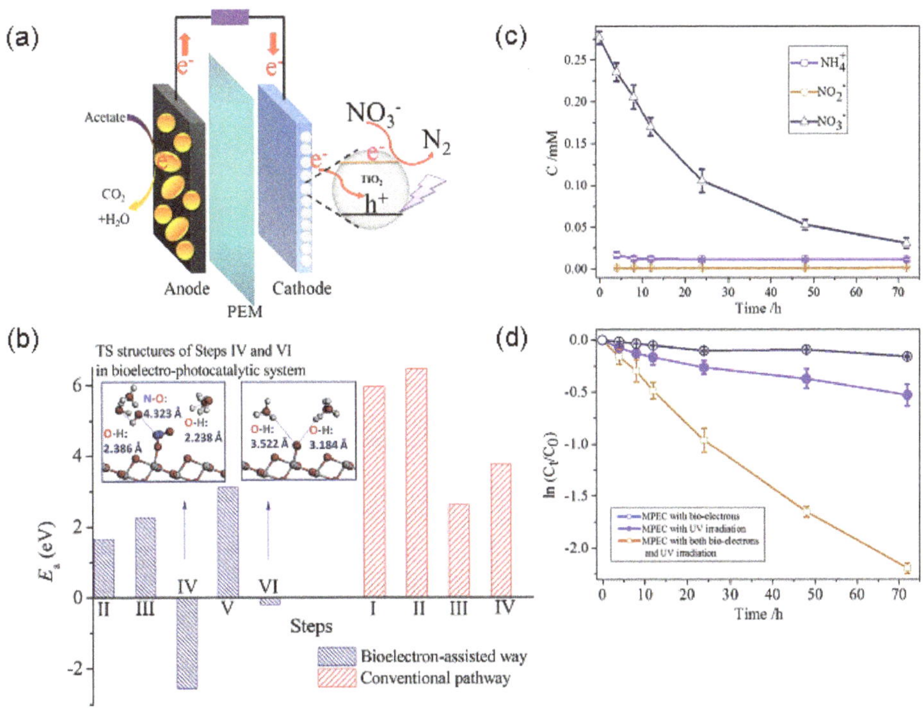

Figure 17. (a) Scheme of the bio electro-photocatalytic denitrification system. (b) Comparison of energy barrier (Ea) for photocatalytic denitrification in a bioelectronic-assisted way and the conventional pathway. The Ea of each step for $NO_3^-$ reduction on the $TiO_2$ (101) surface with the transition state (TS) structures of Steps IV and VI in the bio electro-photocatalytic system. Step I in the bio electro-photocatalytic system is the adsorption of $NO_3^-$ on the $TiO_2$ (101) surface. (c,d) Performance of the bio electro-photocatalytic denitrification system (c) the change of nitrate, nitrite, and ammonium concentrations in the denitrification process; (d) the relative concentration profiles of nitrate during the denitrification process in the bio electro-photocatalytic system (•: with bioanode to supply the bio-electrons but without irradiation; ○: with UV irradiation but without connecting to the bioanode; □: with both UV irradiation and bio-electrons supply from the bioanode). Reproduced with permission from ref. [119]. Copyright 2017 Elsevier.

The photocatalytic denitrification reaction can be described below (Equation (40)):

$$10e_{cb}^- + 2NO_3^- + 12H^+ \rightarrow N_2 + 6H_2O \quad (40)$$

The absence of nitrite generation in the bio electro-photocatalytic denitrification setup is indicative that the reaction pathway is different compared to the conventional denitrification reaction mechanism. The authors provided a valuable solution to increase the efficiency and selectively of photocatalytic denitrification by coupling an electron generation device with a photocatalytic denitrification process and simultaneous activation of nitrate atom pairs for the final formation of $N_2$ from nitrate [119,120].

In the study conducted by Liu et al. [184], a novel two-step reduction process was constructed for the selective removal of nitrate in an aqueous solution of $Na_2SO_3$ using Cu/Fe bimetal photocatalyst. The produced nitrite by the reduction of nitrate on the $Cu^0$ surface could not be converted to ammonia rapidly on the surface of iron oxide layer, leading to the accumulation in time of nitrite. In the next step, the accumulated nitrite was efficiently and easily reduced to nitrogen by $Na_2SO_3$, which worked as an efficient electron donor for nitrite reduction. The selectivity for $N_2$ was over 90%, and the yield of ammonia was below 10% during the two-step reduction process.

Shang et al. [185] investigated the influence of exposed facets of silvered $TiO_2$ photocatalysts on photo denitrification. They found that the nitrate reduction percentage and selectivity to $N_2$ for $Ag_2O/Ag/101\text{-}TiO_2$ reached 99.1 and 81.1%, respectively, due to the formation of the junction at $TiO_2$-metallic $Ag^0$ interfaces and to Z-scheme charge transfer pathway mediated by adjacent Ag.

Recently, Silveira et al. [186] presented a study of the capability of the natural ilmenite ($FeTiO_3$) to reduce nitrate from ultra-pure and mineral water. They claim that natural ilmenite can be a great applicant for reducing $NO_3^-$ in contaminated water. In ultra-pure water, the nitrate is totally converted to $NO_x$ (2%) and $N_2$ (98%) after 210 min. If using oxalate in the mineral water, the nitrate is removed, but $NO_2^-$, $NO_x$, and $N_2$ appear as products. In another study [187], the nitrate reduction in saline waters was explored for the first time employing the $FeTiO_3$/oxalic acid photocatalytic process. A 150 W medium mercury lamp was used. Acidic pH values must be maintained to avoid oxalic acid precipitation by $Ca^{2+}$ present in the water matrix. Under those conditions and compared to ultrapure water, salinity (in the range of 5–33 g/L) has a small influence on nitrate reduction, which is related to the evolution of $C_2O_4^{2-}$ concentration.

Wang et al. [188] prepared a novel $SiW_9/TiO_2/Cu$ composite catalyst and studied the impact of catalyst loading, initial nitrate concentration, polyoxometalate loading, formic acid and $O_2$ on the removal of nitrate under UV light. Nitrate removal up to 76.53% and 82.09% of $N_2$ selectivity was obtained under specific experimental conditions: initial nitrate concentration of 30 mg/L, concentration of formic acid, 30 mmol/L, $SiW_9/Cu$ loading level 8.33%, the catalytic dosage of 0.8 g and presence of $N_2$.

Graphitic carbon nitride ($g-C_3N_4$) has been broadly used in the area of photocatalysis due to its suitable features such as very good stability, graphene-like structure, ease of synthesis and the capability to produce photocarriers. Liu et al. [189] obtained $Ag_yPd_{10-y}/g\text{-}C_xN_4$ Mott–Schottky heterojunction by growing AgPd nanowires (NWs) on the surface of nitrogen-rich $g\text{-}C_xN_4$. Their strategy opens a new way for making photocatalytic hydrogen production in tandem with the reduction of $NO_3^-$ and $NO_2^-$ in water, also extending it to remove metal ions. The $Ag_3Pd_7/g\text{-}C_{1.95}N_4$ catalyst exhibited the highest photocatalytic activity and selectivity for photocatalytic reduction of $NO_3^-$ and $NO_2^-$, and the removal rates of $NO_3^-$ and $NO_2^-$ are 87.4% and 61.8%, respectively, under 365 nm irradiation, at 25 °C.

Soliman et al. [190] studied the reduction of nitrate in water under solar radiation using activated carbon prepared from date palm stone decorated with single and bimetallic nanoparticles. In their work, acetic acid, formic acid, oxalic acid, and ammonium oxalate have been investigated as holes scavengers for nitrate reduction. The obtained results for activated carbon modified with Pd-Ag (using formic acid as a hole scavenger with 0.05 M) showed that the conversion of nitrate (85% after 35 h of natural solar irradiation) takes place mainly through nitrogen gas ($N_2$) rather than nitrite ($NO_2^-$) or ammonium ($NH_4^+$).

The photocatalytic activity strongly depends on the applied experimental conditions, including the mass of the photocatalyst, the incident beam intensity, the type of sacrificial agents, the nitrate concentration in the starting aqueous solution, and so on. Table 4 shows some photocatalytic performance results of different materials reported in the literature, with an emphasis on selectivity towards harmless nitrogen, although an accurate comparison is difficult because of variations in experimental conditions.

**Table 4.** Comparative results of the photocatalytic performance obtained on different materials.

| Photocatalyst | Light Source | Sacrificial Agent | $NO_3^-$ Conversion (%) | $NO_2^-$ Selectivity (%) | $NH_4^+$ Selectivity (%) | $N_2$ Selectivity (%) | Ref. |
|---|---|---|---|---|---|---|---|
| Fresh Ag/P25 | High-pressure Hg lamp 300 W | Formic acid 0.04 M | 99.6 | 2.3 (yield of $NO_2^-$, mgN $L^{-1}$) | 9.3 (yield of $NH_4^+$, mgN $L^{-1}$) | 88.4 | [125] |
| 5% $Ag_2O$/P25 | High-pressure Hg lamp 300 W | Formic acid 0.04 M | 97.2 | 2.4 (yield of $NO_2^-$, mgN $L^{-1}$ | 14.0 (yield of $NH_4^+$, mgN $L^{-1}$) | 83.1 | [125] |
| Fe/$TiO_2$ | High-pressure Hg lamp 110 W | Formic acid 40 mM | 100 | 0 | 13.0 | 87.0 | [127] |
| Cu/$TiO_2$ | High-pressure Hg lamp 110 W | Formic acid 40 mM | 100 | 0 | 37.0 | 63.0 | [127] |
| Cr-$TiO_2$ | Low-pressure Hg lamp 17 W | Formic acid 450 mg $L^{-1}$ (~9.8 mM) | 56.29 | - | - | 98.53 | [139] |
| Zn-$TiO_2$ | Low-pressure Hg lamp 17 W | Formic acid 450 mg $L^{-1}$ (~9.8 mM) | 91.67 | - | - | 95.45 | [139] |
| Ag/$TiO_2$ | High-pressure Hg lamp 125 W | Formic acid 0.04 mol/L | 98.4 | 0 | 0 | 100 | [144] |
| Ag/$TiO_2$ | High-pressure Hg lamp 125 W | Oxalic acid | 16.7 | 2.20 | 0.37 | 84.6 | [144] |
| Pd-Cu/$TiO_2$ | - | Formic acid | 62 | 0 | 6.0 | 94.0 | [164] |
| $LiNbO_3$ | High-pressure Hg lamp 110 W | Humic acid 1.0 mmol/L | 90.1 | - | - | 86.2 | [169] |
| $LiNbO_3$ | High-pressure Hg lamp 110 W | Formic acid 1.0 mmol/L | 98.4 | 0.13 | 1.2 | 95.8 | [169] |
| $Ag_2O$/Ag/101-$TiO_2$ | - | - | 98.57 | 3.55 | 2.42 | 94.03 | [185] |
| $Ag_2O$/Ag/101-$TiO_2$ | - | - | 99.10 | 4.56 | 14.34 | 81.10 | [185] |
| nzv Fe/$TiO_2$ | UV-A lamp 20 W nzv Fe/$TiO_2$ | Formic acid 27 mM | 80.0 | 0 | 39.1 | 60.9 | [191] |
| $SiW_9$/$TiO_2$/Cu | High-pressure mercury lamp 125 W | Formic acid 30 mmol/L | 76.53 | - | - | 82.09 | [188] |
| $Ag_3Pd_7$/g-$C_{1.95}N_4$ | Two light bulbs 40W | - | 87.4 | 61.8 | - | ≈100 | [189] |
| Ag-Pd NPs/activated carbon | Natural solar radiation | Formic acid 0.05 M | 85 | traces | traces | High selectivity to $N_2$ | [190] |

Having a high solubility in water, the nitrate anion is recognized as one of the most widespread contaminants. That is why research is still needed for the development of efficient technologies in the purification of contaminated waters [192].

To completely clarify the nitrate and nitrite photocatalytic reduction mechanism, future studies should quantify the gaseous reaction products in order to elucidate which gaseous species are released during photocatalytic treatment. However, the precise assessment of gases released in the photocatalytic reduction process is extremely challenging from an experimental point of view. It is highly advised to look into reactors that can provide high mass transfer, efficient nitrate reduction, and, on the other hand, a good recovery of N-gases. Combining with other technologies would be a wise choice for improving photocatalytic processes.

## 3. Prospectives and Photocatalytic Approaches in Depollution Technologies

The novelty of this study comes from the fact that this work makes a clear discrimination between oxidative degradation of pollutants, typically leading to the formation of various intermediates, and mineralization, consisting of oxidation to unharmful $CO_2$.

The first condition to apply photocatalysis in depollution technologies is that the organic pollutant should be mineralized to unharmful $CO_2$. The advantage of using solar light in depollution is obvious. Other restrictive conditions hindering of large–scale application of photocatalysis in pollutant removal are the low reaction rates and photocatalytic material-related. Despite the considerable advances in the abatement of numerous recalcitrant compounds, there are remaining challenges to overcome for large-scale practical application. In the meantime, the by-products should be identified and quantified, and their environmental toxicity should be assessed.

The stringent goals in photocatalytic depollution technologies are: (i) high selectivity to $CO_2$; (ii) performant materials characterized by non-toxicity, low band gap, stability, low production costs, high recyclability and (iii) high reaction rates. The efficiency and selectivity of a photocatalyst to carbon dioxide or other unharmful compounds depend on the type of the photocatalytic material and the operating reaction parameters [193]. Another challenge consists in the fact that the majority of photocatalytic systems are based on $TiO_2$ or ZnO, which have large band gap energies (3.1–3.3 eV), thus are mostly active in UV radiation which accounts for only 5% of sunlight [194,195].

### *3.1. Low Reaction Rates*

Due to the low reaction rates, for practical reasons, the photocatalytic technologies should be combined with other techniques such as ozonation [3,196], filtration technology [7], sonication [197–199], thermal activation [200–202]. For instance, Preda et al. [3] investigated the aqueous ammonia oxidation over iron-modified titanate nanorods by using combined treatment with ozone and simulated solar light irradiation. Increasing ammonia conversion was registered relative to the photocatalytic process carried out without ozone, but also the $NO_3^-$ formation was significantly reduced by comparison with the dark ozonation assays. The main achievement of this combined procedure was the increased selectivity of ammonia degradation to gaseous nitrogen-containing end products.

Denny et al. [7] reported the advantages of coupling the photocatalytic and filtration technologies in terms of particulate pollutants elimination together with VOC removal from the air stream. In order to enhance the photodegradation of gas phase ethanol to $CO_2$ and to reduce the production of intermediates such as acetaldehyde, a fluidized bed aerosol generator (FBAG) was adapted to prepare $TiO_2$-loaded ventilation filters as an irradiation source used a UV-light-emitting diode (UV-LED).

According to Adewui [197], photocatalytic processes assisted by ultrasound can be significantly improved and used for the treatment of pollutants in water, sonophotocatalysis (SPC) implying either sequential photocatalytic reaction and ultrasonic irradiation or simultaneous light and ultrasonic irradiation of the investigated system. The main advantages of sonophotocatalysis are clearly emphasized by this review [197]: the better transfer of organic compounds from the bulk solution to catalyst surface, increase the dispersion of chemicals, an extra generation of hydroxyl radicals that are very efficient oxidizing agents as well as the photogenerated holes. The photocatalytic oxidation targeting the mineralization of intermediates such as carboxylic acids can be enhanced by complementary use of ultrasounds [197], many pharmaceuticals and dyes from wastewaters being subjected to sonophotocatalytic studies [198,199].

The "thermo-photo-catalysis" concept, which addresses both environmental and energy fields, was detailed by Nair et al. [200], revealing the main advantages of this dual approach: thermal acceleration of the photocatalytic reactions (known for the limited reaction rates), the occurrence of some photocatalytic processes in the visible-infrared domain (scarcely available in terms of hole/electron photoexcitation), the same efficiency for the pollutant removal as in simple thermocatalytic process but using lower temperatures.

A coupling of advanced oxidation/reduction processes and biological processes for water depollution have been investigated, including ozonation-, Fenton-, electrochemical-biological processes, and also sequential chemical-biological processes [203]. The ozonation in the presence of UV irradiation has become one of the most used advanced oxidation processes for the degradation of organic compounds in general as acids, alcohols, and organochlorines of low molecular weight. Unfortunately, both UV and ozone are quite expensive to generate and need the consumption of large amounts of electric energy. Within the development and application of wastewater treatment technologies, should be taken into account efficiency, cost, and reliability. If the intermediate products obtained require additional removal, then the purification process becomes expensive and complicated. The combination with other technologies, for example, nanotechnology [204], can also be considered.

Andronic et al. [205] tested three different composites in a pilot plant for solar treatment of wastewater using phenol, imidacloprid, and dichloroacetic acid as model pollutants. The investigated photocatalysts were sol-gel $TiO_2$ (as the reference), Vis-active $Cu_xS$ prepared by photochemical precipitation, and highly filterable $TiO_2$-fly ash mixture/composites. The experiments were conducted at a laboratory scale in two solar simulators and under natural solar irradiation at the pilot plant scale, at the Plataforma Solar de Almería, in a Compound Parabolic Collector (CPC) solar radiation system. The research group showed that the high phenol removal efficiency under simulated solar irradiation was attained by the reference $TiO_2$. This behavior was due to titania's large surface area and its anatase/rutile phase composition. Contrarywise, under solar radiation in the CPC reactor pilot, all three pollutants were partially mineralized during the first 40–90 min, but by-products clog the surface, and removal continues without fully oxidizing the organic substrates. After 150 min of solar irradiation at the pilot–plant scale, the difference between the apparent phenol removal (55%) and mineralization (33%) confirms the presence of intermediates at the end of the reaction.

Combined photocatalytic processes proved to be also efficient for the abatement of larger molecule contaminants. Efforts were paid to search for the optimal technology for the removal of wastewater microcontaminants by coupling four different electrochemical processes with a solar CPC reactor at a pilot plant scale [206]. Thus, the anodic oxidation (AO), solar-assisted AO, electro-Fenton (EF), and solar photoelectro-Fenton (SPEF) processes were employed for monitoring the treatment of microcontaminats. This research group selected two different water matrices: one coming from a synthetic retentate with medium content of chlorides (in the range of 550 mg $L^{-1}$, from natural water), while the other one was actual urban wastewater with a higher concentration of chlorides (in the range of 1200–2000 mg $L^{-1}$), spiked with a mix of four microcontaminants (pentachlorophenol, terbutryn, chlorphenvinfos, and diclofenac). These combined technologies successfully removed contaminants of an actual urban wastewater treatment plant secondary effluent up to 80% of the total amount. However, the main disadvantage was that this process was not efficient for dissolved organic carbon (DOC) removal.

Coupling photocatalysis, catalytic, and photocatalytic ozonation, proved to be excellent processes for the abatement of a mixture of seven insecticides at a pilot plant scale [207]. The authors studied a complex mixture of pesticides (e.g., simazine, terbutryn, buprofezin, procymidone, azoxystrobin, imidacloprid, and thiamethoxam), and three sources of $TiO_2$ (N-$TiO_2$, pure $TiO_2$, P25 Degussa) at bench and pilot plant scale to finally combine with ozonation looking for an increase in the degradation efficiency. Ozonation was demonstrated to be the most effective process for the abatement of the targeted mixture of microcontaminants.

### 3.2. Non-Toxicity

Usually, bare and modified semiconductors are used for photocatalysis, especially $TiO_2$-based ones, but unusual photocatalysts such as the highly-defected $SiO_2$ nanotubes proved to be active under solar irradiation [33]. Their activity for methanol and oxalic acid photodegradation in liquid media proved to be close to that registered for $TiO_2$. Therefore, it can be considered a valuable solution for depollution technologies being a low-cost, non-toxic, and environmentally friendly material able to work under solar light, a regenerable energy source.

### 3.3. Industrial Technologies

The establishment photocatalytic depollution technology is a very difficult task due to the fact that real industrial conditions and parameters are less stable and controllable than laboratory conditions, with few companies (Purifics Photo-Cat) applying AOP in industry [194,195]. There is a need to develop pilot scale-up systems that can confirm the technologies presented in this review and eventually be able to use them in industrial applications. Also, another aspect that needs attention is whether photocatalysis will be used alone or as an intermediary step (pre or post-treatment) in a depollution process (maybe used together with ozonation or with physical or biological processes [208].

Conventional photocatalytic nitrate reduction techniques have been developed to produce potable waters, such as adsorption, ion exchange, chemical reduction, membrane filtration, electrochemical, and biological denitrification [209]. These conventional techniques further produce secondary toxic by-products. One of the main problems in the practical application of photocatalysts is the development of methodology for photoreactor scale-up [210]. There are some chemistry, material, and reactor challenges currently limiting large-scale applications of photocatalysis processes of inorganic pollutants [203].

Figure 18 summarizes the challenges and future perspectives to improve photocatalytic remediation technologies.

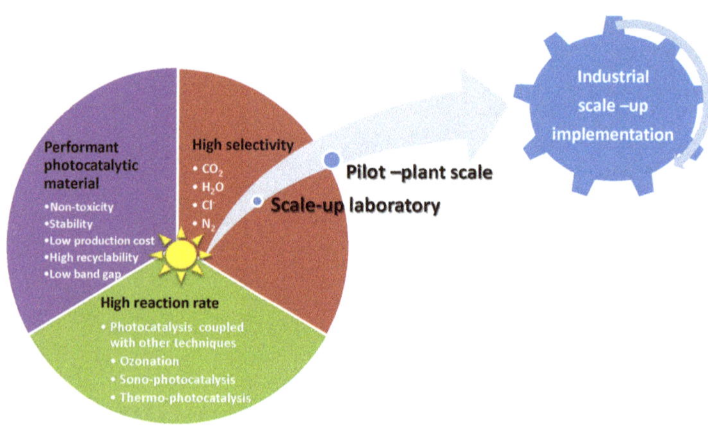


**Figure 18.** Schematic representation of the challenges and photocatalytic approaches in depollution technologies.

## 4. Conclusions

Herein, an extensive screening of the scientific literature related to the photocatalytic removal of various organic and inorganic hazardous compounds is presented. The review focuses on the photomineralization of a few relevant hazardous compounds into $CO_2$ and other harmless products. Specifically, information is provided on the (i) photooxidation of primary alcohols and carboxylic acids in gaseous and liquid media (e.g., methanol, ethanol, oxalic acid) in gaseous and liquid media, (ii) photocatalytic removal of chlorinated and aromatic VOCs from indoor air and water (e.g., trichloroethylene, perchloroethylene, dichloroacetate anion, benzene, toluene, p-xylene, ethylene), (iii) photomineralization of phenol from wastewater, and (iv) efficient removal of nitrate and its conversion, as far as possible, to compounds that do not affect human life and the environment. The degradation of persistent pollutants is critically analyzed, highlighting the main factors affecting the overall process, such as ROS involvement. Particular attention has been paid to the reaction mechanisms established during the photocatalytic removal of inorganic pollutant $NO_3^-$ in connection with the selectivity to harmless nitrogen. The effects of working parameters (e.g., irradiation time, the procedure of operation) on performance are also discussed, along with the intrinsic properties of the applied materials (e.g., surface active sites and structure).

From a large amount of available data, it is clear that further research should be conducted in order to find efficient photocatalysts that are able to mineralize the pollutants into the non-toxic $CO_2$. Only such photocatalysts can be considered for environmental remediation.

Pollutant elimination by only photocatalysis is a highly demanding task and difficult to carry out. Therefore, combining photocatalysis with other techniques is imperative. Photocatalysis as a depollution method has limited efficacy, but from the environmental point of view, it has great significance because the sun is a cheap and endless source. In fact, in nature, all the existent pollutants are exposed to solar irradiation. Thus, photocatalytic studies are highly relevant to the actual environment. Oxide materials (e.g., minerals) are slowly degrading pollutants under solar light, and the fate of intermediates is less known. Thus, the photocatalytic pollutant degradation mechanism in laboratory experiments is very important because it is relevant to the natural depollution processes.

**Author Contributions:** Conceptualization, M.P., C.A., R.-N.S., A.V., F.P. and I.B.; writing—original draft preparation, M.P., C.A., R.-N.S. and A.V.; writing—review and editing, I.B.; supervision, I.B. All authors have read and agreed to the published version of the manuscript.

**Funding:** This research received no external funding.

**Data Availability Statement:** The data that support the plots within this manuscript are available upon reasonable request from the authors.

**Conflicts of Interest:** The authors declare no conflict of interest.

## References

1. Ramalingam, G.; Perumal, N.; Priya, A.K.; Rajendran, S. A review of graphene-based semiconductors for photocatalytic degradation of pollutants in wastewater. *Chemosphere* **2022**, *300*, 134391. [CrossRef] [PubMed]
2. Kang, W.; Chen, S.; Yu, H.; Xu, T.; Wu, S.; Wang, X.; Lu, N.; Quan, X.; Liang, H. Photocatalytic ozonation of organic pollutants in wastewater using a flowing through reactor. *J. Hazard. Mater.* **2021**, *405*, 124277. [CrossRef] [PubMed]
3. Preda, S.; Umek, P.; Zaharescu, M.; Anastasescu, C.; Petrescu, S.V.; Gifu, C.; Eftemie, D.-I.; State, R.; Papa, F.; Balint, I. Iron-modified titanate nanorods for oxidation of aqueous ammonia using combined treatment with ozone and solar light irradiation. *Catalysts* **2022**, *12*, 666. [CrossRef]
4. Mirsadeghi, S.; Zandavar, H.; Rajabi, H.R.; Sajadiasl, F.; Ganjali, M.R.; Pourmortazavi, S.M. Superior degradation of organic pollutants and $H_2O_2$ generation ability on environmentally–sound constructed $Fe_3O_4$-Cu nanocomposite. *J. Mater. Res. Technol.* **2021**, *14*, 808–821. [CrossRef]
5. Qu, Y.; Chen, Z.; Duan, Y.; Liu, L. $H_2O_2$ assisted photocatalysis over Fe-MOF modified BiOBr for degradation of RhB. *J. Chem. Technol. Biotechnol.* **2022**, *97*, 2881–2888. [CrossRef]
6. Binas, V.; Venieri, D.; Kotzias, D.; Kiriakidis, G. Modified $TiO_2$ based photocatalysts for improved air and health quality. *J. Materiomics* **2017**, *3*, 3–16. [CrossRef]
7. Denny, F.; Permana, E.; Scott, J.; Wang, J.; Pui, D.Y.H.; Amal, R. Integrated Photocatalytic Filtration Array for Indoor Air Quality Control. *Environ. Sci. Technol.* **2010**, *44*, 5558–5563. [CrossRef]

8. Shen, M.; Henderson, M.A. Identification of the Active Species in Photochemical Hole Scavenging Reactions of Methanol on $TiO_2$. *J. Phys. Chem. Lett.* **2011**, *2*, 2707–2710. [CrossRef]
9. Tamaki, Y.; Furube, A.; Murai, M.; Hara, K.; Katoh, R.; Tachiya, M. Direct Observation of Reactive Trapped Holes in $TiO_2$ Undergoing Photocatalytic Oxidation of Adsorbed Alcohols: Evaluation of the Reaction Rates and Yields. *J. Am. Chem. Soc.* **2005**, *128*, 416–417. [CrossRef]
10. Mejia, M.I.; Marin, J.M.; Restrepo, G.; Rios, L.A.; Pulgarin, C.; Kiwi, J. Preparation, Testing and Performance of a $TiO_2$/Polyester Photocatalyst for the Degradation of Gaseous Methanol. *Appl. Catal. B.* **2010**, *94*, 166–172. [CrossRef]
11. Setvin, M.; Shi, X.; Hulva, J.; Simschitz, T.; Parkinson, G.S.; Schmid, M.; Valentin, C.D.; Selloni, A.; Diebold, U. Methanol on anatase $TiO_2$ (101): Mechanistic insights into photocatalysis. *ACS Catal.* **2017**, *7*, 7081–7091. [CrossRef]
12. Shen, M.; Acharya, D.P.; Dohnálek, Z.; Henderson, M.A. Importance of diffusion in methanol photochemistry on $TiO_2$ (110). *J. Phys. Chem.* **2012**, *116*, 25465–25469. [CrossRef]
13. Zhou, C.; Ma, Z.; Ren, Z.; Mao, X.; Dai, D.X.; Yang, X. Effect of defects on photocatalytic dissociation of methanol on $TiO_2$ (110). *Chem. Sci.* **2011**, *2*, 1980. [CrossRef]
14. Panayotov, D.A.; Burrows, S.P.; Morris, J.R. Photooxidation mechanism of methanol on rutile $TiO_2$ nanoparticles. *J. Phys. Chem. C.* **2012**, *116*, 6623–6635. [CrossRef]
15. Humayun, M.; Raziq, F.; Khan, A.; Luo, W. Modification strategies of $TiO_2$ for potential applications in photocatalysis: A critical review. *Green Chem. Lett. Rev.* **2018**, *11*, 86–102. [CrossRef]
16. Misra, M.; Kapur, P.; Singla, M.L. Surface plasmon quenched of near band edge emission and enhanced visible photocatalytic activity of Au@ZnO core-shell nanostructure. *Appl. Catal. B Environ.* **2014**, *150*, 605–611. [CrossRef]
17. Chen, W.; Wang, Y.; Liu, S.; Gao, L.; Mao, L.; Fan, Z.; Shangguan, W.; Jiang, Z. Non-noble metal Cu as a cocatalyst on $TiO_2$ nanorod for highly efficient photocatalytic hydrogen production. *Appl. Surf. Sci.* **2018**, *445*, 527–534. [CrossRef]
18. Rosseler, O.; Shankar, M.V.; Karkmaz-Le Du, M.; Schmidlin, L.; Keller, N.; Keller, V. Solar light photocatalytic hydrogen production from water over Pt and Au/$TiO_2$ (anatase/rutile) photocatalysts: Influence of noble metal and porogen promotion. *J. Catal.* **2010**, *269*, 179–190. [CrossRef]
19. Mizukoshi, Y.; Makise, Y.; Shuto, T.; Hu, J.; Tominaga, A.; Shironita, S.; Tanabe, S. Immobilization of noble metal nanoparticles on the surface of $TiO_2$ by the sonochemical method: Photocatalytic production of hydrogen from an aqueous solution of ethanol. *Ult. Sonch.* **2007**, *14*, 387–392. [CrossRef]
20. Gончearenco, E.; Morjan, I.P.; Dutu, E.; Scarisoreanu, M.; Fleaca, C.; Gavrila-Florescu, L.; Dumitrache, F.; Banici, A.M.; Teodorescu, V.S.; Anastasescu, C.; et al. The effect of noble metal addition on the properties of oxide semiconductors nanoparticles. *J. Solid State Chem.* **2022**, *307*, 122817. [CrossRef]
21. Henderson, M.A.; White, J.M.; Uetsuka, H.; Onishi, H. Selectivity changes during organic photooxidation on $TiO_2$: Role of $O_2$ pressure and organic coverage. *J. Catal.* **2006**, *238*, 153–164. [CrossRef]
22. El-Roz, M.; Bazin, P.; Daturi, M.; Thibault-Starzyk, F. On the mechanism of methanol photooxidation to methylformate and carbon dioxide on $TiO_2$: An *operando*-FTIR study. *Phys. Chem. Chem. Phys.* **2015**, *17*, 11277–11283. [CrossRef] [PubMed]
23. DePuccio, D.P.; Landry, C.C. Photocatalytic oxidation of methanol using porous Au/$WO_3$ and visible vight. *Catal. Sci. Technol.* **2016**, *6*, 7512–7520. [CrossRef]
24. Muggli, D.S.; Larson, S.A.; Falconer, J.L. Photocatalytic oxidation of ethanol: Isotopic labeling and transient reaction. *J. Phys. Chem.* **1996**, *100*, 15886–15889. [CrossRef]
25. Muggli, D.S.; McCue, J.T.; Falconer, J.L. Mechanism of the photocatalytic oxidation of ethanol on $TiO_2$. *J. Catal.* **1998**, *173*, 470–483. [CrossRef]
26. Yu, Z.; Chuang, S.S.C. In situ IR study of adsorbed species and photogenerated electrons during photocatalytic oxidation of ethanol on $TiO_2$. *J. Catal.* **2007**, *246*, 118–126. [CrossRef]
27. Fukuhara, D.; Joseph, M.T.; Loumissi, T.; Zhang, C.; Itoi, T.; Zhang, H.; Izumi, Y. Local silver site temperature critically reflected partial and complete photooxidation of ethanol using Ag-$TiO_2$ as revealed by extended X-ray absorption fine structure Debye−Waller factor. *J. Phys. Chem. C.* **2021**, *125*, 14689–14701. [CrossRef]
28. Kawai, T.; Sakata, T. Photocatalytic hydrogen production from liquid methanol and water. *J. Chem. Soc. Chem. Commun.* **1980**, *15*, 694–695. [CrossRef]
29. Villarreal, T.L.; G6mez, R.; Neumann-Spallart, M.; Alonso-Vante, N.; Salvador, P. Semiconductor photooxidation of pollutants dissolved in water: A kinetic model for distinguishing between direct and indirect interfacial hole transfer. I. Photoelectrochemical experiments with polycrystalline anatase electrodes under current doubling and absence of recombination. *J. Phys. Chem. B* **2004**, *108*, 15172–15181. [CrossRef]
30. Haselmann, G.M.; Baumgartner, B.; Wang, J.; Wieland, K.; Gupta, T.; Herzig, C.; Limbeck, A.; Lendl, B.; Eder, D. In situ Pt photodeposition and methanol photooxidation on Pt/$TiO_2$: Pt-loading-dependent photocatalytic reaction pathways studied by liquid-phase infrared spectroscopy. *ACS Catal.* **2020**, *10*, 2964–2977. [CrossRef]
31. Preda, S.; Anastasescu, C.; Balint, I.; Umek, P.; Sluban, M.; Negrila, C.; Angelescu, D.G.; Bratan, V.; Rusu, A.; Zaharescu, M. Charge separation and ROS generation on tubular sodium titanates exposed to simulated solar light. *Appl. Surf. Sci.* **2019**, *470*, 1053–1063. [CrossRef]
32. Papa, F.; Miyazaki, A.; Scurtu, M.; Ianculescu, A.C.; Balint, I. Morphology, chemical state of nanometric-sized Pt–Cu and Pt–Ag particles, and their photocatalytic activity for mineralization of methanol. *J. Nanopart. Res.* **2014**, *16*, 2249. [CrossRef]

33. Anastasescu, C.; Zaharescu, M.; Angelescu, D.; Munteanu, C.; Bratan, V.; Spataru, T.; Negrila, C.; Spataru, N.; Balint, I. Defect-related light absorption, photoluminescence and photocatalytic activity of $SiO_2$ with tubular morphology. *Sol. Energ. Mater. Sol. Cells* **2017**, *159*, 325–335. [CrossRef]
34. Anastasescu, C.; Negrila, C.; Angelescu, D.G.; Atkinson, I.; Anastasescu, M.; Spataru, N.; Zaharescu, M.; Balint, I. Particularities of photocatalysis and formation of reactive oxygen species on insulators and semiconductors: Cases of $SiO_2$, $TiO_2$ and their composite $SiO_2$-$TiO_2$. *Catal. Sci. Technol.* **2018**, *8*, 5657–5668. [CrossRef]
35. Mendive, C.B.; Bredow, T.; Blesabd, M.A.; Bahnemann, D.W. ATR-FTIR measurements and quantum chemical calculations concerning the adsorption and photoreaction of oxalic acid on $TiO_2$. *Phys. Chem. Chem. Phys.* **2006**, *8*, 3232–3247. [CrossRef]
36. Anastasescu, C.; Zaharescu, M.; Balint, I. Unexpected photocatalytic activity of simple and platinum modified tubular $SiO_2$ for the oxidation of oxalic acid to $CO_2$. *Catal. Lett.* **2009**, *132*, 81–86. [CrossRef]
37. Kosanić, M.M. Photocatalytic degradation of oxalic acid over $TiO_2$ powder. *J. Photochem. Photobiol. A Chem.* **1998**, *119*, 119–122. [CrossRef]
38. Craciun, E.; Predoana, L.; Atkinson, I.; Jitaru, I.; Anghel, E.M.; Bratan, V.; Gifu, C.; Anastasescu, C.; Rusu, A.; Raditoiu, V.; et al. $Fe^{3+-}$doped $TiO_2$ nanopowders for photocatalytic mineralization of oxalic acid under solar light irradiation. *J. Photochem. Photobiol. A Chem.* **2018**, *356*, 18–28. [CrossRef]
39. Kiatkittipong, K.; Iwase, A.; Scott, J.; Amal, R. Photocatalysis of heat treated sodium-and hydrogen-titanate nanoribbons for water splitting, $H_2/O_2$ generation and oxalic acid oxidation. *Chem. Eng. Sci.* **2013**, *93*, 341–349. [CrossRef]
40. Cauxa, M.; Fina, F.; Irvine, J.T.S.; Idriss, H.; Howe, R. Impact of the annealing temperature on Pt/g-$C_3N_4$ structure, activity and selectivity between photodegradation and water splitting. *Catal. Today* **2017**, *287*, 182–188. [CrossRef]
41. Karunakaran, C.; Dhanalakshmi, R. Photocatalytic performance of particulate semiconductors under natural sunshine—Oxidation of carboxylic acids. *Sol. Energy Mater. Sol. Cells* **2008**, *92*, 588–593. [CrossRef]
42. Wang, Z.; Xie, X.; Wang, X.; Mahmood, A.; Qiu, H.; Sun, J. Difference of photodegradation characteristics between single and mixed VOC pollutants under simulated sunlight irradiation. *J Photochem. Photobiol. A Chem.* **2019**, *384*, 112029. [CrossRef]
43. Lyu, J.; Gao, J.; Zhang, M.; Fu, Q.; Sun, L.; Hu, S.; Zhong, J.; Wang, S. Construction of homojunction-adsorption layer on anatase $TiO_2$ to improve photocatalytic mineralization of volatile organic compounds. *Appl. Catal. B Environ.* **2017**, *202*, 664–670. [CrossRef]
44. Sansotera, M.; Malek Kheylia, S.G.; Baggioli, A.; Bianchi, C.L.; Pedeferri, P.M.; Diamanti, M.V.; Navarrini, W. Absorption and photocatalytic degradation of VOCs by perfluorinated ionomeric coating with $TiO_2$ nanopowders for air purification. *Chem. Eng. J.* **2019**, *361*, 885–896. [CrossRef]
45. State, R.; Papa, F.; Tabakova, T.; Atkinson, I.; Negrila, C.; Balint, I. Photocatalytic abatement of trichlorethylene over Au and Pd–Au supported on $TiO_2$ by combined photomineralization/hydrodechlorination reactions under simulated solar irradiation. *J. Catal.* **2017**, *346*, 101–108. [CrossRef]
46. Hsu, Y.-C.; Chang, S.-H.; Chung, W.-C.; Chang, M.-B. Photocatalytic removal of trichloroethylene from water with $LaFeO_3$. *Environ. Sci. Pollut. Res.* **2019**, *26*, 26276–26285. [CrossRef]
47. Raciulete, M.; Papa, F.; Negrila, C.; Bratan, V.; Munteanu, C.; Pandele-Cusu, J.; Culita, D.C.; Alkinson, I.; Balint, I. Strategy for modifying layered perovskites toward efficient solar light-driven photocatalysts for removal of chlorinated pollutants. *Catalysts* **2020**, *10*, 637. [CrossRef]
48. Raciulete, M.; Papa, F.; Kawamoto, D.; Munteanu, C.; Culita, D.C.; Negrila, C.; Atkinson, I.; Bratan, V.; Pandele-Cusu, J.; Balint, I. Particularities of trichloroethylene photoatalytic degradation over crystalline $RbLaTa_2O_7$ nanowire bundles grown by solid-state synthesis route. *J. Environ. Chem. Eng.* **2019**, *7*, 102789. [CrossRef]
49. Raciulete, M.; Papa, F.; Culita, D.C.; Munteanu, C.; Atkinson, I.; Bratan, V.; Pandele-Cusu, J.; State, R.; Balint, I. Inpact of $RbLaTa_2O_7$ layered perovskite synthesis conditions on their activity for photocatalytic abatement of trichloroethylene. *Rev. Roum. Chim.* **2018**, *63*, 821–828.
50. Joo, J.C.; Ahn, C.H.; Jang, D.G.; Yoon, Y.H.; Kim, J.K.; Campos, L.; Ahn, H. Photocatalytic degradation of trichloroethylene in aqueous phase using nano-ZNO/Laponite composites. *J. Hazard. Mater.* **2013**, *263*, 569–574. [CrossRef]
51. Friedmann, D.A. General overview of heterogeneous photocatalysis as a remediation technology for wastewaters containing pharmaceutical compounds. *Water* **2022**, *14*, 3588. [CrossRef]
52. Monteiro, R.A.R.; Silva, A.M.T.; Ângelo, J.R.M.; Silva, G.V.; Mendes, A.M.; Boaventura, R.A.R.; Vilar, V.J.P. Photocatalytic oxidation of gaseous perchloroethylene over $TiO_2$ based paint. *J. Photochem. Photobiol. A Chem.* **2015**, *311*, 41–52. [CrossRef]
53. Egerton, T.A.; Mattinson, J.A. The influence of platinum on UV and 'visible' photocatalysis by rutile and Degussa P25. *J. Photochemi. Photobiol. A Chem.* **2008**, *194*, 283–289. [CrossRef]
54. Grzechulska–Damszel, J.; Grześkowiak, M.; Przepiórski, J.; Morawski, A.W. Photocatalytic decomposition of low-concentrated trichloroethylene and tetrachloroethylene in water. *Int. J. Environ. Res.* **2014**, *8*, 347–352. [CrossRef]
55. Suárez, S.; Jansson, I.; Ohtani, B.; Sánchez, B. From titania nanoparticles to decahedral anatase particles: Photocatalytic activity of $TiO_2$/zeolite hybrids for VOCs oxidation. *Catal. Today* **2019**, *326*, 2–7. [CrossRef]
56. Chen, C.-J.; Wu, C.-C.; Hsieh, L.-T.; Chen, K.-C. Treatment of trichloroethylene with photocatalyst-coated optical fiber. *Water* **2019**, *11*, 2391. [CrossRef]
57. Dutschke, M.; Schnabel, T.; Schütz, F.; Springer, C. Degradation of chlorinated volatile organic compounds from contaminated ground water using a carrier-bound $TiO_2/UV/O_3$-system. *J. Environ. Manag.* **2022**, *304*, 114236. [CrossRef]

58. Ji, J.; Xu, Y.; Huang, H.; He, M.; Liu, S.; Liu, G.; Xie, R.; Feng, Q.; Shu, Y.; Zhan, Y.; et al. Mesoporous $TiO_2$ under VUV irradiation: Enhanced photocatalytic oxidation for VOCs degradation at room temperature. *Chem. Eng. J.* **2017**, *327*, 490–499. [CrossRef]
59. Fiorenza, R.; Spitaleria, L.; Perricelli, F.; Nicotra, G.; Fragalà, M.E.; Scirè, S.; Gulino, A. Efficient photocatalytic oxidation of VOCs using ZnO@Au nanoparticles. *J. Photochem. Photobiol. A Chem.* **2023**, *434*, 114232. [CrossRef]
60. Liu, J.; Wang, P.; Qu, W.; Li, H.; Shi, L.; Zhang, D. Nanodiamond-decorated ZnO catalysts with enhanced photocorrosion-resistance for photocatalytic degradation of gaseous toluene. *Appl. Catal. B Environ.* **2019**, *257*, 117880. [CrossRef]
61. Zhang, Y.; Zhao, G.; Chen, Z.; Lian, H.; Gan, L.; Pan, M. Hierarchically nanostructured Ag/ZnO/nBC for VOC photocatalytic degradation: Dynamic adsorption and enhanced charge transfer. *J. Environ. Chem. Eng.* **2022**, *10*, 108690. [CrossRef]
62. Ouachtak, H.; Akhouairi, S.; Haounatic, R.; Addic, A.A.; Jadad, A.; Tahaa, M.L.; Douch, J. 3,4-Dihydroxybenzoic acid removal from water by goethite modified natural sand column fixed-bed: Experimental study and mathematical modeling. *Desalin. Water. Treatm.* **2020**, *194*, 439–449. Available online: https://www.deswater.com/DWT_abstracts/vol_194/194_2020_439.pdf (accessed on 20 January 2023). [CrossRef]
63. Haounati, R.; Alakhras, F.; Ouachtak, H.; Saleh, T.A.; Al-Mazaideh, G.; Alhajri, E.; Jada, A.; Hafid, N.; Addi, A.A. Synthesized of zeolite@$Ag_2O$ nanocomposite as superb stability photocatalysis toward hazardous rhodamine B dye from water. *Arab. J. Sci. Eng.* **2023**, *48*, 169–179. [CrossRef]
64. Raza, W.; Lee, J.; Raza, N.; Luo, Y.; Kim, K.H.; Yang, J. Removal of phenolic compounds from industrial waste water based on membrane-based technologies. *J. Ind. Eng. Chem.* **2019**, *71*, 1–18. [CrossRef]
65. Min, K.; Freeman, C.; Kang, H.; Choi, S.U. The regulation by phenolic compounds of soil organic matter dynamics under a changing environment. *Biomed. Res. Int.* **2015**, *2015*, 825098. [CrossRef]
66. Weber, M.; Weber, M.; Kleine-Boymann, M. Phenol. In *Ullmann's Encyclopedia of Industrial Chemistry*; Wiley-VCH: Weinheim, Germany, 2004; Volume 26, pp. 503–519. [CrossRef]
67. Michałowicz, J.; Duda, W. Phenols—Sources and toxicity. *Pol. J. Environ. Stud.* **2007**, *16*, 347–362.
68. ATSDR—Agency for Toxic Substances and Disease Registry. Medical Management Guidelines for Phenol. 2014. Available online: https://wwwn.cdc.gov/TSP/MMG/MMGDetails.aspx?mmgid=144&toxid=27 (accessed on 14 December 2022).
69. World Health Organization. Phenol Health and Safety Guide. 1994. Available online: http://apps.who.int/iris/bitstream/handle/10665/39958/9241510889-eng.pdf;jsessionid=A1A871567C5E2FEBB6594D0839772049?sequence=1 (accessed on 14 December 2022).
70. Matos, J.; Laine, J.; Herrmann, J.-M. Effect of the type of activated carbons on the photocatalytic degradation of aqueous organic pollutants by UV-irradiated titania. *J. Catal.* **2001**, *200*, 10–20. [CrossRef]
71. Gaya, U.I.; Abdullah, A.H. Heterogeneous photocatalytic degradation of organic contaminants over titanium dioxide: A review of fundamentals, progress and problems. *J. Photochem. Photobiol. C Photochem. Rev.* **2008**, *9*, 1–12. [CrossRef]
72. Dang, T.T.T.; Le, S.T.T.; Channei, D.; Khanitchaidecha, W.; Nakaruk, A. Photodegradation mechanisms of phenol in the photocatalytic process. *Res. Chem. Intermed.* **2016**, *42*, 5961–5974. [CrossRef]
73. Sobczynski, A.; Duczmal, L.; Zmudzinski, W. Phenol destruction by photocatalysis on $TiO_2$: An attempt to solve the reaction mechanism. *J. Mol. Catal. A Chem.* **2004**, *213*, 225–230. [CrossRef]
74. Guo, Z.; Ma, R.; Li, G. Degradation of phenol by nanomaterial $TiO_2$ in wastewater. *Chem. Eng. J.* **2006**, *119*, 55–59. [CrossRef]
75. Wysocka, I.; Kowalska, E.; Trzcinski, K.; Łapinski, M.; Nowaczyk, G.; Zielinska-Jurek, A. UV-Vis-induced degradation of phenol over magnetic photocatalysts modified with Pt, Pd, Cu and Au nanoparticles. *Nanomater.* **2018**, *8*, 28. [CrossRef]
76. Chatterjee, D.; Dasgupta, S. Visible light induced photocatalytic degradation of organic pollutants. *J. Photochem. Photobiol. C Photochem. Rev.* **2005**, *6*, 186–205. [CrossRef]
77. Fu, F.; Shen, H.; Xue, W.; Zhen, Y.; Soomro, A.R.; Yang, X.; Wang, D.; Xu, B.; Chi, R. Alkali-assisted synthesis of direct Z-scheme based $Bi_2O_3/Bi_2MoO_6$ photocatalyst for highly efficient photocatalytic degradation of phenol and hydrogen evolution reaction. *J. Catal.* **2019**, *375*, 399–409. [CrossRef]
78. Zhang, Z.; Wang, W.; Wang, L.; Sun, S. Enhancement of visible-light photocatalysis coupling with narrow-band-bap semiconductor: A case study on $Bi_2S_3/Bi_2WO_6$. *ACS. Appl. Mater. Interfaces* **2012**, *4*, 593–597. [CrossRef]
79. Chien, S.W.; Ng, D.Q.; Kumar, D.; Lam, S.M.; Jaffari, Z.H. Investigating the effects of various synthesis routes on morphological, optical, photoelectrochemical and photocatalytic properties of single-phase perovskite $BiFeO_3$. *J. Phys. Chem. Solid.* **2022**, *160*, 110342. [CrossRef]
80. Jiang, J.; Wang, H.; Chen, X.; Li, S.; Xie, T.; Wang, D.; Lin, Y. Enhanced photocatalytic degradation of phenol and photogenerated charges transfer property over BiOI-loaded ZnO composites. *J. Colloid Interface Sci.* **2017**, *494*, 130–138. [CrossRef]
81. Zhang, Y.; Selvara, R.; Sillanpää, M.; Kim, Y.; Tai, C.-W. The influence of operating parameters on heterogeneous photocatalytic mineralization of phenol over $BiPO_4$. *Chem. Eng. J.* **2014**, *245*, 117–123. [CrossRef]
82. Wang, Y.; Fan, X.; Dong, W.; Zhang, Q.; Liu, J.; Li, R.; Wang, Y.; Zhang, X.; Fan, C. $CeO_2$ nanoparticles decorated $Bi_4O_7$ nanosheets for enhanced photodegradation performance of phenol. *Mater. Lett.* **2022**, *322*, 132465. [CrossRef]
83. Yuan, Y.; Guo, R.-T.; Hong, L.-F.; Lin, Z.-D.; Ji, X.-Y.; Pan, W.-G. Fabrication of a dual S-scheme $Bi_7O9I_3$/g-$C_3N_4$/$Bi_3O_4Cl$ heterojunction with enhanced visible-light-driven performance for phenol degradation. *Chemosphere* **2022**, *287*, 132241. [CrossRef]
84. Jiang, C.; Ge, Y.; Chen, W.; Hua, L.; Li, H.; Zhang, Y.; Cao, S. Hierarchically-structured $TiO_2/MnO_2$ hollow spheres exhibiting the complete mineralization of phenol. *Catalysts* **2019**, *9*, 390. [CrossRef]

85. Sonawane, R.S.; Dongare, M.K. Sol-gel synthesis of Au/$TiO_2$ thin films for photocatalytic degradation of phenol in sunlight. *J. Mol. Catal. A Chem.* **2006**, *243*, 68–76. [CrossRef]
86. Chiou, C.H.; Juang, R.S. Photocatalytic degradation of phenol in aqueous solutions by Pr-doped $TiO_2$ nanoparticles. *J. Hazard. Mater.* **2007**, *149*, 1–7. [CrossRef] [PubMed]
87. Chowdhury, P.; Moreira, J.; Gomaa, H.; Ray, A.K. Visible-solar-light-driven photocatalytic degradation of phenol with dye-sensitized $TiO_2$: Parametric and kinetic study. *Ind. Eng. Chem. Res.* **2012**, *51*, 4523–4532. [CrossRef]
88. Dlamini, M.C.; Dlamini, M.L.; Mente, P.; Tlhaole, B.; Erasmus, R.; Maubane-Nkadimeng, M.S.; Moma, J.A. Photocatalytic abatement of phenol on amorphous $TiO_2$-BiOBr-bentonite heterostructures under visible light irradiation. *J. Ind. Eng. Chem.* **2022**, *111*, 419–436. [CrossRef]
89. Rehman, G.U.; Tahir, M.; Goh, P.S.; Ismail, A.F.; Khan, I.U. Controlled synthesis of reduced graphene oxide supported magnetically separable $Fe_3O_4$@rGO@AgI ternary nanocomposite for enhanced photocatalytic degradation of phenol. *Powder Technol.* **2019**, *356*, 547–558. [CrossRef]
90. Boukhatem, H.; Khalaf, H.; Djouadi, L.; Gonzalez, F.V.; Navarro, R.M.; Santaballa, J.A.; Canle, M. Photocatalytic activity of mont-La (6%)-Cu0.6$Cd_{0.4}$S catalyst for phenol degradation under near UV visible light irradiation. *Appl. Catal. B Environm.* **2018**, *211*, 114–125. [CrossRef]
91. Fan, H.; Yi, G.; Zhang, Z.; Zhang, X.; Li, P.; Zhang, C.; Chen, L.; Zhang, Y.; Sun, Q. Binary $TiO_2$/RGO photocatalyst for enhanced degradation of phenol and its application in underground coal gasification wastewater treatment. *Opt. Mater.* **2021**, *120*, 111482. [CrossRef]
92. Othman, I.; Haija, M.A.; Ismail, I.; Zain, J.H.; Banat, F. Preparation and catalytic performance of $CuFe_2O_4$ nanoparticles supported on reduced graphene oxide $CuFe_2O_4$/rGO) for phenol degradation. *Mater. Chem. Phys.* **2019**, *238*, 121931. [CrossRef]
93. Mohamed, A.; Yousef, S.; Nasser, W.S.; Osman, T.A.; Knebel, A.; Sanchez, E.P.V.; Hashem, T. Rapid photocatalytic degradation of phenol from water using composite nanofibers under UV. *Environ. Sci. Eur.* **2020**, *32*, 160. [CrossRef]
94. Jiang, Z.; Wang, L.; Lei, J.; Liu, Y.; Zhang, J. Photo-Fenton degradation of phenol by CdS/rGO/$Fe^{2+}$ at natural pH with in situ-generated $H_2O_2$. *Appl. Catal. B* **2019**, *241*, 367–374. [CrossRef]
95. Yang, X.; Jia, Q.; Pang, J.; Yang, Y.; Zheng, S.; Jia, J.; Qin, Z. Hierarchical porous N-$TiO_2$/carbon foam composite for enhancement of photodegradation activity under simulated sunlight. *Diam. Relat. Mater.* **2022**, *128*, 109234. [CrossRef]
96. Yuan, Y.; Pan, W.-G.; Guo, R.-T.; Hong, L.-F.; Lin, Z.-D.; Ji, X.-Y. Flower spherical-like $Bi_7O_9$I3/AgI S-scheme heterojunction for phenol photodegradation: The synergetic effect of dual surface plasmon resonance and photothermal property. *Sep. Purif. Technol.* **2022**, *297*, 121538. [CrossRef]
97. Jing, Y.; Yin, H.; Li, C.; Chen, J.; Wu, S.; Liu, H.; Xie, L.; Lei, Q.; Sun, M.; Yu, S. Fabrication of Pt doped $TiO_2$–ZnO@ ZIF-8 core@shell photocatalyst with enhanced activity for phenol degradation. *Environ. Res.* **2022**, *203*, 111819. [CrossRef]
98. Wang, X.; Xu, H.; Luo, X.; Li, M.; Dai, M.; Chen, Q.; Song, H. Enhanced photocatalytic properties of $CeO_2$/$TiO_2$ heterostructures for phenol degradation. *Colloid Interface Sci. Commun.* **2021**, *44*, 100476. [CrossRef]
99. Sharma, N.; Pap, Z.; Baán, K.; Gyulavari, T.; Karacs, G.; Nemeth, Z.; Garg, S.; Hernadi, K. Effective removal of phenol by activated charcoal/BiOCl composite under UV light irradiation. *J. Mol. Struct.* **2022**, *1254*, 132344. [CrossRef]
100. Al-Hamdia, A.M.; Sillanpää, M.; Bora, T.; Duttac, J. Efficient photocatalytic degradation of phenol in aqueous solution by$SnO_2$:Sb nanoparticles. *Appl. Surf. Sci.* **2016**, *370*, 229–236. [CrossRef]
101. Sandulescu, A.; Anastasescu, C.; Papa, F.; Raciulete, M.; Vasile, A.; Spataru, T.; Scarisoreanu, M.; Fleaca, C.; Mihailescu, C.N.; Teodorescu, V.S.; et al. Advancements on basic working principles of photo-driven oxidative degradation of organic substrates over pristine and noble metal-modified $TiO_2$. Model case of phenol photo oxidation. *Catalysts* **2021**, *11*, 487. [CrossRef]
102. Mendoza-Damian, G.; Tzompantzi, F.; Mantilla, A.; Pérez-Hernández, R.; Hernández-Gordillo, A. Improved photocatalytic activity of $SnO_2$–ZnAl LDH prepared by one step $Sn^{4+}$ incorporation. *Appl. Clay Sci.* **2016**, *121*, 127–136. [CrossRef]
103. Raciulete, M.; Anastasescu, C.; Papa, F.; Atkinson, I.; Bradu, C.; Negrila, C.; Eftemie, D.-I.; Culita, D.C.; Miyazaki, A.; Bratan, V.; et al. Band-gap engineering of layered perovskites by Cu spacer insertion as photocatalysts for depollution reaction. *Catalysts* **2022**, *12*, 1529. [CrossRef]
104. Hilsabeck, K.I.; Meiser, J.L.; Sneha, M.; Harrison, J.A.; Zare, R.N. Nonresonant photons catalyze photodissociation of phenol. *J. Am. Chem. Soc.* **2019**, *141*, 1067–1073. [CrossRef] [PubMed]
105. Messele, S.A.; Bengoa, C.; Stüber, F.E.; Giralt, J.; Fortuny, A.; Fabregat, A.; Font, J. Enhanced degradation of phenol by a Fenton-like system (Fe/EDTA/$H_2O_2$) at circumneutral pH. *Catalysts* **2019**, *9*, 474. [CrossRef]
106. Wei, X.Y.; Shao, S.J.; Ding, X.; Jiao, W.Z.; Liu, Y.Z. Degradation of phenol with heterogeneous catalytic ozonation enhanced by high gravity technology. *J. Clean. Prod.* **2019**, *119*, 179–189. [CrossRef]
107. Nawrocki, J.; Kasprzyk-Hordern, B. The efficiency and mechanisms of catalytic ozonation. *Appl. Catal. B Environ.* **2010**, *99*, 27–42. [CrossRef]
108. Lincho, J.; Zaleska-Medynska, A.; Martins, R.C.; Gomes, J. Nanostructured photocatalysts for the abatement of contaminants by photocatalysis and photocatalytic ozonation: An overview. *Sci. Total Environ.* **2022**, *837*, 155776. [CrossRef]
109. Xiao, J.; Xie, Y.; Cao, H. Organic pollutants removal in wastewater by heterogeneous photocatalytic ozonation. *Chemosphere* **2015**, *121*, 1–17. [CrossRef]
110. Xie, Y.; Peng, S.; Feng, Y.; Wu, D. Enhanced mineralization of oxalate by highly active and stable Ce(III)-doped g-$C_3$N4 catalyzed ozonation. *Chemosphere* **2020**, *239*, 124612. [CrossRef]

111. Yu, H.; Wang, M.; Yan, J.; Dang, H.; Zhu, H.; Liu, Y.; Wen, M.; Li, G.; Wu, L. Complete mineralization of phenolic compounds in visible-light-driven photocatalytic ozonation with single-crystal $WO_3$ nanosheets: Performance and mechanism investigation. *J. Hazard. Mater.* **2022**, *433*, 128811. [CrossRef]
112. Nishimoto, S.; Mano, T.; Kameshima, Y.; Miyake, M. Photocatalytic water treatment over $WO_3$ under visible light irradiation combined with ozonation. *Chem. Phys. Lett.* **2012**, *500*, 86–89. [CrossRef]
113. Tawabini, B.; Zubair, A. Bromate control in phenol-contaminated water treated by UV and ozone processes. *Desalination* **2011**, *267*, 16–19. [CrossRef]
114. An, W.; Tian, L.; Hu, J.; Liu, L.; Cui, W.; Liang, Y. Efficient degradation of organic pollutants by catalytic ozonation and photocatalysis synergy system using double-functional $MgO/g\text{-}C_3N_4$ catalyst. *Appl. Surf. Sci.* **2020**, *534*, 147518. [CrossRef]
115. Bhatnagar, A.; Sillanpää, M. A review of emerging adsorbents for nitrate removal from water. *Chem. Eng. J.* **2011**, *168*, 493–504. [CrossRef]
116. Loganathan, P.; Vigneswaran, S.; Kandasamy, J. Enhanced removal of nitrate from water using surface modification of adsorbents—A review. *J. Environ. Manag.* **2013**, *131*, 363–374. [CrossRef]
117. Tugaoen, H.O.N.; Garcia-Segura, S.; Hristovski, K.; Westerhoff, P. Challenges in photocatalytic reduction of nitrate as a water treatment technology. *Sci. Total Environ.* **2017**, *599*, 1524–1551. [CrossRef]
118. Anderson, J.A. Photocatalytic nitrate reduction over $Au/TiO_2$. *Catal. Today* **2011**, *175*, 316–321. [CrossRef]
119. Lin, Z.-Q.; Yuan, S.-J.; Li, W.-W.; Chen, J.-J.; Sheng, G.-P.; Yu, H.-Q. Denitrification in an integrated bioelectro-photocatalytic system. *Water Res.* **2017**, *109*, 88–93. [CrossRef]
120. Yue, M.; Wang, R.; Ma, B.; Cong, R.; Gao, W.; Yang, T. Superior performance of $CuInS_2$ for photocatalytic water treatment: Full conversion of highly stable nitrate ions into harmless $N_2$ under visible light. *Catal. Sci. Technol.* **2016**, *6*, 8300–8308. [CrossRef]
121. Anderson, J.A. Simultaneous photocatalytic degradation of nitrate and oxalic acid over gold promoted titania. *Catal. Today* **2012**, *181*, 171–176. [CrossRef]
122. Soares, O.S.G.P.; Pereira, M.F.R.; Órfão, J.J.M.; Faria, J.L.; Silva, C.G. Photocatalytic nitrate reduction over $Pd\text{–}Cu/TiO_2$. *Chem. Eng. J.* **2014**, *251*, 123–130. [CrossRef]
123. Hirayama, J.; Kamiya, Y. Combining the photocatalyst $Pt/TiO_2$ and the nonphotocatalyst $SnPd/Al_2O_3$ for effective photocatalytic purification of groundwater polluted with nitrate. *ACS Catal.* **2014**, *4*, 2207–2215. [CrossRef]
124. Hirayama, J.; Abe, R.; Kamiya, Y. Combinational effect of $Pt/SrTiO_3$:Rh photocatalyst and $SnPd/Al_2O_3$ non-photocatalyst for photocatalytic reduction of nitrate to nitrogen in water under visible light irradiation. *Appl. Catal. B Environ.* **2014**, *144*, 721–729. [CrossRef]
125. Ren, H.T.; Jia, S.Y.; Zou, J.J.; Wu, S.H.; Han, X. A Facile Preparation of $Ag_2O/P25$ Photocatalyst for Selective Reduction of Nitrate. *Appl. Catal. B Environ.* **2015**, *176*, 53–61. [CrossRef]
126. Oka, M.; Miseki, Y.; Saito, K.; Kudo, A. Photocatalytic reduction of nitrate ions to dinitrogen over layered perovskite $BaLa_4Ti_4O_{15}$ using water as an electron donor. *Appl. Catal. B Environ.* **2015**, *179*, 407–411. [CrossRef]
127. Sá, J.; Alcaraz Agüera, C.; Gross, S.; Anderson, J.A. Photocatalytic nitrate reduction over metal modified $TiO_2$. *Appl. Catal. B Environ.* **2009**, *85*, 192–200. [CrossRef]
128. Doudrick, K.; Yanga, T.; Hristovski, K.; Westerhoff, P. Photocatalytic nitrate reduction in water: Managing the hole scavenger and reaction by-product selectivity. *Appl. Catal. B Environ.* **2013**, *136*, 40–47. [CrossRef]
129. Gekko, H.; Hashimotoa, K.; Kominami, H. Photocatalytic reduction of nitrite to dinitrogen in aqueous suspensions of metal-loaded titanium (IV) oxide in the presence of a hole scavenger: An ensemble effect of silver and palladium co-catalysts. *Phys. Chem. Chem. Phys.* **2012**, *14*, 7965–7970. [CrossRef]
130. Adachi, M.; Kudo, A. Effect of Surface Modification with layered double hydroxide on reduction of nitrate to nitrogen over $BaLa_4Ti_4O_{15}$ photocatalyst. *Chem. Lett.* **2012**, *41*, 1007–1008. [CrossRef]
131. Shi, H.; Li, C.; Wang, L.; Wang, W.; Meng, X. Selective reduction of nitrate into $N_2$ by novel Z-scheme $NH_2$-MIL-101(Fe)/$BiVO_4$ heterojunction with enhanced photocatalytic activity. *J. Hazard. Mater.* **2022**, *424*, 127711. [CrossRef]
132. Ruiz-Beviá, F.; Fernández-Torres, M.J. Effective catalytic removal of nitrates from drinking water: An unresolved problem? *J. Clean. Prod.* **2019**, *217*, 398–408. [CrossRef]
133. Zazo, J.A.; García-Muñoz, P.; Pliego, G.; Silveira, J.E.; Jaffe, P.; Casas, J.A. Selective reduction of nitrate to $N_2$ using ilmenite as a low cost photo-catalyst. *Appl. Catal. B Environ.* **2020**, *273*, 118930. [CrossRef]
134. Silveira, J.E.; Ribeiro, A.R.; Carbajo, J.; Pliego, G.; Zazo, J.A.; Casas, J.A. The photocatalytic reduction of $NO_3^-$ to $N_2$ with ilmenite ($FeTiO_3$): Effects of groundwater matrix. *Water Res.* **2021**, *200*, 117250. [CrossRef]
135. Gomathi Devi, L.; Kavitha, R. A review on non metal ion doped titania for the photocatalytic degradation of organic pollutants under UV/solar light: Role of photogenerated charge carrier dynamics in enhancing the activity. *Appli. Catal. B Environ.* **2013**, *140*, 559–587. [CrossRef]
136. Girish Kumar, S.; Koteswara Rao, K.S.R. Comparison of modification strategies towards enhanced charge carrier separation and photocatalytic degradation activity of metal oxide semiconductors ($TiO_2$, $WO_3$ and ZnO). *Appl. Surf. Sci.* **2017**, *391*, 124–148. [CrossRef]
137. Lazar, M.A.; Varghese, S.; Nair, S.S. Photocatalytic water treatment by titanium dioxide: Recent updates. *Catalysts* **2012**, *2*, 572–601. [CrossRef]

138. Zhang, L.; Mohamed, H.H.; Dillert, R.; Bahnemann, D. Kinetics and mechanisms of charge transfer processes in photocatalytic systems: A review. *J. Photochem. Photobiol. C Photochem. Rev.* **2012**, *13*, 263–276. [CrossRef]
139. Luiz, D.B.; Andersen, S.L.F.; Berger, C.; José, H.J.; Moreira, R.F.P.M. Photocatalytic reduction of nitrate ions in water over metal-modified $TiO_2$. *J. Photochem. Photobiol. A Chem.* **2012**, *246*, 36–44. [CrossRef]
140. Doudrick, K.; Monzón, O.; Mangonon, A.; Hristovski, K.; Westerhoff, P. Nitrate reduction in water using commercial titanium dioxide photocatalysts (P25, P90, and Hombikat UV100). *J. Environ. Eng.* **2012**, *138*, 852–861. [CrossRef]
141. Kominami, H.; Nakaseko, T.; Shimada, Y.; Furusho, A.; Inoue, H.; Murakami, S.Y.; Kera, Y.; Ohtani, B. Selective photocatalytic reduction of nitrate to nitrogen molecules in an aqueous suspension of metal-loaded titanium (IV) oxide particles. *Chem. Commun.* **2005**, *23*, 2933–2935. [CrossRef]
142. Li, L.; Xu, Z.; Liu, F.; Shao, Y.; Wang, J.; Wan, H.; Zheng, S. Photocatalytic nitrate reduction over Pt–Cu/$TiO_2$ catalysts with benzene as hole scavenger. *J. Photochem. Photobio. A Chem.* **2010**, *212*, 113–121. [CrossRef]
143. Li, Y.; Wasgestian, F. Photocatalytic reduction of nitrate ions on $TiO_2$ by oxalic acid. *J. Photochem. Photobiol. A Chem.* **1998**, *112*, 255–259. [CrossRef]
144. Zhang, F.; Jin, R.; Chen, J.; Shao, C.; Gao, W.; Li, L.; Guan, N. High photocatalytic activity and selectivity for nitrogen in nitrate reduction on Ag/$TiO_2$ catalyst with fine silver clusters. *J. Catal.* **2005**, *232*, 424–431. [CrossRef]
145. Nakamura, K.; Yoshida, Y.; Mikami, I.; Okuhara, T. Selective hydrogenation of nitrate in water over Cu–Pd/mordenite. *Appl. Catal. B Environ.* **2006**, *65*, 31–36. [CrossRef]
146. Prüsse, U.; Hähnlein, M.; Daum, J.; Vorlop, K.D. Improving the catalytic nitrate reduction. *Catal. Today* **2000**, *55*, 79–90. [CrossRef]
147. Prüsse, U.; Vorlop, K.D. Supported bimetallic palladium catalysts for water-phase nitrate reduction. *J. Molec. Catal. A Chem.* **2001**, *173*, 313–328. [CrossRef]
148. Compton, J.E.; Harrison, J.A.; Dennis, R.L.; Greaver, T.L.; Hill, B.H.; Jordan, S.J.; Walker, H.; Campbell, H.V. Ecosystem services altered by human changes in the nitrogen cycle: A new perspective for US decision making. *Ecol. Lett.* **2011**, *14*, 804–815. [CrossRef]
149. Garcia-Segura, S.; Mostafa, E.; Baltruschat, H. Could $NO_x$ be released during mineralization of pollutants containing nitrogen by hydroxyl radical? Ascertaining the release of N-volatile species. *Appl. Catal. B Environ.* **2017**, *207*, 376–384. [CrossRef]
150. Kumar, S.G.; Rao, K.K. Tungsten-based nanomaterials ($WO_3$ & $Bi_2WO_6$): Modifications related to charge carrier transfer mechanisms and photocatalytic applications. *Appl. Surf. Sci.* **2015**, *355*, 939–958. [CrossRef]
151. Ranjit, K.T.; Viswanathan, B. Photocatalytic reduction of nitrite and nitrate ions to ammonia on M/$TiO_2$ catalysts. *J. Photochem. Photobiol. A Chem.* **1997**, *108*, 73–78. [CrossRef]
152. Park, H.; Park, Y.; Kim, W.; Choi, W. Surface modification of $TiO_2$ photocatalyst for environmental applications. *J. Photochem. Photobiol. C Photochem. Rev.* **2013**, *15*, 1–20. [CrossRef]
153. Yang, J.; Dai, J.; Li, J. Visible-light-induced photocatalytic removal of aqueous nitrate with Nd, N-codoped titania nanoparticles. *Sci. Adv. Mater.* **2013**, *5*, 1013–1023. [CrossRef]
154. Hamanoi, O.; Kudo, A. Reduction of nitrate and nitrite ions over Ni-ZnS photocatalyst under visible light irradiation in the presence of a sacrificial reagent. *Chem. Lett.* **2002**, *31*, 838–839. [CrossRef]
155. Miyazaki, A.; Matsuda, K.; Papa, F.; Scurtu, M.; Negrila, C.; Dobrescu, G.; Balint, I. Impact of particle size and metal–support interaction on denitration behavior of well-defined Pt–Cu nanoparticles. *Catal. Sci. Technol.* **2015**, *5*, 492–503. [CrossRef]
156. State, R.; Scurtu, M.; Miyazaki, A.; Papa, F.; Atkinson, I.; Munteanu, C.; Balint, I. Influence of metal-support interaction on nitrate hydrogenation over Rh and Rh-Cu nanoparticles dispersed on $Al_2O_3$ and $TiO_2$ supports. *Arab. J. Chem.* **2017**, *10*, 975–984. [CrossRef]
157. Vasile, A.; Papa, F.; Bratan, V.; Munteanu, C.; Teodorescu, M.; Atkinson, I.; Anastasescu, M.; Kawamoto, D.; Negrila, C.; Ene, C.D.; et al. Water denitration over titania-supported Pt and Cu by combined photocatalytic and catalytic processes: Implications for hydrogen generation properties in a photocatalytic system. *J. Environ. Chem. Eng.* **2022**, *10*, 107129. [CrossRef]
158. Chelu, M.; State, R.; Munteanu, C.; Atkinson, I.; Rusu, A.; Bratan, V.; Musuc, A.; Balint, I.; Papa, F. Enhanced photocatalytic activity of ZnO nanoparticles obtained by "green" synthesis with well dispersed Pd-Au bimetallic nanoparticles. *Rev. Roum. Chim.* **2018**, *63*, 837–845. Available online: https://revroum.lew.ro/wp-content/uploads/2018/09/Art%2007.pdf (accessed on 19 December 2022).
159. Park, C.; Kwak, H.; Moon, G.H.; Kim, W. Biomimetic photocatalysts for the conversion of aqueous-and gas-phase nitrogen species to molecular nitrogen via denitrification and ammonia oxidation. *J. Mater. Chem. A* **2021**, *9*, 19179–19205. [CrossRef]
160. Bems, B.; Jentoft, F.; Schlögl, C.R. Photoinduced decomposition of nitrate in drinking water in the presence of titania and humic acids. *Appl. Catal. B Environ.* **1999**, *20*, 155–163. [CrossRef]
161. Orth, W.S.; Gillham, R.W. Dechlorination of trichloroethene in aqueous solution using $Fe^0$. *Environ. Sci. Technol.* **1995**, *30*, 66–71. [CrossRef]
162. Alowitz, M.J.; Scherer, M.M. Kinetic of nitrate, nitrite and Cr(VI) reduction by iron metal. *Environ. Sci. Technol.* **2002**, *36*, 299–306. [CrossRef]
163. Hou, Z.; Chen, F.; Wang, J.; François-Xavier, C.P.; Wintgens, T. Novel Pd/$GdCrO_3$ composite for photo-catalytic reduction of nitrate to $N_2$ with high selectivity and activity. *Appl. Catal. B Environ.* **2018**, *232*, 124–134. [CrossRef]
164. Wehbe, N.; Jaafar, M.; Guillard, C.; Herrmann, J.-M.; Miachon, S.; Puzenat, E.; Guilhaume, N. Comparative study of photocatalytic and non-photocatalytic reduction of nitrates in water. *Appl. Catal. A Gen.* **2009**, *368*, 1–8. [CrossRef]

165. Gao, W.; Jin, R.; Chen, J.; Guan, X.; Zeng, H.; Zhang, F. Titania-supported bimetallic catalysts for photocatalytic reduction of nitrate. *Catal. Today* **2004**, *90*, 331–336. [CrossRef]
166. Mikami, I.; Sakamoto, Y.; Yoshinaga, Y.; Okuhara, T. Kinetic and adsorption studies on the hydrogenation of nitrate and nitrite in water using Pd-Cu on active carbon support. *Appl. Catal. B Environ.* **2003**, *44*, 79–86. [CrossRef]
167. Devadas, A.; Vasudevan, S.; Epron, F. Nitrate reduction in water: Influence of the addition of a second metal on the performances of the Pd/$CeO_2$ catalyst. *J. Hazard. Mater.* **2012**, *669*, 1412–1417. [CrossRef]
168. Hirayama, J.; Kondo, H.; Miura, Y.K.; Abe, R.; Kamiya, Y. Highly effective photocatalytic system comprising semiconductor photocatalyst and supported bimetallic non-photocatalyst for selective reduction of nitrate to nitrogen in water. *Catal. Commun.* **2012**, *20*, 99–102. [CrossRef]
169. Liu, G.; You, S.; Ma, M.; Huang, H.; Ren, N. Removal of nitrate by photocatalytic denitrification using nonlinear optical material. *Environ. Sci. Techn.* **2016**, *50*, 11218–11225. [CrossRef]
170. Sato, T.; Sato, K.; Fujishiro, Y.; Yoshioka, T.; Okuwaki, A. Photochemical reduction of nitrate to ammonia using layered hydrous titanate/cadmium sulphide nanocomposites. *J. Chem. Technol. Biotechnol.* **1996**, *67*, 345–349. [CrossRef]
171. Ketir, W.; Bouguelia, A.; Trari, M. $NO_3^-$ removal with a new delafossite $CuCrO_2$ photocatalyst. *Desalination* **2009**, *244*, 144–152. [CrossRef]
172. Ranjit, K.T.; Viswanathan, B. Photocatalytic reduction of nitrite and nitrate ions over doped $TiO_2$ catalysts. *J. Photochem. Photobiol. A Chem.* **1997**, *107*, 215–220. [CrossRef]
173. Kato, H.; Kudo, A. Photocatalytic reduction of nitrate ions over tantalate photocatalysts. *Phys. Chem. Chem. Phys.* **2002**, *4*, 2833–2838. [CrossRef]
174. Kudo, A.; Domen, K.; Maruya, K.; Onishi, T. Photocatalytic reduction of $NO_3^-$ to form $NH_3$ over Pt–$TiO_2$. *Chem. Lett.* **1987**, *6*, 1019–1022. [CrossRef]
175. Bahadori, E.; Tripodi, A.; Ramis, G.; Rossetti, I. Semi-batch photocatalytic reduction of nitrates: Role of process conditions and co-catalysts. *ChemCatChem* **2019**, *11*, 4642–4652. [CrossRef]
176. Osterloh, F.E. Inorganic nanostructures for photoelectrochemical and photocatalytic water splitting. *Chem. Soc. Rev.* **2013**, *42*, 2294–2320. [CrossRef]
177. Wang, Z.; Li, C.; Domen, K. Recent developments in heterogeneous photocatalysts for solar-driven overall water splitting. *Chem. Soc. Rev.* **2019**, *48*, 2109–2125. [CrossRef]
178. Kudo, A.; Miseki, Y. Heterogeneous photocatalyst materials for water splitting. *Chem. Soc. Rev.* **2009**, *38*, 253–278. [CrossRef]
179. Kamat, P.V. Quantum dot solar cells. Semiconductor nanocrystals as light harvesters. *J. Phys. Chem. C* **2008**, *112*, 18737–18753. [CrossRef]
180. Linic, S.; Aslam, U.; Boerigter, C.; Morabito, M. Photochemical transformations on plasmonic metal nanoparticles. *Nature Mater.* **2015**, *14*, 567–576. [CrossRef]
181. Wei, L.; Adamson, M.A.; Vela, J. $Ni_2P$-modified $Ta_3N_5$ and TaON for photocatalytic nitrate reduction. *Chem. Nano Mat.* **2020**, *6*, 1179–1185. [CrossRef]
182. Wei, L.; Liu, D.J.; Rosales, B.A.; Evans, J.W.; Vela, J. Mild and selective hydrogenation of nitrate to ammonia in the absence of noble metals. *ACS Catal.* **2020**, *10*, 3618–3628. [CrossRef]
183. Hou, Z.; Chu, J.; Liu, C.; Wang, J.; Li, A.; Lin, T.; François-Xavier, C.P. High efficient photocatalytic reduction of nitrate to $N_2$ by Core-shell Ag/$SiO_2$@ c$TiO_2$ with synergistic effect of light scattering and surface plasmon resonance. *Chem. Eng. J.* **2021**, *415*, 128863. [CrossRef]
184. Liu, Y.; Gong, X.; Yang, W.; Wang, B.; Yang, Z.; Liu, Y. Selective reduction of nitrate into nitrogen using Cu/Fe bimetal combined with sodium sulfite. *Ind. Eng. Chem. Res.* **2019**, *58*, 5175–5185. [CrossRef]
185. Shang, Z.; An, X.; Liu, H.; Hu, C.; Qu, J. Silvered $TiO_2$ for facet-dependent photocatalytic denitrification. *ACS Appl. Nano Mater.* **2021**, *4*, 13534–13542. [CrossRef]
186. Silveira, J.E.; De Souza, A.S.; Pansini, F.N.; Ribeiro, A.R.; Scopel, W.L.; Zazo, J.A.; Casas, J.A.; Paz, W.S. A comprehensive study of the reduction of nitrate on natural $FeTiO_3$: Photocatalysis and DFT calculations. *Sep. Purif. Technol.* **2023**, *306*, 122570. [CrossRef]
187. Silveira, J.E.; Garcia-Costa, A.L.; Carbajo, J.; Ribeiro, A.R.; Pliego, G.; Paz, W.S.; Zazo, J.A.; Casas, J.A. Nitrate removal in saline water by photo-reduction using natural $FeTiO_3$ as catalyst. *Chem. Eng. J. Adv.* **2022**, *12*, 100387. [CrossRef]
188. Wang, L.; Fu, W.; Zhuge, Y.; Wang, J.; Yao, F.; Zhong, W.; Ge, X. Synthesis of polyoxometalates (POM)/$TiO_2$/Cu and removal of nitrate nitrogen in water by photocatalysis. *Chemosphere* **2021**, *278*, 130298. [CrossRef]
189. Liu, X.; Liu, H.; Wang, Y.; Yang, W.; Yu, Y. Nitrogen-rich g-$C_3N_4$@ AgPd Mott-Schottky heterojunction boosts photocatalytic hydrogen production from water and tandem reduction of $NO_3^-$ and $NO_2^-$. *J. Colloid Interface Sci.* **2021**, *581*, 619–626. [CrossRef]
190. Soliman, A.M.; Alshamsi, D.; Murad, A.A.; Aldahan, A.; Ali, I.M.; Ayesh, A.I.; Elhaty, I.A. Photocatalytic removal of nitrate from water using activated carbon-loaded with bimetallic Pd-Ag nanoparticles under natural solar radiation. *J. Photochem. Photobiol. A Chem.* **2022**, *433*, 114175. [CrossRef]
191. Liu, Y.; Lee, J.; Zhao, Y.; Zhang, M.; Wang, L.; Duan, Q. A novel preparation approach and denitrification performance of $TiO_2$/$Fe^0$ photocatalysts. *Desalination Water Treat.* **2016**, *57*, 3125–3131. [CrossRef]
192. Ahmed, M.J.; Hameed, B.H.; Khan, M.A. Recent advances on activated carbon-based materials for nitrate adsorption: A review. *J. Anal. Appl. Pyrolysis* **2022**, *169*, 105856. [CrossRef]

193. Cheng, Z.W.; Jiang, Y.F.; Zhang, L.L.; Chen, J.M.; Wei, Y.Y. Conversion characteristics and kinetic analysis of gaseous $\alpha$-pinene degraded by a VUV light in various reaction media. *Sep. Purif. Technol.* **2011**, *77*, 26–32. [CrossRef]
194. Geng, Q.; Wang, H.; Chen, R.; Chen, L.; Li, K.; Dong, F. Advances and challenges of photocatalytic technology for air purification. *Natl. Sci. Open* **2022**, *1*, 20220025. [CrossRef]
195. Younis, S.A.; Kim, K.H. Heterogeneous photocatalysis scalability for environmental remediation: Opportunities and challenges. *Catalysts* **2020**, *10*, 1109. [CrossRef]
196. Figueredo, M.; Rodríguez, E.M.; Fernando, J.R.; Beltrán, J. Photocatalytic ozonation in water treatment: Is there really a synergy between systems? *Water Res.* **2021**, *206*, 117727. [CrossRef]
197. Adewui, Y.G. Sonochemistry in environmental remediation. 2. Heterogeneous sonophotocatalytic oxidation processes for the treatment of pollutants in water. *Environ. Sci. Technol.* **2005**, *39*, 8557–8570. [CrossRef]
198. Khan, M.F.; Bakhtiar, S.H.; Zada, A.; Raziq, F.; Saleemi, H.A.; Khan, M.S.; Ismail, P.M.; Alguno, A.C.; Capangpangan, R.Y.; Ali, A.; et al. Ag modified ZnO microsphere synthesis for efficient sonophotocatalytic degradation of organic pollutants and $CO_2$ conversion. *Environ. Nanotechnol. Monit. Manag.* **2022**, *18*, 100711. [CrossRef]
199. Paustian, D.; Franke, M.; Stelter, M.; Braeutigam, P. Sonophotocatalysis—Limits and possibilities for synergistic effects. *Catalysts* **2022**, *12*, 754. [CrossRef]
200. Nair, V.; Muñoz-Batista, M.J.; Fernández-García, M.; Luque, R.; Colmenares, J.C. Thermo-photocatalysis: Environmental and energy applications. *ChemSusChem* **2019**, *12*, 2098–2116. [CrossRef]
201. Muñoz-Batista, M.J.; Eslava-Castillo, A.M.; Kubacka, A.; Fernández-García, M. Thermo-photo degradation of 2-propanol using a composite ceria-titania catalyst: Physico-chemical interpretation from a kinetic model. *Appl. Catal. B Environ.* **2018**, *225*, 298–306. [CrossRef]
202. Wang, G.; Huang, B.; Lou, Z.; Wang, Z.; Qin, X.; Zhang, X.; Dai, Y. Valence state heterojunction $Mn_3O_4/MnCO_3$: Photo and thermal synergistic catalyst. *Appl. Catal. B Environ.* **2016**, *180*, 6–12. [CrossRef]
203. Zhang, C.; Li, Y.; Shen, H.; Shuai, D. Simultaneous coupling of photocatalytic and biological processes: A promising synergistic alternative for enhancing decontamination of recalcitrant compounds in water. *Chem. Eng. J.* **2021**, *403*, 126365. [CrossRef]
204. Tyagi, S.; Rawtani, D.; Khatri, N.; Tharmavaram, M. Strategies for nitrate removal from aqueous environment using nanotechnology: A review. *J. Water Process. Eng.* **2018**, *21*, 84–95. [CrossRef]
205. Andronic, L.; Isac, L.; Miralles-Cuevas, S.; Visa, M.; Oller, I.; Duta, A.; Malato, S. Pilot-plant evaluation of $TiO_2$ and $TiO_2$-based hybrid photocatalysts for solar treatment of polluted water. *J. Hazard. Mater.* **2016**, *320*, 469–478. [CrossRef] [PubMed]
206. Salmerón, I.; Rivas, G.; Oller, I.; Martínez-Piernas, A.; Agüera, A.; Malato, S. Nanofiltration retentate treatment from urban wastewater secondary effluent by solar electrochemical oxidation processes. *Sep. Purif. Technol.* **2021**, *254*, 117614. [CrossRef]
207. Gomes, J.; Roccamante, M.; Contreras, S.; Medina, F.; Oller, I.; Martin, R.C. Scale-up impact over solar photocatalytic ozonation with benchmark-P25 and N-$TiO_2$ for insecticides abatement in water. *J. Environm. Chem. Eng.* **2021**, *9*, 104915. [CrossRef]
208. Augugliaro, V.; Litter, M.; Palmisano, L.; Soria, J. The combination of heterogeneous photocatalysis with chemical and physical operations: A tool for improving the photoprocess performance. *J. Photochem. Photobiol. C Photochem. Rev.* **2006**, *7*, 127–144. [CrossRef]
209. Ajmal, Z.; Naciri, Y.; Hsini, A.; Bresolin, B.M.; Qadeer, A.; Nauman, M.; Arif, M.; Irshad, M.K.; Khan, K.A.; Djellabi, R.; et al. Prospects of Photocatalysis in the Management of Nitrate Contamination in Potable Water. In *Progress and Prospects in the Management of Oxyanion Polluted Aqua Systems. Environmental Contamination Remediation and Management*; Oladoja, N.A., Unuabonah, E.I., Eds.; Springer: Cham, Switzerland, 2021; Chapter 7; pp. 185–217. [CrossRef]
210. Sacco, O.; Vaiano, V.; Sannino, D. Main parameters influencing the design of photocatalytic reactors for wastewater treatment: A mini review. *J. Chem. Technol. Biotechnol.* **2020**, *95*, 2608–2618. [CrossRef]

catalysts

MDPI

*Review*

# A Review: Photocatalysts Based on BiOCl and g-$C_3N_4$ for Water Purification

**Qiang Ren [1], Juming Liu [2], Qi Yang [1,*] and Wei Shen [3,*]**

[1] School of Water Resources and Environment, China University of Geosciences (Beijing), Beijing 100084, China; mars103@126.com
[2] School of Chemical Engineering, Inner Mongolia University of Technology, Huhhot 010051, China; liujuming@imut.edu.cn
[3] China Waterborne Transport Research Institute, Ministry of Transport of the People's Republic of China, Beijing 100088, China
* Correspondence: yq@cugb.edu.cn (Q.Y.); shenwei@wti.ac.cn (W.S.)

**Abstract:** Many organic pollutants are discharged into the environment, which results in the frequent detection of organic pollutants in surface water and underground water. Some of the organic pollutants can stay for a long time in the environment due to their recalcitrance. Advanced oxidation processes (AOPs) can effectively treat the recalcitrant organic compounds in water. Photocatalysis as one of the AOPs has attracted a lot of interest. BiOCl and g-$C_3N_4$ are nice photocatalysts. However, their catalytic activity should be further improved for industrial utilization. The construction of heterojunction between the two different components is deemed as an efficient strategy for developing a highly efficient photocatalyst. As a typical type-II heterojunction, g-$C_3N_4$/BiOCl heterojunctions showed better photocatalytic performance. To date, the g-$C_3N_4$/BiOCl composites were mainly studied in the field of water purification. The photoactivity of the pristine catalysts was greatly enhanced by the combination of the two materials. However, three kinds of proposed mechanisms were used to explain the improvement of the g-$C_3N_4$/BiOCl heterojunctions. But few researchers tried to explain why there were three different scenarios employed to explain the charge transfer. According to the articles reviewed, no direct evidence could indicate whether the band structures of the heterojunctions based on BiOCl and g-$C_3N_4$ were changed. Therefore, many more studies are needed to reveal the truth. Having a clearer understanding of the mechanism is beneficial for researchers to construct more efficient photocatalysts. This article is trying to start a new direction of research to inspire more researchers to prepare highly effective photocatalysts.

**Keywords:** BiOCl; $C_3N_4$; photocatalysis; mechanism

**Citation:** Ren, Q.; Liu, J.; Yang, Q.; Shen, W. A Review: Photocatalysts Based on BiOCl and g-$C_3N_4$ for Water Purification. *Catalysts* **2021**, *11*, 1084. https://doi.org/10.3390/catal11091084

Academic Editors: Ioan Balint and Monica Pavel

Received: 19 August 2021
Accepted: 6 September 2021
Published: 8 September 2021



## 1. Introduction

With the development of human civilization, many refractory pollutants were discharged to the environment, which are hard to be degraded by traditional purification methods. For example, pollutants like tetracycline [1], bisphenol A [2], Astrazone Black [3], estriol [4], and tetraethylated rhodamine [5] can hardly be degraded by normal waste-water treatment plants. Generally, the degradation of recalcitrant organic pollutants relies on the consumption of energy, such as the Fenton process. Photocatalysis has drawn much attention because of the utilization of solar energy and friendliness to the environment.

In photocatalysis, the heterojunction of two different materials is deemed as an efficient strategy to develop diverse hybrid composites with multiple functionalities [6]. Photocatalysts normally possess some defects that restrict the utilization of the material, for example, ultrafast recombination of photo-induced charge carriers [7], low efficient utilization of sunlight, and wide bandgap [8]. Hybridization of two different catalysts would be an excellent way to improve the photoactivity. Lots of photocatalysts were synthesized in order to enhance the photoactivity of the catalysts, such as $MoS_2$/g-$C_3N_4$ [9], $CuInS_2$/g-$C_3N_4$ [10], $Ag_2O$/g-$C_3N_4$ [11], $Ag_2O/TiO_2$ [12], AgI/$CuBi_2O_4$ [13], CuS/$BiVO_4$ [14], $Ag_3PO_4/MoS_2$ [15], g-$C_3N_4/SiO_2$ [16], $ZnFe_2O_4/TiO_2$ [17], $LaFeO_3/SnS_2$ [18], $Bi_2O_3$/g-$C_3N_4$ [19], and $Ag_2O/Bi_5O_7I$ [20]. Recently, heterojunctions based on BiOCl and g-$C_3N_4$ drew much attention because of the abundance of the materials in the environment.

Graphitic carbon nitride is regarded as a nice photocatalyst because of its nontoxicity, stability in pH over a broad range (0–14), easy to prepare, and the narrow bandgap [21]. After being first synthesized in 1834 [22], carbon nitride has been used in many areas, such as virus inactivation [23], activation of benzene [24], $H_2$ revolution [25,26], fuel cells [27], $CO_2$ reduction [28], and organic pollutants degradation [29]. However, because of its high recombination rate of the photogenerated charge carriers and low BET surface area, the application of g-$C_3N_4$ is restricted. Recently, many studies focused on building heterojunctions to improve its photoactivity, such as the system of $WO_3$/g-$C_3N_4$ [30], $ZnWO_4$/g-$C_3N_4$ [31], and $In_2S_3$/g-$C_3N_4$ [32]. According to these studies, coupling graphitic carbon nitride with other kinds of semiconductors could construct better photocatalysts by reducing the recombination rate of the photogenerated charge carriers or increasing the surface area.

On the contrary, layered structure of BiOCl facilitates the photogenerated charge carriers' separation and endows it with a strong ability to degrade organic pollutants [33]. Morphology control was employed by many researchers to improve the pristine catalyst [34]. For example, according to E. Ramírez Meneses and co-workers [35], the addition of capping agents could affect the morphology of BiOCl. However, the as-prepared catalysts were unable to be excited by visible light. In order to expand the light absorption range of BiOCl, many researchers synthesized heterostructures like carbon dots/BiOCl [36], BiOCl/g-$C_3N_4$ [37], $WO_3$/BiOCl [38], $Bi_2MoO_6$-BiOCl [33], BiOCl/BiOBr [39], m-$Bi_2O_4$/BiOCl [40], BiOCl/$BiVO_4$ [41], $Bi_2O_2CO_3$/BiOCl [42], and BiOI/BiOCl [43].

Among them, the composition of BiOCl and g-$C_3N_4$ is considered as an excellent combination. The heterojunction could enhance the separation of the photo-induced charge carriers and enable the catalyst to respond to visible light [43,44]. Noble metal doping is also considered as a good method to improve the semiconductor. However, the high cost of noble metal doping restricts its utilization. If a noble metal doping catalyst is used repeatedly, the catalyst will be eroded, and perhaps generate new pollutants [45]. Heterojunctions of semiconductors is friendly to the environment, stable, and abundant in nature. Especially, the system of g-$C_3N_4$/BiOCl could be used repeatedly and facile to be produced.

Some researchers also found the photoactivity of g-$C_3N_4$/BiOCl heterojunctions could be further enhanced by combining them with other materials. For example, the systems of $Bi_2S_3$/BiOCl/g-$C_3N_4$ [46], BiOCl/g-$C_3N_4$/kaolinite [47], and g-$C_3N_4$/CDs (carbon dots)/BiOCl [48]. Notably, through the addition of mediators, Z-scheme catalysts can be synthesized, such as the systems of g-$C_3N_4$/Au/BiOCl [49] and BiOCl/RGO/protonated g-$C_3N_4$ [50]. However, few researchers have focused on figuring out which method could prompt the photoactivity of the binary heterojunction. According to all the articles reviewed here, analysis of the proposed mechanism was an important section. Based on the adopted characterizations and experiments, the mechanism was discussed to help readers to understand the whole photocatalysis process.

First, in this article, the methods of preparation and the applications of the g-$C_3N_4$/BiOCl heterojunctions are reviewed. Then, the binary heterojunctions mentioned in this article were classified into three types according to the proposed mechanisms. The major difference between them is whether the alignment of band structures was taken into account after the syntheses. Many researchers carried out some experiments to prove their mechanisms. For instance, ESR and trapping experiments could demonstrate the main reactive species during the reaction. DFT (density functional theory) calculation was also adopted to anticipate the band structures and main reactive species [51]. However, all the methods seemed still not enough to directly prove the mechanism. Some of the studies seem controversial to each other. All three types of proposed mechanisms will be discussed in this article while trying to find some patterns. Though dye sensitization happened when the heterojunctions were used to degrade some pollutant, this article tries to discuss whether the alignment of the band structure should be taken into account. At last, in the section of summary and outlook, the direction of the study on the mechanism is proposed.

## 2. Synthesis of g-$C_3N_4$/BiOCl Heterojunction

In the field of photocatalysis, g-$C_3N_4$ has been the focus of research in the past decade. Because the photo-induced electrons of g-$C_3N_4$ recombines with the hole very quickly. Though visible light accounts for more than 40% of solar light, this feature limits its utilization as a perfect photocatalyst.

The indirect band structure of BiOCl means photo-induced electron-hole pairs cannot recombine very quickly [8]. Furthermore, its layered structure facilitates the generation of the reactive species. The biggest problem is that this material only responds to UV-light, which merely accounts for 5% of solar light. This means BiOCl cannot make use of solar energy sufficiently.

In order to overcome the shortcomings of the semiconductors mentioned above, metal and carbonaceous materials were introduced [52]. Doping of those materials could improve charge separation and light absorption of the pristine catalysts. For example, noble metal with high electric conductivity and different Fermi level energy were doped into the semiconductor. The metal particles acted as the electron reservoirs or acceptors on the surface of the catalyst. However, holes generated on the semiconductor cannot migrate to noble metal particles because of the existence of the Schottky barrier. Thus, charge separation is realized through elemental doping [22]. However, doping of other materials has its limitations as mentioned above. None-mental heterojunctions are becoming more attractive. The semiconductor-semiconductor hybridization is generally supposed to be an effective way to improve the photoactivity of the semiconductors. Composite of BiOCl and g-$C_3N_4$ is proved to be an excellent form of a heterojunction photocatalyst.

As it is known to us all, efficient methods to enhance the activity of photocatalyst can be divided into two types: (1) facilitating photo-induced charge separation; (2) improving the efficiency of solar energy utilization. There is no doubt that making a heterojunction between two different semiconductors is an excellent way to realize the two aims at the same time. Compositing BiOCl and g-$C_3N_4$ can not only improve charge carriers' separation, but also make the composite respond to visible light.

Heterojunctions of g-$C_3N_4$/BiOCl could be synthesized by many methods. According to the articles reviewed, there are three different methods overall employed by researchers, hydrothermal, deposition–precipitation, and solvent thermal. It is hard to say which one of them is the best. Generally, catalysts with larger surface area and thinner morphology could facilitate the degradation of pollutants. All the heterojunctions reviewed showed enhanced visible light absorption. Thus, it is reasonable to say the photoactivity of the catalysts was improved through the construction of the heterojunction.

Graphitic carbon nitride synthesized through calcination of urea showed higher surface area and better photoactivity [22]. Considering the theoretical specific surface area of perfect monolayer g-$C_3N_4$ is 2500 $m^2\ g^{-1}$, there is still a long distance to cover [53]. However, it is not the main reason that explains the photocatalytic improvement of the composite based on the system of g-$C_3N_4$/BiOCl. The BET surface area of pristine g-$C_3N_4$ could be 150.10 $m^2\ g^{-1}$, but the photoactivity is much lower than the hybrid. The surface area of BiOCl-CNs-3% is much lower than BiOCl-CNs-5%, whereas the photoactivity is much higher [6].

The mass ratio of the heterojunction is a very important factor. There is no doubt that the addition of BiOCl can prompt the photo-induced charge separation, and $C_3N_4$ can enhance the absorption of visible light in the systems. Theoretically, higher mass ratio of BiOCl means more chances for charge carriers to be separated, which could improve the photoactivity of the composite. When the mass ratio of one certain component increases, the photoactivity of the composite will not stop being enhanced until the generation of and separation of photo-induced charge carriers reaches a certain balance. However, higher ratio of BiOCl does not always mean higher photoactivity. In the system of (20%) g-$C_3N_4$/BiOCl when the mass ratio of g-$C_3N_4$ reached 20%, the catalyst showed the best photocatalysis performance [21]. When the mass ratio was below 20%, increasing the ratio of g-$C_3N_4$ led to the photocatalytic enhancement of the heterojunction. If the mass ratio increased further, the activity started to decrease. Because too much g-$C_3N_4$ tends to agglomerate with itself, the contact between BiOCl and g-$C_3N_4$ was weakened [54]. The same thing happened when the mass ratio of BiOCl increased.

Generally, the morphology of g-$C_3N_4$ is hard to control unless using a certain template or different precursor [55]. Direct calcination of the precursors could only produce bulk g-$C_3N_4$. In order to improve the connection between the g-$C_3N_4$ and the BiOCl, exfoliation of the bulk g-$C_3N_4$ was adopted by many researchers. There are mainly three exfoliation methods that were adopted, chemical blowing, thermal, and liquid exfoliation. Previous researchers have proved that g-$C_3N_4$ exfoliated by thermal exfoliation method exhibited better photocatalytic performance than others through comparative experiments [56]. According to the articles reviewed, g-$C_3N_4$ commonly synthesized before the construction of the binary heterojunctions, so the construction of the heterojunction could not affect the morphology of g-$C_3N_4$. However, the presence of g-$C_3N_4$ could influence the morphology of BiOCl in the binary heterojunction. Therefore, the morphology control of BiOCl should be a very important factor that influences the photocatalytic performance of the binary heterojunction system under visible light irradiation.

Moreover, facet control of BiOCl is also a very efficient method to enhance photoactivity of the heterojunction. BiOCl with exposed 001 and 010 facets could be synthesized through PH value adjustment during the process of the construction route [57]. The better visible-light photocatalytic performance of BiOCl-010 could be attributed to the larger surface area, which could also beneficial for the construction of the heterojunction.

### 2.1. Hydrothermal Method

Generally, in the hydrothermal method, g-$C_3N_4$ powder-prepared through polymerization-was dispersed in deionized water, and then the solution was stirred in order to prepare a suspension. Commonly, Bi $(NO_3)_3 \cdot 5H_2O$ and KCl were added into the as-prepared suspension. Subsequently, the mixed suspension was transferred to an autoclave and heated to gain the BiOCl/g-$C_3N_4$. To date, three kinds of BiOCl/g-$C_3N_4$ heterojunction were prepared through the hydrothermal method (listed in Table 1).

**Table 1.** G-$C_3N_4$/BiOCl heterojunctions synthesized by hydrothermal method.

| Catalyst (Mass Ratio %) | Template | Morphology | Size | Year | Ref. |
|---|---|---|---|---|---|
| BiOCl/g-$C_3N_4$ (50/50) | SDBS | nanoplate | <5 nm | 2017 | [58] |
| g-$C_3N_4$/BiOCl (23.03/76.97) | - | nanodisc | 35–50 nm | 2017 | [59] |
| BiOCl/g-$C_3N_4$ (40/60) | - | - | 2 μm | 2015 | [60] |

Just as analyzed above, morphology of BiOCl is the most important variable in the process. Generally speaking, BiOCl of the BiOCl/g-$C_3N_4$ system with larger surface area is expected to show better visible light absorption. So, the main purpose of this section is to find out which method could make the morphology of BiOCl become thinner.

L. Song and co-workers employed $NH_4Cl$ as blowing agent to make ultrathin g-$C_3N_4$ nanosheets (prepared by polymerization of melamine) in order to enhance the contact between the two materials [59]. In that article, two-dimensional g-$C_3N_4$/BiOCl heterojunctions were prepared through a facile hydrothermal method. The width of Pristine BiOCl is less than 3 μm and thicknesses is 50–60 nm. Notably, the construction of the binary heterojunction affected the morphology of the BiOCl. The thickness of nanodisc-like BiOCl was around 35–50 nm after loaded by ultrathin graphitic carbon nitride.

Yifan Yang and coworkers also adopted a hydrothermal method to synthesize the heterojunction [60]. Interestingly, before the construction of the g-$C_3N_4$/BiOCl heterojunction, they synthesized g-$C_3N_4$ and BiOCl separately. After the formation of the heterojunction, it was observed that the width of BiOCl was 2 μm and the thickness was about 200 nm. Obviously, the surface area of the heterojunction should be smaller than that synthesized in the study mentioned above, though it showed better photoactivity than pristine catalysts.

According to Yongkui Huang and colleagues, they used $BiCl_3$ as a precursor to synthesize the heterojunction [58]. It was reported that nanoplate-shaped BiOCl decorated on the surface of g-$C_3N_4$ with thickness below 5 nm and widths of 20~30 nm. However, the result of X-ray powder diffraction (XRD) showed that the 001-facet exposed BiOCl and was the dominate, which is supposed to beneficial for the UV-light photocatalytic reaction.

Especially, when it comes to visible-light, among all the articles reviewed in this section, though heterojunction synthesized by using $BiCl_3$ as precursor showed the thinnest structure, method employed by L. Song and co-workers should be the better way to synthesize g-$C_3N_4$/BiOCl heterojunction.

### 2.2. Deposition-Precipitation Method

Deposition-precipitation is another facile way to synthesize g-$C_3N_4$/BiOCl binary heterojunction. According to the studies reviewed, most researchers adopted this method to fabricate the heterojunction as listed in Table 2. Qingbo Li and co-workers proved that facet control was still the main factor that affected the photoactivity of the heterojunction under visible light illumination, though proper way of exfoliation could also improve the photocatalytic performance [61]. They synthesized g-$C_3N_4$, BiOCl-010, and BiOCl-001, separately, before the construction of the heterojunction. It was obvious that the BiOCl-010 possessed shaper edges and smaller size, according to Figure 1, which means its surface area was larger.

**Figure 1.** SEM images of (**a**) g-$C_3N_4$/BiOCl-001 and (**b**) g-$C_3N_4$/BiOCl-010. Reproduced with permission from Iqbal W et al, Catalysis Science & Technology; published by Royal Society of Chemistry, 2018.

Furthermore, the unique hierarchical flowerlike morphology of BiOCl could improve the photoactivity of the heterojunction because of its enlarged surface area [6]. This was proved by Liwen Lei and coworkers through synthesizing flower-like BiOCl by using Arabic gum (AG) as a template [34]. The heterojunction using AG showed better photocatalytic performance than that without using the template. Weidong Hou and coworkers also synthesized a flower-like g-$C_3N_4$/BiOCl heterojunction employing a microwave-assisted method [62]. However, the function of microwave was to accelerate the reaction process and to enhance the purity of the heterojunction [63,64]. Therefore, the utilization of microwave could not control the morphology of the composite. Compared to the study completed by Weidong Hou and colleagues, the presence of ethylene glycol could facilitate the formation of flower-like heterojunction. Tiekun Jia and colleagues also constructed a flower-like heterojunction by using ethylene glycol and glycerine [65]. The presence of glycerine increased the surface area of the composite.

**Table 2.** G-$C_3N_4$/BiOCl heterojunctions synthesized by deposition–precipitation method.

| Catalyst (Mass Ratio %) | Template | Morphology | Size | BET Surface Area | Year | Ref. |
|---|---|---|---|---|---|---|
| BiOCl/g-$C_3N_4$ (97/3) | - | Hierarchical flowerlike | 0.15 μm/10 nm | 19.04 | 2014 | [6] |
| g-$C_3N_4$/BiOCl (20/80) | - | Nanoplate | 1 μm | - | 2014 | [21] |
| $C_3N_4$/BiOCl (20/80) | Arabic gum | Flower-like | 200 nm/5–8 nm | 49.37 | 2014 | [34] |
| ng-CN/BOC-010 (70/30) | - | Nanoparticle-nanosheet | - | 18.10 | 2015 | [61] |
| (OV)BiOCl/g-$C_3N_4$-10 | - | Flower-like | 2 μm/33.7 nm | 11.66 | 2020 | [62] |
| BiOCl/g-$C_3N_4$ (10/90) | - | Sheet-like+ microplate | 51.8 nm | - | 2019 | [63] |
| g-$C_3N_4$ /BiOCl (55/45) | - | Hierarchical flower-like | 1 μm/10 nm | 44.2 | 2017 | [65] |
| BiOCl-g-$C_3N_4$ | - | Two-dimensional structure | 10 nm | - | 2014 | [66] |
| BiOCl-g-$C_3N_4$ (50/50) | CTAC | Wrinkle two-dimensional structure | 10 nm | - | 2014 | [67] |
| g-$C_3N_4$/BiOCl (20/80) | - | Nanoplate + sheets | 1 μm | - | 2015 | [68] |
| BiOCl/(0.1g)g-$C_3N_4$ | - | nanosheet | 20 nm | 6.60 | 2017 | [69] |

Though it was observed that the presence of g-$C_3N_4$ during the synthesis of the BiOCl would make the morphology of BiOCl become thinner, the addition of some template could construct even thinner morphology, which means enlarged surface area. Yang Bai and coworkers synthesized a g-$C_3N_4$/BiOCl heterojunction by using cetyltrimethylammonium chloride (CTAC) as the template [66]. The size of the composite was about 10 nm. The template facilitated the formation of BiOCl nano-dots deposited on the surface of g-$C_3N_4$ as reported by Chun-zhi Zheng and colleagues [67]. However, if the heterojunction was synthesized without using any template, the morphology of the composite was supposed to be larger. Jiangbo Sun and coworkers constructed a g-$C_3N_4$/BiOCl-010 heterojunction just by adjusting the PH value [68]. The size of it was about 5 μm.

According to Lingjun Song and colleagues, the thickness of BiOCl decreased from 40 to 20 nm after combined with g-$C_3N_4$ [69]. The same phenomenon was observed by Weidong Hou and coworkers, the as-prepared flower-like BiOCl became thinner as the content of g-$C_3N_4$ increased [62]. The width and thickness of pristine BiOCl became smaller than 2 μm and 33.7 nm after the construction of the binary heterojunction, as demonstrated in Figure 2. Notably, among all the articles reviewed in this section, Shan Shi and colleagues employed $NaBiO_3$ to synthesize the g-$C_3N_4$/BiOCl heterojunction instead of $Bi(NO_3)_3{\cdot}5H_2O$ [21]. According to the study, though its size was about 1 μm, the BiOCl in the as-prepared heterojunction was 001 facets exposed, which could facilitate response to the UV light illumination. When it came to visible light photocatalytic reaction, the facile route adopted by Liwen Lei and coworkers could synthesize the heterojunction with larger surface area among all the deposition–precipitation methods reviewed.

### 2.3. Solvent-Thermal Method

The solvent thermal method is similar to the aforementioned hydrothermal method. The main difference is the starting materials of solvent thermal method are dissolved in some organic solvent while those of hydrothermal method dissolved in deionized water. All the g-$C_3N_4$/BiOCl heterojunctions synthesized by solvent thermal method are listed in Table 3.

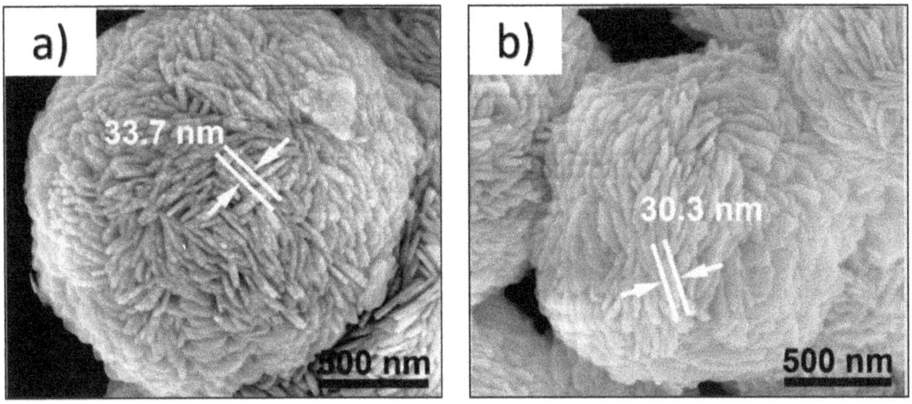


**Figure 2.** SEM images of (**a**) flower-like BiOCl and (**b**) g-$C_3N_4$/BiOCl. Reproduced with permission from Hou W et al, ChemistrySelect; published by John Wiley and Sons, 2020.

**Table 3.** G-$C_3N_4$/BiOCl heterojunctions synthesized by solvent thermal method.

| Catalyst (Mass Ratio %) | Template | g-$C_3N_4$ Precursor | Morphology | Diameter or Thickness | BET Surface Area ($m^2$/g) | Year | Ref. |
|---|---|---|---|---|---|---|---|
| BiOCl-$C_3N_4$ (50/50) | IL: [HMIm]Cl | melamine | nanoflowers | - | 24.26 | 2013 | [44] |
| BiOCl/g-$C_3N_4$ (85/15) | CTAB | melamine | rolled flake+ lamellar | 70 nm | 47.1 | 2019 | [70] |
| (OV)BiOCl-g-$C_3N_4$ (50/50) | PVP | urea | Ultrathin nanosheet | ~4.3 nm | 62.0 | 2017 | [71] |
| BOC/CN (60/40) | - | melamine | ultrathin layered structure | around 3.5 nm | 68.5 | 2019 | [72] |
| BiOCl/$C_3N_4$ | - | urea | Nanoplate+ rough slice | - | 47.1 | 2017 | [73] |
| g-$C_3N_4$/BiOCl (1/99) | [C16mim]Cl | Dicyandiamide′ | three-dimensional spherical structure | 1 μm | 22.58 | 2016 | [74] |

Notably, the ionic liquid 1-hexyl-3-methylimidazolium chloride (IL: [HMIm]Cl) was used as one of the starting materials. It was used as a template and the source of chloride. For example, Xiao-jing Wang and colleagues prepared a novel g-$C_3N_4$/BiOCl heterojunction through a facile solvent thermal route [44]. According to the study, ionic liquid (IL: [HMIm]Cl) not only played an important role as the source of Cl, but also as a template to direct the growth of the nanoplate into flowerlike nanoparticles. The as-prepared g-$C_3N_4$/BiOCl composite was synthesized with larger surface area than the pristine catalysts and better photocatalytic performance. Ionic liquid was also adopted by other researchers, S. Yin and colleagues prepared a g-$C_3N_4$/BiOCl composite that employed the ionic liquid [$C_{16}$mim]Cl as the source of Cl [74]. Like the study mentioned above, $Bi(NO_3)_3 \cdot 5H_2O$ and ethylene glycol were used as starting materials. Similarly, the two composites showed a flower-like microsphere structure instead of the nano-slice structure. It is reasonable to believe that the presence of ionic liquid could direct the morphology of the g-$C_3N_4$/BiOCl heterojunction in these two studies. According to the XRD patterns of them, the BiOCl of the two heterojunctions were not 001 facet dominant like showed in Figure 3. This might mean the heterojunction could respond better to visible light than UV-light.

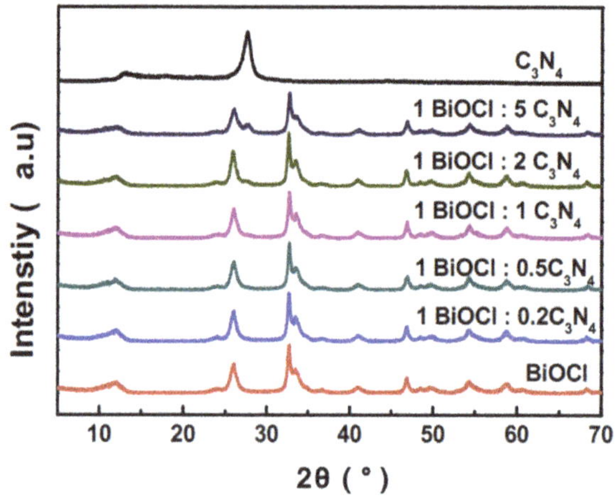


**Figure 3.** XRD patterns of BiOCl, $C_3N_4$, and g-$C_3N_4$/BiOCl heterojunction in different compositions. Reproduced with permission from Wang XJ et al, Chemical Engineering Journal; published by Elsevier BV, 2013.

Among all the heterojunctions synthesized through the solvent thermal method, only ionic liquid assisted routes could prepare flower-like g-$C_3N_4$/BiOCl heterojunctions. There were other templates that researchers employed to regulate the morphology of the heterojunction. For instance, cetyltrimethyl ammonium bromide (CTAB) and urea were used by Wei Cai and colleagues to prepare a nano-sliced g-$C_3N_4$/BiOCl heterojunction [70]. The construction of the heterojunction did not change the morphology of the BiOCl obviously. However, the presence of the template could not just direct the morphology. According to Qiao Wang and coworkers, the presence of the template could facilitate the generation of oxygen vacancies [71]. Unlike the aforementioned studies, NaCl was adopted as the source of Cl, and Polyvinylpyrrolidone (PVP) was used as the template. A facile solvent thermal method was adopted to prepare an ultrathin g-$C_3N_4$/BiOCl heterojunction. Just like reported by Xianlong Zhang and colleagues [72], the formation of the heterojunction could make the morphology of the BiOCl become thinner. It was proved that the absence of PVP lowered the photoactivity of the heterojunction. Oxygen vacancies were supposed to act as the trap of electrons and facilitated the generation of superoxide radicals. The result of the density-functional calculation indicated that the presence of the oxygen vacancies narrowed the bandgap, thus improving the absorption of light.

Though the addition of some templates could help to prepare heterojunctions with thinner structure and enhance the photoactivity, the presence of them might cause some environmental problems. So, scientists adopted some template-free solvent thermal methods to prepare the binary heterojunction. Xianlong Zhang and colleagues synthesized a 2 D/2 D g-$C_3N_4$/BiOCl composite via a solvent thermal method without the presence of any templates [72]. According to the study, HCl was used as the source of the Cl. The study indicated that ultrathin g-$C_3N_4$/BiOCl nanosheets were prepared without the presence of any templates. The two pristine catalysts were self-assembled to form a face-to-face contact, which facilitated the charge transfer and the light absorption. Wenwen Liu and coworkers prepared a square-like nanoplates heterojunction without using templates [73]. NaCl aqueous solution was the source of the Cl. Compared to the study mentioned before, HCl might help to enlarge the surface area of the heterojunction.

Therefore, the heterojunction prepared through the template-free solvent thermal method adopted by Xianlong Zhang and colleagues showed the thinnest structure among all the solvent thermal methods. However, the as-prepared BiOCl was 001 facet dominant. It is reasonable to believe that the as-prepared catalyst would exhibit better photoactivity under UV-light illumination. According to the articles reviewed, all the heterojunction were not 001 facet exposed, except that prepared by Xianlong Zhang and colleagues. So, when it comes to visible light illumination, heterojunction synthesized by Qiao Wang and coworkers could exhibit better photocatalytic performance than other catalysts because of its enlarged surface area.

### 2.4. Calcination Method

Besides the methods mentioned above, Wenjie Shan and colleagues synthesized the g-$C_3N_4$/BiOCl heterojunction through an in-situ calcination method [75]. They obtained a composite with enhanced photoactivity through the direct calcination of $Bi_2O_3$ and guanidine hydrochloride. Despite the different starting materials, the as-prepared composite showed superior catalytic performance and enlarged surface area (25.50 $m^2$ $g^{-1}$), similar to the flower-like structured g-$C_3N_4$/BiOCl heterojunctions synthesized through other methods. Notably, the 001 facet of as-prepared catalyst showed very strong signal. The heterojunction might show better photocatalytic performance under UV-light illumination.

Therefore, the methods employed to synthesize the g-$C_3N_4$/BiOCl binary photocatalyst mainly focused on making a thinner and hierarchical structure. Facilitating the charge transfer between the two components is one of the main purposes. The larger surface area might be beneficial for the enlarged interfacial contact area, which is why the thinner structured heterojunction exhibited obviously enhanced photoactivity compared to the pristine catalysts. It is believed larger surface area of one certain material could provide more active sites for the absorption of organic matters and photocatalytic reactions [45,76,77]. On one hand, higher surface area means higher mass to surface area rate and more area for other catalysts to deposit on it. On the other hand, enlarged surface area endows photocatalysts to absorb visible light more efficiently [41]. Among the articles reviewed, the exfoliation and introduction of organic template produced the thinner structured BiOCl heterojunction that possessed larger face-to-face interface with g-$C_3N_4$ and facilitated the efficient charge separation of light-induced charge carriers of the heterojunction system [65,71]. A similar pattern could be observed in other heterojunction systems [7,34,44,49,78]. However, whether a certain structure is beneficial for photoactivity generally depends on the circumstance [79]. The function of morphology still needs to be studied further to reveal a clearer pattern.

## 3. Applications of g-$C_3N_4$/BiOCl Heterojunction

To date, the g-$C_3N_4$/BiOCl heterojunction was mainly applied in the degradation of organic dyes, residual pharmaceutical agents, and plasticizers in aqueous solution. The heterojunction was not applied in fields of $H_2$ generation, $CO_2$ reduction, water splitting, or disinfection yet.

### 3.1. Dye Degradation

Organic dyes are widely used, particularly azo dyes, and are very difficult to be degraded by conventional biological treatment because of the complex aromatic structure, which is highly hazardous to the human race [80,81]. For example, RhB is almost unable to be degraded under irradiation of visible light without the addition of a photocatalyst [34]. Degradation of rhodamine B was employed repeatedly to test the photocatalytic performance of the g-$C_3N_4$/BiOCl composite. Methyl orange (MO) and methylene blue (MB) were also used to test photoactivity of the binary heterojunction. All the heterojunctions applied to dye degradation were listed in Table 4.

**Table 4.** G-$C_3N_4$/BiOCl heterojunctions applied to dye degradation.

| Catalyst (Mass Ratio%) | Application | Efficiency/Time | Light Source | Concentration of the Pollutant | Main Reactive Species | Stability and Reusability | Ref. |
|---|---|---|---|---|---|---|---|
| BiOCl/g-$C_3N_4$ (97/3) | Degradation of RhB | 100%/40 min | 400 W halogen lamp sodium nitrite solution (2 M) to eliminate UV light ($\lambda <$ 400 nm) and thermal effect | 50 mg/L | $\bullet O_2^-$ | stable after 7 irradiation cycles | [6] |
| g-$C_3N_4$/BiOCl (20/80) (001) | Degradation of RhB | 100%/35 min | 500 W Xenon lamp with a light filter 400–800 nm | 7 mg/L | - | 65% after 10 irradiation cycles | [21] |
| $C_3N_4$/BiOCl (20/80) | Degradation of RhB | ~100%/20 min | 500 W Xe arc lamp UV-cut off filter ($\lambda >$ 420 nm) | 20 mg/L | $\bullet O_2^-$, hole | 90% after 5 irradiation cycles | [34] |
| BiOCl-$C_3N_4$ (50/50) (001) | Degradation methyl orange (MO) | 95%/80 min | 300 W xenon arc lamp 400 nm cutoff filter | 10 mg/L | hole | stable after 6 irradiation cycles | [44] |
| BiOCl/g-$C_3N_4$ (50/50) (001) | Degradation of RhB | >90%/50 min | 300 W Xelamp 400 nm cutoff filter | 10 mg/L | - | - | [58] |
| g-$C_3N_4$/BiOCl (23.03/76.97) | Degradation of RhB | 100%/30 min | 300 W Xe arc lamp 400 nm cutoff filter 35 mW/$cm^2$ | 10 mg/L | $\bullet O_2^-$, hole | stable after 4 irradiation cycles | [59] |
| BiOCl/g-$C_3N_4$ (40/60) (001) | Degradation of methylene-blue (MB) | 80%120 min | 500 W Xenon lamp 420 nm cutoff filter | 5–10 mol/L | - | stable after 5 irradiation cycles | [60] |
| ng-CN/BOC-010 (70/30) | Degradation methyl orange (MO) | >90%/150 min | 300 W metal-halide lamp 420 nm cutoff filter | 20 mg/L | $\bullet O_2^-$, hole | - | [61] |
| g-$C_3N_4$ /BiOCl (55/45) | Degradation of methylene blue (MB) | 100%/30 min | daylight lamp 60 W, $\lambda \geq$ 400 nm | 5–10 mol/L | hole | ~99% after 5 irradiation cycles | [65] |
| BiOCl-g-$C_3N_4$ | Degradation of RhB | 99%/60 min | 300 W xenon lamp incident lightpower: 6 W | 10 mg/L | $\bullet OH$, $\bullet O_2^-$ | - | [66] |
| BiOCl-g-$C_3N_4$ (50/50) | Degradation of RhB | 99%/35 min | 300 W xenon lamp | 10 mg/L | - | - | [67] |
| g-$C_3N_4$/BiOCl (20/80) | Degradation of RhB | 100%/20 min | 500-W Xe lamp with a light filter 400–800 nm | 7 mg/L | - | 56% after 5 irradiation cycles | [68] |

**Table 4.** *Cont.*

| Catalyst (Mass Ratio%) | Application | Efficiency/Time | Light Source | Concentration of the Pollutant | Main Reactive Species | Stability and Reusability | Ref. |
|---|---|---|---|---|---|---|---|
| BiOCl/(0.1 g)g-$C_3N_4$ (001) | Degradation of RhB | 100%/50 min | 300 W Xe arc lamp<br>400 nm cutoff filter<br>35 mW/$cm^2$ | 10 mg/L | •$O_2^-$, hole | stable after 5 irradiation cycles | [69] |
| BiOCl/g-$C_3N_4$ (85/15) | Degradation of RhB | 90%/30 min | 300 W Xe lamp<br>400 nm cutoff filter | 25 mg/L | •$O_2^-$ | - | [70] |
| BOC/CN (60/40) | Degradation of RhB | 95.93%/80 min | 300 W Xe lamp<br>420 nm cutoff filter | 10 mg/L | •$O_2^-$, hole | 89% after 5 irradiation cycles | [72] |
| BiOCl/$C_3N_4$ | Degradation of methyl orange (MO) | 84.28%/180 min | 300 W Xe lamp<br>0.5 mol·$L^{-1}$ $Na_2SO_4$<br>Solution ($\lambda \geq 420$ nm) | 10 mg/L | •$O_2^-$, hole | stable after 4 irradiation cycles | [73] |
| g-$C_3N_4$/BiOCl (1/99) | Degradation of RhB | 94 %/30 min | 300 W Xe lamp<br>400 nm cutoff filter | 10 mg/L | •$O_2^-$, hole | - | [74] |
| g-$C_3N_4$/BiOCl (001) | degradation of methylorange (MO) | 70%/300 min | 500 W halogen tungsten lamp<br>420 nm cut-off filter | 20 mg/L | - | stable after 4 irradiation cycles | [75] |

The main defect that influences the photoactivity of g-$C_3N_4$ is the low separation efficiency of the photo-induced charge carriers. It is believed the layered structure of $[Bi_2O_2]^{2+}$ blocks of bismuth oxyhalides (BiOX) could enhance the separation of the charge carriers [68]. This is why BiOX attracts the attentions of many researchers. Therefore, the introduction of BiOX could improve the photocatalytic performance of the binary heterojunction. Among heterojunctions of g-$C_3N_4$/BiOCl, g-$C_3N_4$/BiOBr, and g-$C_3N_4$/BiOI, g-$C_3N_4$/BiOCl composite showed the best visible light photocatalytic performance tested by degradation of RhB, according to J. Sun and colleagues [68]. So, the compositing of g-$C_3N_4$ and BiOCl could be a very efficient way to synthesize a better photocatalyst.

Obviously, the degradation rate of organic dyes was accelerated under visible-light illumination through the combination of the two pristine catalysts. Because of the dye sensitization, though RhB could be degraded over BiOCl under visible light irradiation, the introduction of g-$C_3N_4$ further enhanced the photocatalytic performance of the composite [67]. It was reported that the degradation efficiency could be about 25 times higher than pure g-$C_3N_4$ [6]. Y. Yang and colleagues introduced g-$C_3N_4$ into pristine BiOCl, then the degradation rate of MB was increased about 5.9 times higher under the illumination of visible light [60]. L. Song and coworkers demonstrated that the degradation rate of RhB reached 89 and 50% over pristine BiOCl and g-$C_3N_4$, respectively, while that of the as-prepared g-$C_3N_4$/BiOCl heterojunction reached almost 100% within 30 min of visible light irradiation [59].

As mentioned above, facet control could improve the photoactivity of the binary heterojunction greatly. The degradation rate of MO under visible light irradiation over ng-CN/BiOC-010 was about two times that over ng-CN/BiOC-001 [61,69]. Though prepared by different researchers, heterojunctions synthesized by 010 dominant BiOCl (010HB) showed improved photoactivity compared to 001 dominant BiOCl (001HB). According to the first ten heterojunctions listed in Table 4, though the concentration of RhB was different, 001HB could completely degrade RhB in about 50 min. Those prepared by 010HB showed better photoactivity. RhB could be degraded in about 30 min. The degradation rate of MB over 001HB was inferior to 010HB. 001HB could only remove MB in 2 h [60], while MB was degraded in 30 min over 010HB [65].

Just as discussed above, surface area is another crucial factor that influences the photoactivity of the heterojunction under visible light illumination. Yang Bai, Xianlong Zhang, and coworkers also synthesized 010HB [66,72], but the degradation of the model pollutant (RhB 10mg l-1) lasted for about more than an hour. Whereas, the flower-like structured heterojunction prepared by Liwen Lei and colleagues could remove RhB (20 mg/L) in 20 min [34]. The as-prepared catalyst showed the best photoactivity among all the heterojunctions applied to RhB degradation. Compared to the former two studies, the major difference was its enlarged surface area. Therefore, it is reasonable to believe that larger surface area could enhance the degradation of pollutant under visible light illumination.

The similar results could be observed by other researchers. According to Qingbo Li, Wenwen Liu, and their colleagues [61,73], the as-prepared 010HB could remove MO in about three hours under visible light irradiation. Whereas, 001HB synthesized by Wenjie Shan and coworkers degraded MO in 5 h under similar conditions [75]. However, surface area might be the main factor that influences the degradation of MO under visible light illumination. The flower-like 001HB constructed by Xiaojing Wang and colleagues could degrade MO in 80 min [44], though it was 001 facet dominant. As all the experiments were conducted under similar conditions, it is reasonable to believe the enlarged surface area improved the photocatalytic performance of 001HB greatly.

Furthermore, the photodegradation rate of MB over pure g-$C_3N_4$ was higher than pure BiOCl, according to T. Jia and colleagues [65]. The removal efficiency of MB over as-prepared heterojunction was about two times higher than the pristine catalysts. The main reason might be that MB could not be photosensitized. The light absorption rate of the catalysts might play a more important role. S. Shi and colleagues also observed a similar phenomenon, and the MB degradation rate over the as-prepared heterojunction of the two catalysts was about four times higher than the pristine materials [21].

Therefore, according to the articles discussed in this section, g-$C_3N_4$/BiOCl heterojunctions with 001 facet dominant BiOCl and enlarged surface area could appear superior in terms of dye degradation.

### 3.2. Other Applications

Besides dye degradation, other applications of the g-$C_3N_4$/BiOCl composite were seldom reported, as according to Table 5.

**Table 5.** Other applications of g-$C_3N_4$/BiOCl heterojunctions.

| Catalyst (Mass Ratio %) | Application | Light Source | Efficiency/Time | Main Reactive Species | Stability and Reusability | Ref. |
|---|---|---|---|---|---|---|
| (OV)BiOCl/g-$C_3N_4$-10 | Degradation of carbamazepine | Visible light | 49%/240 min | •$O_2^-$, hole | ~50% after 5 irradiation cycles | [62] |
| BiOCl/g-$C_3N_4$(10/90) | Degradation of nizatidine | LED (365 nm) | 96%/30 min | •$O_2^-$, hole | ~92% after 5 irradiation cycles | [63] |
| (OV)BiOCl-g-$C_3N_4$(50/50) | Degradation of 4-chlorophenol | Short-arc xenon lamp 420 nm cutoff filter | 95 %/2 h | •$O_2^-$, hole | 81% after 4 irradiation cycles | [71] |
| g-$C_3N_4$/BiOCl | degradation of dibutyl phthalate and methyl orange | 500 W halogen tungsten lamp 420 nm cutoff filter | 60%/300 min(DBP) 70%/300 min(MO) | - | stable after 4 irradiation cycles | [75] |

One of the most important applications is the removal of recalcitrant industrial materials. 4-chlorophenol (4-CP) is an important material widely used in many areas. It could be used to manufacture sanitizers, germicides, precursors of pesticides, and dyes [82]. Just like other endocrine disruptors (bisphenol A (BPA), bisphenol S (BPS), and bisphenol F (BPF)), which are hazardous to the environment, they were employed by Q. Wang and co-workers to test the photocatalytic performance of the g-$C_3N_4$/BiOCl heterojunction [71]. It was indicated that the refractory pollutants could be mostly degraded within 2 h, while these endocrine disruptors were impossible to remove by conventional wastewater treatment.

As one of the plasticizers, about 60% of Dibutyl phthalate (DBP) was photodegraded over the g-$C_3N_4$/BiOCl catalyst within 300 min of visible light irradiation, according to W. Shan and colleagues [75]. Plasticizers, such as DBP, could adversely affect the neurodevelopment of infant and child, which is very hard to be degraded by normal wastewater treatment plants [83]. Photocatalysis is supposed to be an effective way to remove DBP from aqueous solution.

Degradation of the residual pharmaceutical agents is another application of the g-$C_3N_4$/BiOCl binary heterojunction. For example, carbamazepine, as a psychotropic and antiepileptic drug, is a recalcitrant and toxic chemical, which was adopted by researchers to test the photoactivity of the g-$C_3N_4$/BiOCl heterojunction [62]. It was indicated that the binary heterojunction showed more superior photoactivity than pure catalysts.

The contamination of antibiotics has drawn lots of attention in recent years. The amine-based pharmaceutical nizatidine, which could cause environmental problems, was applied in the field of photocatalysis by some researchers [63]. The result indicated that the degradation rate of the recalcitrant pollutant was accelerated over the g-$C_3N_4$/BiOCl photocatalyst. Application of antibiotic degradation, as one of the important applications of photocatalysis, is expected to have a bright future.

Moreover, according to B. Zhang and co-workers [84], g-$C_3N_4$/BiOCl could be used to modify an ITO electrode. That is the only application reported recently other than the degradation of recalcitrant pollutants.

However, it is not clear that if the same factors influenced the dye degradation could improve the removal of other chemicals, because there is still not enough research to be compared with.

According to Tables 4 and 5, the g-$C_3N_4$/BiOCl heterojunctions reviewed showed excellent stability and reusability. Almost every as-prepared catalyst exhibited stable photoactivity within at least five recycles. Furthermore, XRD patterns of the catalysts after several cycles showed that the crystal phase of the materials still stayed intact [21,58,61,71].

## 4. Mechanisms of the BiOCl/g-$C_3N_4$ Heterojunctions

P-type photocatalyst BiOCl combined with typical n-type photocatalyst g-$C_3N_4$ could form a conventional type-II photocatalyst with a staggered-gap band structure. There are three different types of semiconductor heterojunctions overall as shown in Figure 4. In a type-I heterojunction, conduction band (CB) and valence band (VB) of the semiconductor (SC1) are higher and lower than that of the other semiconductor (SC2), respectively. When SC1 and SC2 construct type-II heterojunction, CB and VB of SC1 are higher than of SC2. Because of the built-in electric field formed inside of the composite, photo-induced electrons tend to migrate to the CB of SC2. At the same time, holes accumulate on VB of SC1 rapidly. Because the electrons and holes migrate to different semiconductors, charge separation is enhanced. The pattern of charge carriers' movement in type-III heterojunction is the same as in type-II heterojunction. The difference of the band structures of the two semiconductors is even larger than in the type-II heterojunction [52].

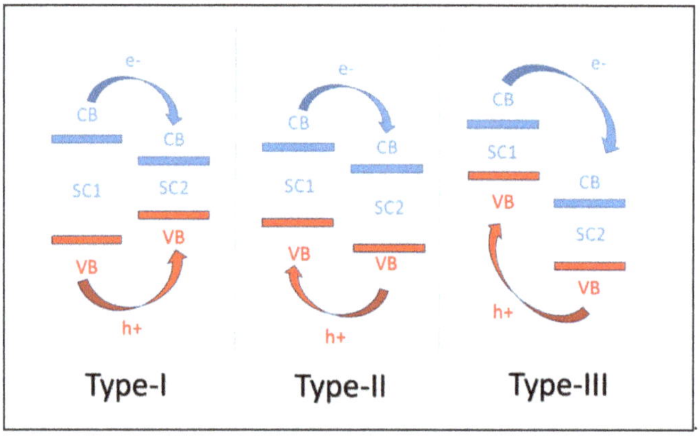


**Figure 4.** Three different types of semiconductor heterojunction.

Separation of electron-hole pairs could be prompted by the construction of the g-$C_3N_4$/BiOCl heterojunction. Solar energy utilization of the heterojunction was also more efficient since the wavelength of photo-response was broadened [8]. According to the articles reviewed, an interesting phenomenon is observed, some researchers believed that band structures of the two catalysts stayed unchanged after the combination of the two materials, whereas others thought that CB and VB of g-$C_3N_4$/BiOCl changed to align Fermi energy levels.

Theoretically, the Fermi energy level of an n-type catalyst is close to the bottom of CB. The Fermi energy level of a p-type catalysts is close to the top of VB [85]. After construction of heterojunction, the CB and VB of n-type catalyst tend to move downward, while those of p-type catalyst are moving upward. This kind of heterojunction based on g-$C_3N_4$ and BiOCl is labeled as PCNB in this article. The CB and VB of some g-$C_3N_4$/BiOCl heterojunctions stayed the same after the combination. This kind of heterojunction is denoted as the system of CNB here.

### 4.1. CNB Heterojunction

Generally, the proposed mechanism for the generation of reactive radicals on the surface of CNB heterojunction is shown in Figure 5. Photoexcited electrons firstly generated in the conduction band of g-$C_3N_4$ by irradiation of visible light because of its relatively mild band gap (2.7 eV). When it comes to dye degradation, photo-induced charge carriers also generated through dye sensitization. Then, electrons transferred to the conduction band of BiOCl because the conduction band of BiOCl is less negative than that of g-$C_3N_4$. Photogenerated electrons tend to transfer to a less negative conduction band. Electrons could react with $O_2$ on the surface of CNB to generate superoxide radicals. At the same time, holes remaining in the valence band of g-$C_3N_4$ react with surface-absorbed $H_2O$ to generate hydroxyl radicals, so that separation of photo-generated charge carriers is improved and the catalyst can response to visible light. However, the redox ability of the heterojunction was sacrificed when photoactivity is improved, because the holes accumulated on VB of g-$C_3N_4$.

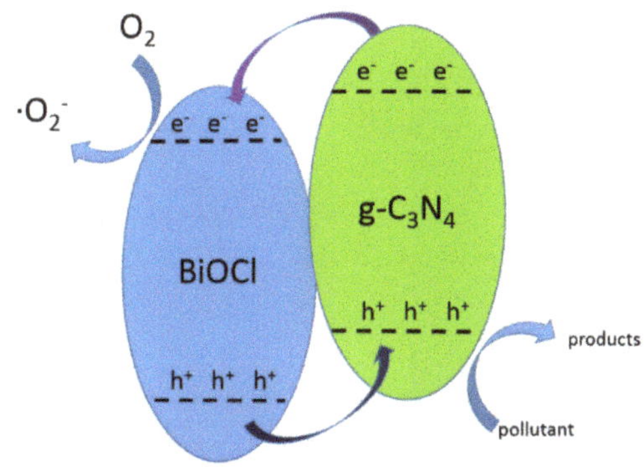


**Figure 5.** The mechanism for the generation of reactive radicals over CNB.

The mechanism mentioned above was adopted by Faisal Al Marzouqi and co-workers to explain the degradation of nizatidine over the BiOCl/g-$C_3N_4$ heterojunction [63]. The degradation efficiency of nizatidine was improved under the irradiation of visible light. According to the XRD pattern, the as-prepared catalyst was constructed by pure BiOCl and g-$C_3N_4$. The construction of the heterojunction was verified. As shown in the UV-vis diffuse reflectance spectra, the absorption edge of BiOCl was about 364 nm (in the UV range), and that of g-$C_3N_4$ was about 450 nm (in the visible range). After being combined, the absorption band edge of the heterojunction could be up to 476 nm. The photoactivity of the heterojunction was improved. The bandgap value for 10% BiOCl/g-$C_3N_4$ sample was 2.6 eV, which endowed the catalyst with the highest photoactivity among all the as-prepared samples. Therefore, the bandgap of the composite was narrowed by combination of the two components. The degradation rate of nizatidine was enhanced by the construction of the heterojunction as shown in Figure 6b. This improvement was explained by the double-charge transfer mechanism as proposed in Figure 5. Obviously, the CB and VB of both pristine catalysts did not change. The generation of reactive radicals depicted in the article was the same as that in Figure 5. However, the article provided no further evidence to prove the main reactive radicals. The presence of hydroxyl radicals was supposed to be the main cause of the degradation of nizatidine in the article. But the study did not exclude the possibility that the hydroxyl radicals could be generated from superoxide radicals. Y. Yang and colleagues demonstrated hydroxyl and superoxide radicals were the main species during the photocatalytic oxidation of MB, too [60]. Hydroxyl radicals were supposed to be produced in the VB of g-$C_3N_4$.

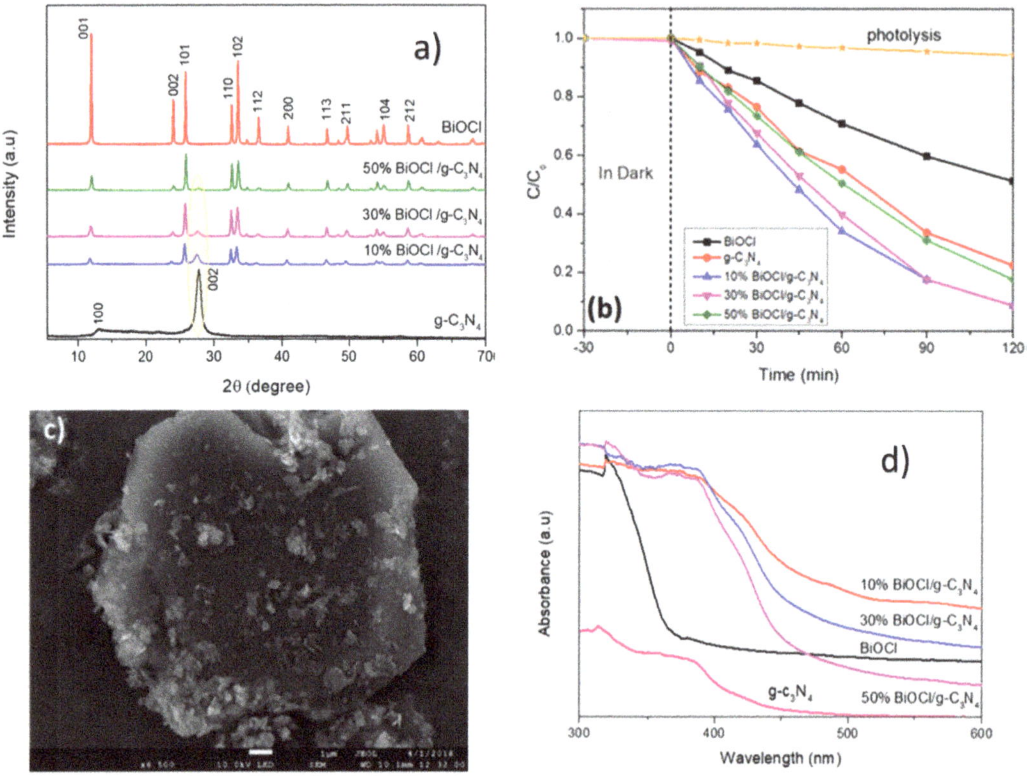


**Figure 6.** (**a**) XPR pattern of as-prepared BiOCl/g-$C_3N_4$ samples; (**b**) Degradation rate of nizatidine at an initial concentration of 5 mg/L and pH = 5.6 with all the prepared samples; (**c**) SEM image of 10% BiOCl/g-$C_3N_4$ sample; (**d**) UV-vis diffuse reflectance spectra of the obtained samples. Reproduced with permission from Al Marzouqi F et al, ACS Omega; published by American Chemical Society, 2013.

To date, lots of CNB heterojunctions were reported. Wenwen Liu and colleagues constructed a 2-dimensional layered BiOCl/g-$C_3N_4$ composite, and the photodegradation of MO was greatly improved through constructing a CNB heterojunction [73]. When the mass ratio of BiOCl reached 70%, BiOCl/g-$C_3N_4$ heterojunction showed the highest photocatalytic performance. EIS images and PL spectra were carried out to prove that better charge separation was realized. The proposed mechanism was similar to that shown in Figure 5. Electrons generated in the conduction band of g-$C_3N_4$, and then transferred to the conduction band of BiOCl. As a result, superoxide radicals generated on the surface of the heterojunction. Holes in the valence band of $C_3N_4$ were accumulated to participate in the degradation of MO degradation. Trapping experiments exhibited $\bullet O_2^-$ and holes were the main reactive species in the degradation of MO, which could be the evidence of the proposed mechanism. In this study, the VB and CB positions of BiOCl and g-$C_3N_4$ were determined by the Mott-Schottky curve. The alignment of band edges during the combination of the two materials was not taken into consideration, though the researchers did not directly adopt the standard values. The presence of the main reactive species was consistent with the proposed mechanism. Liwen Lei and co-workers prepared another heterostructure photocatalyst by combining BiOCl and g-$C_3N_4$ [34]. Arabic gum (AG) was added while synthesizing the heterojunction. They also proved that the superoxide and holes are the main reactive species through trapping experiments. The mechanism shown in Figure 5 was also adopted to explain the degradation of RhB over the composite.

However, the $BiOCl/g\text{-}C_3N4$ heterojunction prepared by Xiaojing Wang and colleagues showed a different result [44]. Like the studies mentioned above [73], XPR, FT-IR spectroscopy, and PL emission spectra were carried out to demonstrate the formation of the heterojunction. The light response wavelength of BiOCl was broadened, while the charge separation was enhanced. Trapping experiments were also carried out to detect the main reactive species in the photocatalytic process. It turned out that $\bullet O_2^-$ was not the main reactive species, whereas holes played an important role during the degradation of MO.

Why the hydroxyl radicals were not generally supposed to generate during the photocatalytic reaction was not mentioned in the above studies. Zhang Sai and co-workers explained the reason in their study [86], the standard CB and VB potentials of $g\text{-}C_3N_4$ are approximately $-1.3$ and 1.40 eV, respectively. The standard redox potential of $\bullet O_2^-/O_2$ is $-0.13$ eV (vs. NHE), which is more positive than the CB potential of $g\text{-}C_3N_4$. So, it is very easy for e- on the CB of $g\text{-}C_3N_4$ to generate superoxide radicals. The VB potential of $g\text{-}C_3N_4$ is less positive than the standard potential of $\bullet OH/OH^-$, which is +1.99 eV (vs. NHE). This makes holes on the VB of the catalyst and cannot be captured and to produce $\bullet OH$ radicals. If the CB and VB of the $g\text{-}C_3N_4/BiOCl$ catalysts stay unchanged after the construction of the type-II heterojunction, electrons accumulate on the CB of BiOCl ($-1.1$ eV) [8] to form $\bullet O_2^-$. Holes migrate to the VB of $g\text{-}C_3N_4$, but cannot generate hydroxyl radicals. Therefore, superoxide radicals and holes are the main reactive species in the systems of $BiOCl/g\text{-}C_3N_4$. This theory is consistent with the results mentioned above. The work of L. Song and co-workers also suggested that the standard redox potential of the VB of $g\text{-}C_3N_4$ was not positive enough to generate $\bullet OH$ groups [69]. J. Sun and colleagues directly used the standard potentials of the pristine catalysts to describe the mechanism without taking the alignment of the Fermi energy level into account [68].

Q. Li and co-workers employed the result of X-ray photoelectron spectroscopy (VB XPS) spectra to determine the VB of pure $g\text{-}C_3N_4$, which was 1.44 eV NHE [61]. Compared to the standard potential of $\bullet OH/OH^-$, the generation of $\bullet OH$ was not expected to happen on the VB of $g\text{-}C_3N_4$. The result of trapping experiments suggested that $\bullet O_2^-$ and holes were the dominant reactive species during the degradation of MO.

Some other researchers did not only adopt trapping experiments to determine the main species, for example, L. Song and co-workers also adopted ESR spectra and trapping experiments to find out the main reactive species [59]. The presence of superoxide radicals was directly proved by the ESR test. The generation of hydroxyl radicals was not detected. Trapping experiments proved holes also played an important role during the oxidation of RhB.

Just like the aforementioned study of Xiaojing Wang and colleagues [44], T. Jia and colleagues determined the CB and VB potentials of BiOCl and $g\text{-}C_3N_4$ by using theoretical calculation, then holes were proved to be the main reactive species during the oxidation of MB through trapping experiments [65].

There are some other studies that adopted a similar mechanism to explain the degradation of pollutants over ternary catalysts based on the system of $BiOCl/g\text{-}C_3N_4$, like systems of $BiOCl/g\text{-}C_3N_4$/kaolinite [47], $g\text{-}C_3N_4$/CDs/BiOCl [48], $BiOCl/CdS/g\text{-}C_3N_4$ [87], and $BiOI\text{-}BiOCl/C_3N_4$ [88].

However, Xiaojuan Bai and colleagues demonstrated that hydroxyl radicals were still produced, though the VB of $g\text{-}C_3N_4$ was not positive enough [89]. They synthesized a kind of photocatalyst by modifying $g\text{-}C_3N_4$ with fullerene. After the modification, the degradation rate of MB was improved. Trapping and ESR experiments proved that holes and $\bullet OH$ were the main reactive species in the photodegradation of MB. After the modification, the VB of $C_{60}/g\text{-}C_3N_4$ was more positive by 0.17 eV. Considering the theory depicted above, that was not positive enough to generate $\bullet OH$ directly on the VB of $g\text{-}C_3N_4$. The mechanism was further researched by adding $N_2$ to create an anoxic suspension. The degradation of MB was almost unchanged in the presence of $N_2$, which indicated that the $\bullet OH$ was generated on the surface of the composite, but not through the reaction induced by electrons on the CB of g-$C_3N_4$. This study seems contradictory to the theory described above that the VB of g-$C_3N_4$ was not positive enough to produce •OH [86].

The CNB system is a typical type-II heterojunction due to the band structures of the two materials. Trapping experiments were carried out to clarify the main reactive species, which proved to be superoxide radicals and holes. However, according to the study discussed above [86], there is still something unclear about the mechanism depicted in this section. Some more works are required to elucidate the reaction that happened over the heterojunction of CNB.

*4.2. PCNB Heterojunction*

According to the theory of semiconductor physics about p-n junction, n-type semiconductor g-$C_3N_4$ combines with p-type semiconductor BiOCl to form one composite, which tends to have one single Fermi energy level under one certain circumstance. As mentioned above, the Fermi level of g-$C_3N_4$ is supposed to be higher than that of BiOCl. After the construction of PCNB heterojunction, band structures of the two materials were expected to be changed to align the Fermi levels (EF). The photocatalytic mechanism of PCNB system is shown in Figure 7.

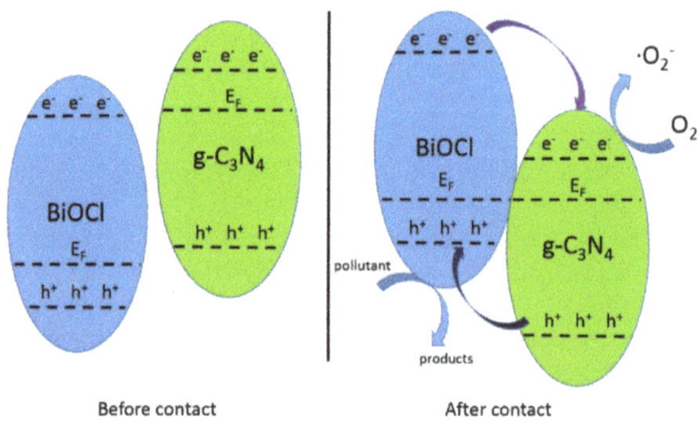


**Figure 7.** The schematic mechanism of g-$C_3N_4$/BiOCl p-n junction.

Hybrid density-functional theory (DFT) calculation was used to anticipate the properties of the g-$C_3N_4$/BiOCl composite [51]. G-$C_3N_4$ and BiOCl could form a stable composite with a narrower bandgap (2.1 eV), then the absorption of visible light was enhanced. According to the values of work function (WF), g-$C_3N_4$ and BiOCl were supposed to be positively and negatively charged after the contact, respectively, then the built-in electric field between the two materials was formed. Photo-induced charge carriers' separation was greatly improved since the lifetime of them was prolonged, and the recombination of electrons and holes was hindered in the system of g-$C_3N_4$/BiOCl. By the values of the valence and conduction band offset (VBO and CBO), which were 0.69 and 1.78 eV, respectively. The band structures of g-$C_3N_4$ and BiOCl were changed after contact as shown in Figure 7. Just as depicted above, theoretically, p-n junction possesses one single Fermi energy level, so energy bands of p-type semiconductor tend to move upward, whereas that of n-type semiconductor tend to move downward. Then, the as-prepared heterojunction can still be defined as a type-II heterojunction. Compared to pure catalysts, the photo response of PCNB heterojunction can be expanded to visible light region, and separation of photo-induced charge carriers could be promoted.

Though the author of the article mentioned above anticipated the properties of the binary heterojunction by theoretical calculation, many studies of other researchers could provide proof of the results. For example, Xianlong Zhang and co-workers synthesized a PCNB heterojunction by using g-$C_3N_4$ and BiOCl in the absence of the surfactant [72]. XRD data of the samples indicated that the formation of heterojunction did not change the structure of pristine g-$C_3N_4$ and BiOCl as shown in Figure 8a. XPS analysis was carried out to prove the strong interaction between g-$C_3N_4$ and BiOCl in the heterojunction. UV-Vis DRS spectrum (Figure 8b) indicated the wavelength of light response was expanded to visible light region. The recombination of electrons and holes was inhibited by the construction of heterojunction, which could be inferred from the photoluminescence spectra shown in Figure 8d. The results of the trapping experiments indicated that the dominant reactive species in the degradation of RhB are $\bullet O_2^-$ and $h^+$. The proposed degradation mechanism of RhB over the as-prepared catalysts was similar to that shown in Figure 7 which was different from that in Figure 5. The conduction band edge and valance band edge of BiOCl moved upward, while the band edges of g-$C_3N_4$ moved downward after the contact of the two materials. The experiments and tests conducted above verified the results of hybrid density-functional theory (DFT) calculation.

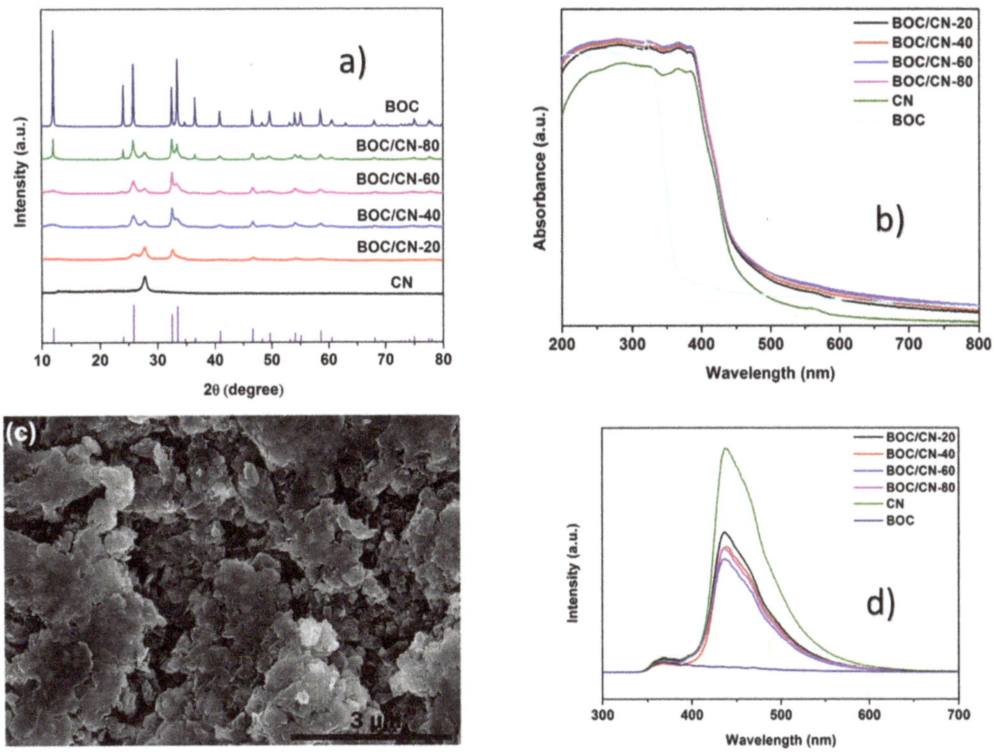


**Figure 8.** (**a**) XRD data, (**b**) UV-Vis DRS spectrum, (**c**) SEM, (**d**) photoluminescence spectra of all the samples. Reproduced with permission from Zhang X et al, Applied Surface Science; published by Elsevier BV, 2019.

Oxygen vacancies of PCNB were also introduced, which enhanced the photoactivity. Generally, oxygen vacancy is a common defect on oxide surfaces [90]. Oxygen vacancies could be introduced by certain methods. For example, removal of the surface oxygen atoms and elimination of the surface hydroxyl groups could be realized simultaneously through microwave irradiation and reaction between ethylene glycol and BiOCl [91]. However, there still is not a proper way to detect OVs quantitatively at present. The understanding of OVs is still infant. There are still many steps to take to elucidate the function of oxygen vacancy.

Qiao Wang and co-workers synthesized an oxygen vacancy-rich 2 D/2 D BiOCl/g-$C_3N_4$ p-n junction [71]. The efficiency of dechlorination and hydroxylation of 4-chlorophenol over the as-prepared heterojunction were improved under the illumination of visible light. Notably, photoactivity of the catalyst was promoted greatly by introducing oxygen vacancies through the addition of the template of PVP. The presence of OVs could be detected by electron spin resonance (ESR) spectroscopy. Oxygen vacancy is a very common surface defect that exists on the surface of BiOCl, which could produce a new state in the bandgap and localize the photo-induced electrons [92]. The light absorption of BiOCl photocatalyst was expended to the visible light region in the presence of oxygen vacancies. Furthermore, OVs could facilitate the generation of superoxide radicals. Therefore, the introduction of OVs could improve the photoactivity of the catalyst. Other researchers used multiple methods to detect OVs. W. Hou and colleagues also synthesized another oxygen vacancy-rich PCNB heterojunction [62]. They not only employed ESR spectroscopy to prove the presence of OVs on the surface of the catalyst, but also adopted $O_2$-TPD profiles (Figure 9) to further detect the OVs. ESR and trapping experiments were also conducted to determine the main reactive species in the study, simultaneously.

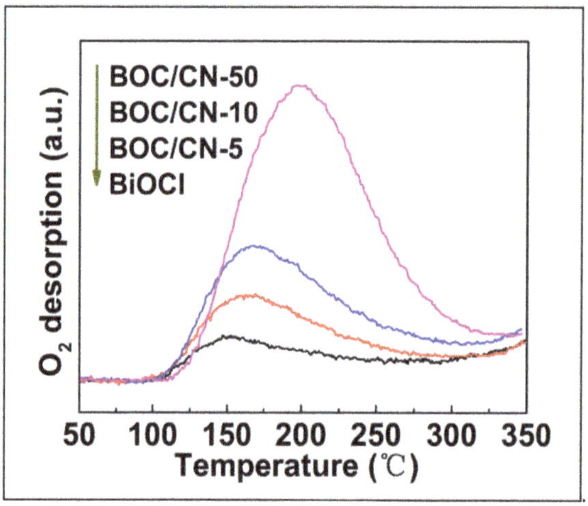


**Figure 9.** The $O_2$-TPD profiles of BiOCl, BOC/CN-5, BOC/CN-10, and BOC/CN-50. Reproduced with permission from Hou W et al, ChemistrySelect; published by John Wiley and Sons, 2020.

However, some researchers just adopted the mechanism depicted in Figure 7 to explain the photocatalytic reaction that happened over the g-$C_3N_4$/BiOCl heterojunction. For instance, Sheng Yin and co-workers synthesized a p-n junction g-$C_3N_4$/BiOCl with the assistance of ionic liquid [C16mim]Cl [74]. The characterization of as-prepared heterojunction showed the two materials constructed a stable heterojunction and stayed intact during the synthesis process. The photocatalytic performance of the composite was enhanced. However, the researchers did not use other methods, like trapping experiments, to prove the proposed mechanism.

In recent years, researchers have employed some advanced techniques to figure out the photocatalytic mechanism of the g-$C_3N_4$/BiOCl heterojunction. According to Z. Chen and co-workers, ultrafast transient absorption (TA) spectroscopy was adopted to test the mechanism [93]. TA spectroscopy provided more direct evidence of the PCNB heterojunction mechanism. Representative TA kinetic profiles indicated that the photo-induced electrons transferred to the CB of g-$C_3N_4$ in the binary heterojunction. If the band structures of the two photocatalysts stayed the same, the electrons were supposed to accumulate in the CB of BiOCl because of the more positive redox potential. It is reasonable to believe the results of TA spectroscopy could be the evidence of the PCNB mechanism.

Just like the study mentioned above [61], W. Cai and colleagues also adopted the valence band X-ray photoelectron spectroscopy (VB XPS) to determine the VB of BiOCl and g-$C_3N_4$ [70]. However, the same technique was employed by both articles, which obtained different results. It was indicated that the alignment of the Fermi levels happened while preparing the g-$C_3N_4$/BiOCl composite. The photo-induced electrons gathered in the CB of g-$C_3N_4$, while the accumulation of the holes happened on the VB of BiOCl. The potential of the VB (1.58 eV) was not positive enough to produce •OH. It was reasonable to deduce that the superoxide radicals were the main reactive species.

### 4.3. Z-Scheme Heterojunction

Normally, whether the band structures of heterojunctions changed after contact with each other, these systems cannot have high-efficient charge separation and strong redox ability at the same time [94]. Many researchers tried to find a different way to produce a new form of photocatalyst, which possessed high-efficient charge separation without sacrificing the redox ability of pristine catalyst since the construction of heterojunction is a perfect way to improve the photoactivity. Z-scheme photocatalyst is such an ideal system. By combining two or more kinds of materials, the new composite could respond to visible light and the charge separation can be improved, but the redox abilities of these catalysts were unharmed, like the system of g-$C_3N_4$/Au/BiOCl [49] and BiOCl/RGO/protonated g-$C_3N_4$ [50]. Z-scheme photocatalysts could be synthesized by not only chemical methods but also mechanical force, just like the system of $WO_3$/$NaNbO_3$ [95]. According to Yang Bai and co-workers, directly combining g-$C_3N_4$ and BiOCl could construct a Z-scheme catalyst through a facile method [66]. Typically, Z-scheme catalysts can be divided into three types, which could be labeled as PS-C-PS, PS-PS, and PS-A/D-PS [96–98]. In the PS-A/D-PS system, there exists an acceptor/donor pair as a common electron mediator. The schematic diagram of Z-scheme electron transfer in the PS-A/D-PS system is shown in Figure 10(a).

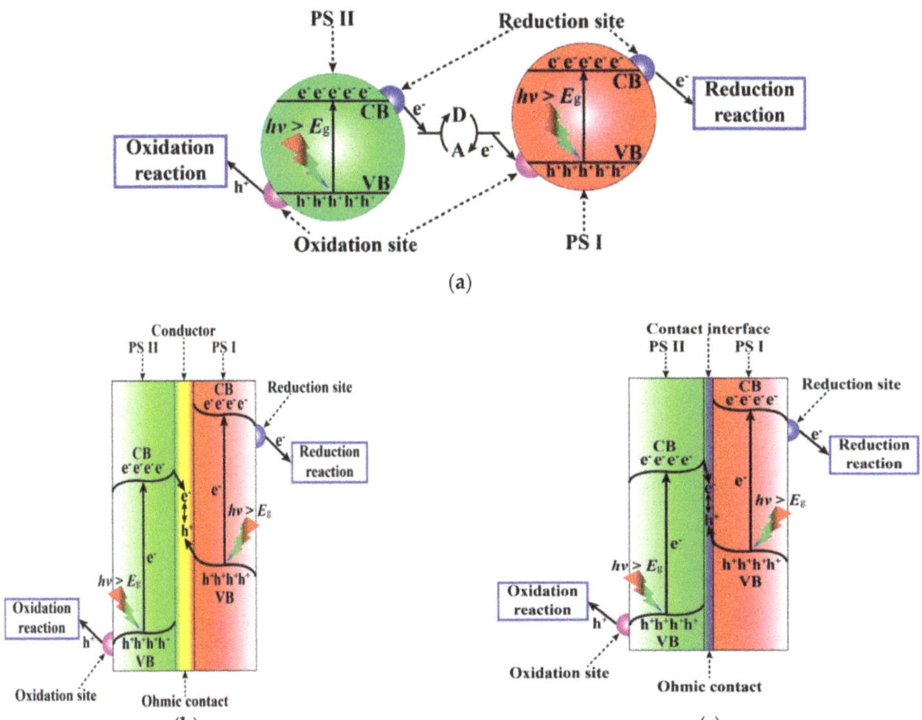


**Figure 10.** Schematic diagram of Z-scheme electron transfer in (**a**) PS-A/D-PS, (**b**) PS-C-PS, and (**c**) PS-PS system. Reproduced with permission from Zhou P et al, Advanced Materials; published by John Wiley and Sons, 2014.

Because the system of PS-C-PS is more stable than PS-A/D-PS, the PS-C-PS catalyst can be used in many different circumstances. Some researchers found the system of PS-A/D-PS was eroded after being used several times [99,100]. The PS-C-PS and PS-PS systems have a wider application range. These two systems also called all-solid-state Z-scheme photocatalysts. The schematic diagrams of Z-scheme electron transfer in the system of PS-C-PS and PS-PS are shown in Figure 10b,c.

The g-$C_3N_4$/Au/BiOCl heterojunction could be classified as a typical PS-C-PS catalyst, Au was employed as the mediator. Z-scheme catalyst of g-$C_3N_4$/BiOCl can be classified as a PS-PS type catalyst. It is believed that the conductive mediator of the PS-C-PS system could block the visible light to lower the efficiency of light energy absorption due to the surface plasmon resonance effect [101]. In this point of view, the system of PS-PS Z-scheme photocatalyst turns out to exhibit better photocatalysis performance. Because of the better photoactivity, Z-scheme catalysts drew a lot of attention, recently. Like the systems discussed in the former two sections, the enhancement of the Z-scheme catalysts based on g-$C_3N_4$ and BiOCl could be ascribed to better separation of charge carriers and wavelength expansion of light absorption.

To date, only one kind of Z-scheme g-$C_3N_4$/BiOCl photocatalyst was reported [66]. The proposed catalytic mechanism of the BiOCl/g-$C_3N_4$ system is shown in Figure 11. The variation of the band structure during the synthesis of the heterojunction was not taken into account. Results of the trapping experiments indicated that the main reactive species were hydroxyl and superoxide radicals. This was considered as evidence of the formation of the Z-scheme heterojunction, according to the study.

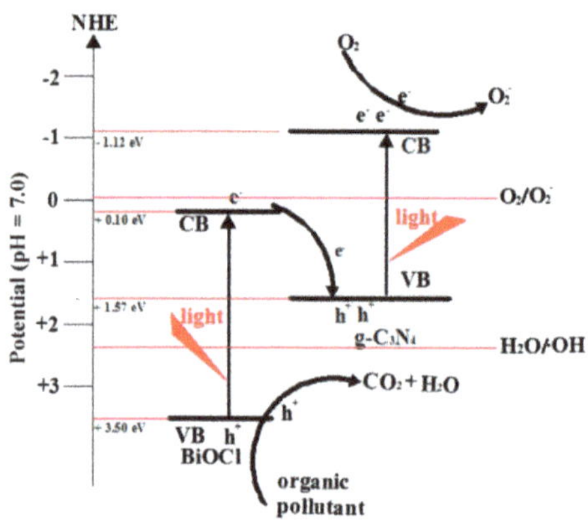


**Figure 11.** Direct Z-scheme photocatalytic mechanism of BiOCl/g-$C_3N_4$. Reproduced with permission from Bai Y et al, RSC Advances; published by Royal Society of Chemistry, 2014.

When the heterojunction was exposed under the illumination of visible light, the generation of •OH decreased. At the same time, the generation of superoxide radicals was not affected. The theoretical values of the band's redox potentials were directly adopted without being verified further.

The g-$C_3N_4$/Au/BiOCl Z-scheme system also adopted a similar theory that considered that the band structures of the three materials stayed the same after combined. The main reactive species during the degradation of RhB was the photo-induced holes. The aforementioned study of BiOCl/RGO/protonated g-$C_3N_4$ also expected the band structures stayed the same during the formation of the heterojunction. The main reactive species of the antibiotic TC degradation were holes and $\bullet O_2^-$, and the presence of them was taken as the evidence of the Z-scheme mechanism.

Therefore, all the Z-scheme photocatalysts mentioned above did not take the alignment of Fermi energy levels during the synthesis of the heterojunctions into account. Then, theoretical values of the redox potentials were adopted to explain the proposed mechanisms. The presence of specific reactive species was used as proof of the proposed mechanism.

It is common sense that the enhancement caused by the construction of the heterojunction is due to better charge separation and expended wavelength of light absorption. However, when it comes to the explanation of the mechanism, there are three different scenarios. As far as we know, the CB of BiOCl and the VB of g-$C_3N_4$ is not negative and positive enough to generate superoxide and hydroxyl radicals, respectively. However, just like the study mentioned above, though the VB of $C_{60}$/g-$C_3N_4$ was not supposed to be positive enough to produce hydroxyl radicals, the presence of hydroxyl radicals was still detected in an anoxic environment, which excluded the possibility that hydroxyl radicals might be produced from superoxide radicals. According to the CNB heterojunctions mentioned above, a similar phenomenon was observed. The ionic-liquid-assisted solvent-thermal route synthesized BiOCl/g-$C_3N_4$ generated hydroxyl radicals under the illumination of visible light [44], but the radicals might be produced in the VB of g-$C_3N_4$. The authors carried out no further tests to figure out the source of the radicals.

Additionally, some studies mentioned above might be explained by more than one mechanism. The improvement of the heterojunctions could be translated by all the three mechanisms. For example, the Z-scheme mechanism was used to describe photocatalytic reaction over the g-$C_3N_4$/BiOCl heterojunction [66], but it is still plausible to explain the process by the mechanism of CNB heterojunction. Photo-induced electrons and holes accumulated on the CB of BiOCl and the VB of g-$C_3N_4$, respectively. So, it is reasonable to think that the main species were superoxide radicals and holes according to the mechanism showed in Figure 5. The generation of •OH could be ascribed to $•O_2^-$. The presence of 420 nm cutoff filter meant that BiOCl cannot be photo-excited, so the concentration of $•O_2^-$ decreased, then the chain of the reaction generated •OH was cut off. The results of the experiments could be evidence of the relation between those two reactive species.

The photocatalytic reaction could be also explained by the mechanism of PCNB heterojunction. The band structures changed after being combined. The electrons and holes accumulated on the CB of g-$C_3N_4$ and the VB of BiOCl just as shown in Figure 6. The results of trapping experiments could be evidence of the PCNB mechanism. The generation of •OH could be ascribed to $•O_2^-$, either.

Therefore, it is difficult to testify the charge transfer mechanism over the g-$C_3N_4$/BiOCl photocatalysts. However, there is still no direct evidence of the mechanisms mentioned above. All the theories used to explain them seem very reasonable. Although trapping experiments, ESR tests, and other methods could detect the presence of the main reactive species, we still cannot figure out exactly where they come from. So, there are still many steps to take in order to precisely describe the mechanism.

## 5. Other Methods to Improve Photoactivity of Catalysts Based on CN and BOC

There are many studies focused on how to further enhance the photoactivity of the g-$C_3N_4$/BiOCl system. For example, Chengyun Zhou and colleagues used carbon-doped g-$C_3N_4$ (CCN) to combine with BiOCl. Then, the as-prepared p-n heterojunction showed improved photoactivity through degradation of tetracycline (TC) [7]. Notably, the bandgap of BiOCl was modulated by adjusting the ratio of Cl and O to synthesize a catalyst with a mild bandgap (2.33 eV), which was denoted as $Bi_{12}O_{17}Cl_2$. The as-prepared $Bi_{12}O_{17}Cl_2$ was able to respond to visible light due to its narrowed band gap. Through integrating with carbon-doped g-$C_3N_4$, a heterojunction of CCN/ $Bi_{12}O_{17}Cl_2$ was composited. CCN was integrated with $Bi_{12}O_{17}Cl_2$ to form a stable heterojunction, and each of the elements were uniformly distributed on the surface of the photocatalyst. The absorption of visible light and separation of photo-induced charge carriers was enhanced by the coupling of the two materials. The degradation of TC over the heterojunction was improved greatly compared to pure catalysts. The results of ESR proved the presence of •OH and $•O_2^-$ during the degradation. Trapping experiments proved the presence of •OH, $•O_2^-$, and holes affecting the degradation process. This study also adopted the PCNB mechanism to explain the mineralization process. Those two materials formed a p-n junction after being integrated. The band structures were changed because of the construction of the composite. There are also many other ternary systems based on $BiOCl/g\text{-}C_3N_4$ that were synthesized, like $BiOCl/Bi_2MoO_6/g\text{-}C_3N_4$ [102], $BiOCl/TiO_2\text{-}C_3N_4$ [103], $g\text{-}C_3N_4@BiOCl/Bi_{12}O_{17}Cl_2$ [104], $g\text{-}C_3N_4$/oxygen-deficient BiOCl nanocomposite/graphene quantum dots [105], $BiOCl/CdS/g\text{-}C_3N_4$ [87], $g\text{-}C_3N_4/BiOCl_xI_{1\text{-}x}$ [106], $g\text{-}C_3N_4/BiOCl_xBr_{1\text{-}x}$ [107], $BiOI/BiOCl/g\text{-}C_3N_4$ [88], and $Bi_2S_3/BiOCl/g\text{-}C_3N_4$ [46]. Unlike the CCN/ $Bi_{12}O_{17}Cl_2$ system, all the mechanisms of ternary systems mentioned above did not take the alignment of Fermi levels into account.

Ajay Kumara and co-workers also synthesized a new kind of heterostructure through constructing quaternary magnetic $BiOCl/g\text{-}C_3N_4/Cu_2O/Fe_3O_4$ nano-heterojunction [108]. According to the vibrating sample magnetometry (VSM) studies, the addition of $Fe_3O_4$ endowed the heterojunction to be separated from liquid magnetically. The bandgap of the heterojunction was 2.58 eV, which indicated that the as-prepared catalyst could respond to visible light. The recombination of the photo-induced charge carriers was greatly inhibited. The degradation rate over the quaternary heterojunction was about 2.7 and 2.4 times higher as BiOCl and $g\text{-}C_3N_4$, respectively. It was even 0.5 times higher than $BiOCl/g\text{-}C_3N_4$ binary heterojunction. The main reactive species in the photocatalytic process were •OH and $•O_2^-$, according to the results of trapping experiments. The mechanism was as depicted in Figure 12. The alignment of the Fermi energy levels was considered to happen during the preparation. P-n junctions were considered to be formed at the $g\text{-}C_3N_4$/BiOCl interface and the $g\text{-}C_3N_4/Cu_2O$ interface, respectively.

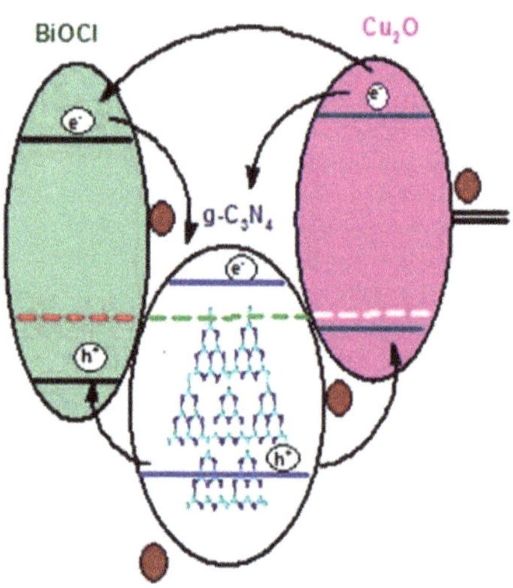


**Figure 12.** Proposed mechanism of quaternary heterojunction of $BiOCl/g\text{-}C_3N_4/Cu_2O/Fe_3O_4$. Reproduced with permission from Kumar A et al, Chemical Engineering Journal; published by Elsevier BV, 2018.

## 6. Summary and Outlook

According to the articles reviewed above, facet control and morphology of BiOCl were very important to the photoactivity of the heterojunctions. $G\text{-}C_3N_4$/BiOCl heterojunction with enlarged surface area accelerated the degradation of the azo dye. 010 facets exposed BiOCl of the heterojunction could enhance the absorption of the visible light.

BiOCl and $g\text{-}C_3N_4$ are both excellent photocatalysts despite their disadvantages, like fast photo-induced charge carriers' recombination and low efficient solar energy absorption. By coupling those two catalysts, we could get CNB heterojunction, PCNB heterojunction, and Z-scheme heterojunction. A single catalyst cannot have all the advanced features simultaneously. The construction of catalysts based on g-$C_3N_4$/BiOCl is a good strategy to fabricate a perfect photocatalyst. Combining g-$C_3N_4$/BiOCl heterojunction with other materials could provide more active sites, and further improve its capability to respond to visible light or make the composite magnetic recyclable.

To date, the heterojunction based on BiOCl and g-$C_3N_4$ mainly used in the purification of water, according to the articles reviewed. The mechanisms used to explain the photocatalytic processes could be divided into three different scenarios. Though there were some researchers that employed certain advanced techniques to prove the source of the reactive species and the charge transfer over the as-prepared catalysts, there is still not enough direct evidence of the mechanisms. Gaining a clearer understanding of the charge transfer is very important for researchers to prepare better photocatalysts. The industrial application of photocatalysis will benefit from this direction of research. Nowadays, few researchers have focused on this direction. This article intends to inspire more studies to clarify the route of charge transfer.

**Author Contributions:** Conceptualization, investigation, resources, writing—original draft preparation, editing, Q.R.; writing—review, J.L.; supervision, Q.Y. and W.S. All authors have read and agreed to the published version of the manuscript.

**Funding:** This research received no external funding.

**Conflicts of Interest:** The authors declare no conflict of interest.

## References

1. Pan, M.; Wong, C.K.C.; Chu, L.M. Distribution of Antibiotics in Wastewater-Irrigated Soils and Their Accumulation in Vegetable Crops in the Pearl River Delta, Southern China. *J. Agric. Food Chem.* **2014**, *62*, 11062–11069. [CrossRef] [PubMed]
2. Sharma, V.; Anquandah, G.A.K.; Yngard, R.A.; Kim, H.; Fekete, J.; Bouzek, K.; Ray, A.K.; Golovko, D. Nonylphenol, octylphenol, and bisphenol-A in the aquatic environment: A review on occurrence, fate, and treatment. *J. Environ. Sci. Health Part A* **2009**, *44*, 423–442. [CrossRef]
3. Yesilada, O.; Asma, D.; Cing, S. Decolorization of textile dyes by fungal pellets. *Process. Biochem.* **2003**, *38*, 933–938. [CrossRef]
4. Ahern, J.; Fairchild, R.; Thomas, J.S.; Carr, J.; Patterson, H.H. Characterization of BiOX compounds as photocatalysts for the degradation of pharmaceuticals in water. *Appl. Catal. B Environ.* **2015**, *179*, 229–238. [CrossRef]
5. Sun, S.; Wang, W.; Zhang, L.; Zhou, L.; Yin, W.; Shang, M. Visible Light-Induced Efficient Contaminant Removal by $Bi_5O_7I$. *Environ. Sci. Technol.* **2009**, *43*, 2005–2010. [CrossRef]
6. Chang, F.; Xie, Y.; Zhang, J.; Chen, J.; Li, C.; Wang, J.; Luo, J.; Deng, B.; Hu, X. Construction of exfoliated g-$C_3N_4$ nanosheets–BiOCl hybrids with enhanced photocatalytic performance. *RSC Adv.* **2014**, *4*, 28519–28528. [CrossRef]
7. Zhou, C.; Lai, C.; Xu, P.; Zeng, G.; Huang, D.; Li, Z.; Zhang, C.; Cheng, M.; Hu, L.; Wan, J.; et al. Rational Design of Carbon-Doped Carbon Nitride/$Bi_{12}O_{17}Cl_2$ Composites: A Promising Candidate Photocatalyst for Boosting Visible-Light-Driven Photocatalytic Degradation of Tetracycline. *ACS Sustain. Chem. Eng.* **2018**, *6*, 6941–6949. [CrossRef]
8. Yang, Y.; Zhang, C.; Lai, C.; Zeng, G.; Huang, D.; Cheng, M.; Wang, J.; Chen, F.; Zhou, C.; Xiong, W. BiOX (X = Cl, Br, I) photocatalytic nanomaterials: Applications for fuels and environmental management. *Adv. Colloid Interface Sci.* **2018**, *254*, 76–93. [CrossRef] [PubMed]
9. Liu, Y.; Zhang, H.; Ke, J.; Zhang, J.; Tian, W.; Xu, X.; Duan, X.; Sun, P.H.; Tade, M.; Wang, S. 0D ($MoS_2$)/2D (g-$C_3N_4$) heterojunctions in Z-scheme for enhanced photocatalytic and electrochemical hydrogen evolution. *Appl. Catal. B Environ.* **2018**, *228*, 64–74. [CrossRef]
10. Guo, F.; Shi, W.; Li, M.; Shi, Y.; Wen, H. 2D/2D Z-scheme heterojunction of $CuInS_2$/g-$C_3N_4$ for enhanced visible-light-driven photocatalytic activity towards the degradation of tetracycline. *Sep. Purif. Technol.* **2019**, *210*, 608–615. [CrossRef]
11. Liang, S.; Zhang, D.; Pu, X.; Yao, X.; Han, R.; Yin, J.; Ren, X. A novel $Ag_2O$/g-$C_3N_4$ p-n heterojunction photocatalysts with enhanced visible and near-infrared light activity. *Sep. Purif. Technol.* **2019**, *210*, 786–797. [CrossRef]
12. Liu, G.; Wang, G.; Hu, Z.; Su, Y.; Zhao, L. $Ag_2O$ nanoparticles decorated $TiO_2$ nanofibers as a p-n heterojunction for enhanced photocatalytic decomposition of RhB under visible light irradiation. *Appl. Surf. Sci.* **2019**, *465*, 902–910. [CrossRef]
13. Guo, F.; Shi, W.; Wang, H.; Han, M.; Guan, W.; Huang, H.; Liu, Y.; Kang, Z. Study on highly enhanced photocatalytic tetracycline degradation of type II AgI/$CuBi_2O_4$ and Z-scheme AgBr/$CuBi_2O_4$ heterojunction photocatalysts. *J. Hazard. Mater.* **2018**, *349*, 111–118. [CrossRef]
14. Lai, C.; Zhang, M.; Li, B.; Huang, D.; Zeng, G.; Qin, L.; Liu, X.; Yi, H.; Cheng, M.; Li, L.; et al. Fabrication of CuS/$BiVO_4$ (0 4 0) binary heterojunction photocatalysts with enhanced photocatalytic activity for Ciprofloxacin degradation and mechanism insight. *Chem. Eng. J.* **2019**, *358*, 891–902. [CrossRef]

15. Zhu, C.; Zhang, L.; Jiang, B.; Zheng, J.; Hu, P.; Li, S.; Wu, M.; Wu, W. Fabrication of Z-scheme $Ag_3PO_4/MoS_2$ composites with enhanced photocatalytic activity and stability for organic pollutant degradation. *Appl. Surf. Sci.* **2016**, *377*, 99–108. [CrossRef]
16. Hao, Q.; Niu, X.; Nie, C.; Hao, S.; Zou, W.; Ge, J.; Chen, D.; Yao, W. A highly efficient g-$C_3N_4/SiO_2$ heterojunction: The role of $SiO_2$ in the enhancement of visible light photocatalytic activity. *Phys. Chem. Chem. Phys.* **2016**, *18*, 31410–31418. [CrossRef]
17. Nguyen, T.B.; Huang, C.; Doong, R.-A. Photocatalytic degradation of bisphenol A over a $ZnFe_2O_4/TiO_2$ nanocomposite under visible light. *Sci. Total Environ.* **2019**, *646*, 745–756. [CrossRef]
18. Luo, J.; Li, R.; Chen, Y.; Zhou, X.; Ning, X.; Zhan, L.; Ma, L.; Xu, X.; Xu, L.; Zhang, L. Rational design of Z-scheme $LaFeO_3/SnS_2$ hybrid with boosted visible light photocatalytic activity towards tetracycline degradation. *Sep. Purif. Technol.* **2019**, *210*, 417–430. [CrossRef]
19. He, R.; Zhou, J.; Fu, H.; Zhang, S.; Jiang, C. Room-temperature in situ fabrication of $Bi_2O_3$/g-$C_3N_4$ direct Z-scheme photocatalyst with enhanced photocatalytic activity. *Appl. Surf. Sci.* **2018**, *430*, 273–282. [CrossRef]
20. Chen, Y.; Zhu, G.; Hojamberdiev, M.; Gao, J.; Zhu, R.; Wang, C.; Wei, X.; Liu, P. Three-dimensional $Ag_2O/Bi_5O_7I$ p–n heterojunction photocatalyst harnessing UV–vis–NIR broad spectrum for photodegradation of organic pollutants. *J. Hazard. Mater.* **2018**, *344*, 42–54. [CrossRef]
21. Shi, S.; Gondal, M.; Al-Saadi, A.; Fajgar, R.; Kupcik, J.; Chang, X.; Shen, K.; Xu, Q.; Seddigi, Z. Facile preparation of g-$C_3N_4$ modified BiOCl hybrid photocatalyst and vital role of frontier orbital energy levels of model compounds in photoactivity enhancement. *J. Colloid Interface Sci.* **2014**, *416*, 212–219. [CrossRef] [PubMed]
22. Ong, W.-J.; Tan, L.-L.; Ng, Y.H.; Yong, S.-T.; Chai, S.-P. Graphitic Carbon Nitride (g-$C_3N_4$)-Based Photocatalysts for Artificial Photosynthesis and Environmental Remediation: Are We a Step Closer to Achieving Sustainability? *Chem. Rev.* **2016**, *116*, 7159–7329. [CrossRef]
23. Li, Y.; Zhang, C.; Shuai, D.; Naraginti, S.; Wang, D.; Zhang, W. Visible-light-driven photocatalytic inactivation of $MS_2$ by metal-free g-$C_3N_4$: Virucidal performance and mechanism. *Water Res.* **2016**, *106*, 249–258. [CrossRef]
24. Goettmann, F.; Fischer, A.; Antonietti, M.; Thomas, A. Metal-free catalysis of sustainable Friedel–Crafts reactions: Direct activation of benzene by carbon nitrides to avoid the use of metal chlorides and halogenated compounds. *Chem. Commun.* **2006**, 4530–4532. [CrossRef] [PubMed]
25. Wang, X.; Maeda, K.; Thomas, A.; Takanabe, K.; Xin, G.; Carlsson, J.M.; Domen, K.; Antonietti, M. A metal-free polymeric photocatalyst for hydrogen production from water under visible light. *Nat. Mater.* **2009**, *8*, 76–80. [CrossRef] [PubMed]
26. Liu, J.; Liu, Y.; Liu, N.; Han, Y.; Zhang, X.; Huang, H.; Lifshitz, Y.; Lee, S.-T.; Zhong, J.; Kang, Z. Metal-free efficient photocatalyst for stable visible water splitting via a two-electron pathway. *Science* **2015**, *347*, 970–974. [CrossRef]
27. Zheng, Y.; Liu, J.; Liang, J.; Jaroniec, M.; Qiao, S.Z. Graphitic carbon nitride materials: Controllable synthesis and applications in fuel cells and photocatalysis. *Energy Environ. Sci.* **2012**, *5*, 6717–6731. [CrossRef]
28. Pawar, R.; Kang, S.; Ahn, S.H.; Lee, C.S. Gold nanoparticle modified graphitic carbon nitride/multi-walled carbon nanotube (g-$C_3N_4$/CNTs/Au) hybrid photocatalysts for effective water splitting and degradation. *RSC Adv.* **2015**, *5*, 24281–24292. [CrossRef]
29. Lan, M.; Fan, G.; Yang, L.; Li, F. Enhanced visible-light-induced photocatalytic performance of a novel ternary semiconductor coupling system based on hybrid Zn–In mixed metal oxide/g-$C_3N_4$ composites. *RSC Adv.* **2014**, *5*, 5725–5734. [CrossRef]
30. Huang, L.; Xu, H.; Li, Y.; Li, H.; Cheng, X.; Xia, J.; Xu, Y.; Cai, G. Visible-light-induced $WO_3$/g-$C_3N_4$ composites with enhanced photocatalytic activity. *Dalton Trans.* **2013**, *42*, 8606–8616. [CrossRef]
31. Wang, Y.; Wang, Z.; Muhammad, S.; He, J. Graphite-like $C_3N_4$ hybridized ZnWO4 nanorods: Synthesis and its enhanced photocatalysis in visible light. *CrystEngComm* **2012**, *14*, 5065–5070. [CrossRef]
32. Xing, C.; Wu, Z.; Jiang, D.; Chen, M. Hydrothermal synthesis of $In_2S_3$/g-$C_3N_4$ heterojunctions with enhanced photocatalytic activity. *J. Colloid Interface Sci.* **2014**, *433*, 9–15. [CrossRef]
33. Han, G.; Li, D.Y.; Zheng, Y.F.; Song, X.C. Enhanced Visible-Light-Responsive Photocatalytic Properties of $Bi_2MoO_6$-BiOCl Nanoplate Composites. *J. Nanosci. Nanotechnol.* **2018**, *18*, 5575–5581. [CrossRef]
34. Lei, L.; Gao, D.; Jin, H.; Zhang, Q.; Xu, J.; Fu, Z. A novel enhanced visible-light-driven photocatalyst via hybridization of nanosized BiOCl and graphitic $C_3N_4$. *Dalton Trans.* **2015**, *44*, 795–803. [CrossRef] [PubMed]
35. Ramírez-Meneses, E.; Valencia-Barrón, J.P.; Hernández-Pérez, M.A.; Domínguez-Crespo, M.A.; Torres, A.; Palacios, E. Synthesis and Characterization of BiOCl Powders with Soft Templates. *J. Inorg. Organomet. Polym. Mater.* **2018**, *28*, 2350–2364. [CrossRef]
36. Lin, W.; Yu, X.; Shen, Y.; Chen, H.; Zhu, Y.; Zhang, Y.; Meng, H. Carbon dots/BiOCl films with enhanced visible light photocatalytic performance. *J. Nanoparticle Res.* **2017**, *19*, 56. [CrossRef]
37. Singh, S.; Sharma, R.; Khanuja, M. A review and recent developments on strategies to improve the photocatalytic elimination of organic dye pollutants by BiOX (X=Cl, Br, I, F) nanostructures. *Korean J. Chem. Eng.* **2018**, *35*, 1955–1968. [CrossRef]
38. Yu, C.L.; Chen, J.C.; Zhou, W.Q.; Wei, L.F.; Fan, Q.Z. Grinding calcination preparation of $WO_3$/BiOCl heterostructures with enhanced visible light photocatalytic activity. *Mater. Res. Innov.* **2014**, *19*, 54–59. [CrossRef]
39. Cui, Z.; Song, H.; Ge, S.; He, W.; Liu, Y. Fabrication of BiOCl/BiOBr hybrid nanosheets with enhanced superoxide radical dominating visible light driven photocatalytic activity. *Appl. Surf. Sci.* **2019**, *467*, 505–513. [CrossRef]
40. Junxiu, W.; Zhenzong, Z.; Xi, W.; Yi, S.; Yongfu, G.; Keung, W.P.; Renbi, B. Synthesis of novel p-n heterojunction m-$Bi_2O_4$/BiOCl nanocomposite with excellent photocatalytic activity through ion-etching method. *Chin. J. Catal.* **2018**, *39*, 1792–1803.

41. Song, L.; Pang, Y.; Zheng, Y.; Chen, C.; Ge, L. Design, preparation and enhanced photocatalytic activity of porous BiOCl/$BiVO_4$ microspheres via a coprecipitation-hydrothermal method. *J. Alloys Compd.* **2017**, *710*, 375–382. [CrossRef]
42. Yu, L.; Zhang, X.; Li, G.; Cao, Y.; Shao, Y.; Li, D. Highly efficient $Bi_2O_2CO_3$/BiOCl photocatalyst based on heterojunction with enhanced dye-sensitization under visible light. *Appl. Catal. B Environ.* **2016**, *187*, 301–309. [CrossRef]
43. Zhong, Y.; Liu, Y.; Wu, S.; Zhu, Y.; Chen, H.; Yu, X.; Zhang, Y. Facile Fabrication of BiOI/BiOCl Immobilized Films with Improved Visible Light Photocatalytic Performance. *Front. Chem.* **2018**, *6*, 58. [CrossRef]
44. Wang, X.J.; Wang, Q.; Li, F.-T.; Yang, W.-Y.; Zhao, Y.; Hao, Y.-J.; Liu, S.-J. Novel BiOCl–$C_3N_4$ heterojunction photocatalysts: In situ preparation via an ionic-liquid-assisted solvent-thermal route and their visible-light photocatalytic activities. *Chem. Eng. J.* **2013**, *234*, 361–371. [CrossRef]
45. Zhang, C.; Li, Y.; Shuai, D.; Shen, Y.; Xiong, W.; Wang, L. Graphitic carbon nitride (g-$C_3N_4$)-based photocatalysts for water disinfection and microbial control: A review. *Chemosphere* **2019**, *214*, 462–479. [CrossRef] [PubMed]
46. Liu, S.; Liu, Y.; Dai, G.; Bao, X.; Huang, N.; Peng, R.; Zhou, Y. Synthesis and characterization of novel $Bi_2S_3$/BiOCl/g-$C_3N_4$ composite with efficient visible-light photocatalytic activity. *Mater. Lett.* **2019**, *241*, 190–193. [CrossRef]
47. Dong, X.; Sun, Z.; Zhang, X.; Li, C.; Zheng, S. Construction of BiOCl/g-$C_3N_4$/kaolinite composite and its enhanced photocatalysis performance under visible-light irradiation. *J. Taiwan Inst. Chem. Eng.* **2018**, *84*, 203–211. [CrossRef]
48. Asadzadeh-Khaneghah, S.; Habibi-Yangjeh, A.; Yubuta, K. Novel g-$C_3N_4$ nanosheets/CDs/BiOCl photocatalysts with exceptional activity under visible light. *J. Am. Ceram. Soc.* **2018**, *102*, 1435–1453. [CrossRef]
49. Zhao, S.; Zhang, Y.; Zhou, Y.; Fang, J.; Wang, Y.; Zhang, C.; Chen, W. Fabrication of sandwich-structured g-$C_3N_4$/Au/BiOCl Z-scheme photocatalyst with enhanced photocatalytic performance under visible light irradiation. *J. Mater. Sci.* **2018**, *53*, 6008–6020. [CrossRef]
50. Xue, J.; Li, X.; Ma, S.; Xu, P.; Wang, M.; Ye, Z. Facile fabrication of BiOCl/RGO/protonated g-$C_3N_4$ ternary nanocomposite as Z-scheme photocatalyst for tetracycline degradation and benzyl alcohol oxidation. *J. Mater. Sci.* **2018**, *54*, 1275–1290. [CrossRef]
51. Yao, W.; Zhang, J.; Wang, Y.; Ren, F. Hybrid density functional study on the mechanism for the enhanced photocatalytic properties of the ultrathin hybrid layered nanocomposite g-$C_3N_4$/BiOCl. *Appl. Surf. Sci.* **2018**, *435*, 1351–1360. [CrossRef]
52. Marschall, R. Semiconductor Composites: Strategies for Enhancing Charge Carrier Separation to Improve Photocatalytic Activity. *Adv. Funct. Mater.* **2014**, *24*, 2421–2440. [CrossRef]
53. Fu, J.; Yu, J.; Jiang, C.; Cheng, B. g-$C_3N_4$-Based Heterostructured Photocatalysts. *Adv. Energy Mater.* **2018**, *8*, 1701503. [CrossRef]
54. Xu, H.; Yan, J.; Xu, Y.; Song, Y.; Li, H.; Xia, J.; Huang, C.; Wan, H. Novel visible-light-driven AgX/graphite-like $C_3N_4$ (X=Br, I) hybrid materials with synergistic photocatalytic activity. *Appl. Catal. B Environ.* **2013**, *129*, 182–193. [CrossRef]
55. Iqbal, W.; Yang, B.; Zhao, X.; Rauf, M.; Waqas, M.; Gong, Y.; Zhang, J.; Mao, Y. Controllable synthesis of graphitic carbon nitride nanomaterials for solar energy conversion and environmental remediation: The road travelled and the way forward. *Catal. Sci. Technol.* **2018**, *8*, 4576–4599. [CrossRef]
56. Rono, N.; Kibet, J.K.; Martincigh, B.S.; Nyamori, V.O. A comparative study between thermal etching and liquid exfoliation of bulk graphitic carbon nitride to nanosheets for the photocatalytic degradation of a model environmental pollutant, Rhodamine B. *J. Mater. Sci. Mater. Electron.* **2021**, *32*, 687–706. [CrossRef]
57. Jiang, J.; Zhao, K.; Xiao, X.; Zhang, L. Synthesis and Facet-Dependent Photoreactivity of BiOCl Single-Crystalline Nanosheets. *J. Am. Chem. Soc.* **2012**, *134*, 4473–4476. [CrossRef] [PubMed]
58. Huang, Y. BiOCl Nanoplates Decorated on g-$C_3N_4$ for Enhanced Photocatalytic Activities. *Int. J. Electrochem. Sci.* **2017**, 4351–4359. [CrossRef]
59. Song, L.; Pang, Y.; Zheng, Y.; Ge, L. Hydrothermal synthesis of novel g-$C_3N_4$/BiOCl heterostructure nanodiscs for efficient visible light photodegradation of Rhodamine B. *Appl. Phys. A* **2017**, *123*, 500. [CrossRef]
60. Yang, Y.; Zhou, F.; Zhan, S.; Liu, Y.; Yin, Y. Enhanced Photocatalytic Activity of BiOCl Hybridized with g-$C_3N_4$. *J. Inorg. Organomet. Polym. Mater.* **2016**, *26*, 91–99. [CrossRef]
61. Li, Q.; Zhao, X.; Yang, J.; Jia, C.-J.; Jin, Z.; Fan, W. Exploring the effects of nanocrystal facet orientations in g-$C_3N_4$/BiOCl heterostructures on photocatalytic performance. *Nanoscale* **2015**, *7*, 18971–18983. [CrossRef]
62. Hou, W.; Deng, C.; Xu, H.; Li, D.; Zou, Z.; Xia, H.; Xia, D. n–p BiOCl@g-$C_3N_4$ Heterostructure with Rich-oxygen Vacancies for Photodegradation of Carbamazepine. *ChemistrySelect* **2020**, *5*, 2767–2777. [CrossRef]
63. AlMarzouqi, F.; Al Farsi, B.; Kuvarega, A.T.; Al Lawati, H.A.J.; Al Kindy, S.M.Z.; Kim, Y.; Selvaraj, R. Controlled Microwave-Assisted Synthesis of the 2D-BiOCl/2D-g-$C_3N_4$ Heterostructure for the Degradation of Amine-Based Pharmaceuticals under Solar Light Illumination. *ACS Omega* **2019**, *4*, 4671–4678. [CrossRef]
64. Tompsett, G.A.; Conner, W.C.; Yngvesson, K.S. Microwave Synthesis of Nanoporous Materials. *Chemphyschem Eur. J. Chem. Phys. Phys. Chem.* **2006**, *7*, 296–319. [CrossRef]
65. Jia, T.; Li, J.; Long, F.; Fu, F.; Zhao, J.; Deng, Z.; Wang, X.; Zhang, Y. Ultrathin g-$C_3N_4$ Nanosheet-Modified BiOCl Hierarchical Flower-Like Plate Heterostructure with Enhanced Photostability and Photocatalytic Performance. *Crystals* **2017**, *7*, 266. [CrossRef]
66. Bai, Y.; Wang, P.-Q.; Liu, J.-Y.; Liu, X.-J. Enhanced photocatalytic performance of direct Z-scheme BiOCl–g-$C_3N_4$ photocatalysts. *RSC Adv.* **2014**, *4*, 19456–19461. [CrossRef]
67. Zheng, C.-Z.; Zhang, C.-Y.; Zhang, G.-H.; Zhao, D.-J.; Wang, Y.-Z. Enhanced photocatalytic performance of g-$C_3N_4$ with BiOCl quantum dots modification. *Mater. Res. Bull.* **2014**, *55*, 212–215. [CrossRef]

68. Sun, J.; Song, J.; Gondal, M.A.; Shi, S.; Lu, Z.; Xu, Q.; Chang, X.; Xiang, D.; Shen, K. Preparation of g-$C_3N_4$/BiOX (X = Cl, Br, I) composites, and their photocatalytic activity under visible light irradiation. *Res. Chem. Intermed.* **2014**, *41*, 6941–6955. [CrossRef]
69. Song, L.; Zheng, Y.; Chen, C. Sonication-assisted deposition–precipitation synthesis of graphitic $C_3N_4$/BiOCl heterostructured photocatalysts with enhanced rhodamine B photodegradation activity. *J. Mater. Sci. Mater. Electron.* **2017**, *28*, 15861–15869. [CrossRef]
70. Cai, W.; Tang, J.; Shi, Y.; Wang, H.; Jiang, X. Improved in Situ Synthesis of Heterostructured 2D/2D BiOCl/g-$C_3N_4$ with Enhanced Dye Photodegradation under Visible-Light Illumination. *ACS Omega* **2019**, *4*, 22187–22196. [CrossRef] [PubMed]
71. Wang, Q.; Wang, W.; Zhong, L.; Liu, D.; Cao, X.; Cui, F. Oxygen vacancy-rich 2D/2D BiOCl-g-C3N4 ultrathin heterostructure nanosheets for enhanced visible-light-driven photocatalytic activity in environmental remediation. *Appl. Catal. B Environ.* **2018**, *220*, 290–302. [CrossRef]
72. Zhang, X.; An, D.; Feng, D.; Liang, F.; Chen, Z.; Liu, W.; Yang, Z.; Xian, M. In situ surfactant-free synthesis of ultrathin BiOCl/g-$C_3N_4$ nanosheets for enhanced visible-light photodegradation of rhodamine B. *Appl. Surf. Sci.* **2019**, *476*, 706–715. [CrossRef]
73. Liu, W.; Qiao, L.; Zhu, A.; Liu, Y.; Pan, J. Constructing 2D BiOCl/$C_3N_4$ layered composite with large contact surface for visible-light-driven photocatalytic degradation. *Appl. Surf. Sci.* **2017**, *426*, 897–905. [CrossRef]
74. Yin, S.; Di, J.; Li, M.; Sun, Y.; Xia, J.; Xu, H.; Fan, W.; Li, H. Ionic liquid-assisted synthesis and improved photocatalytic activity of p-n junction g-C3N4/BiOCl. *J. Mater. Sci.* **2016**, *51*, 4769–4777. [CrossRef]
75. Shan, W.; Hu, Y.; Bai, Z.; Zheng, M.; Wei, C. In situ preparation of g-$C_3N_4$/bismuth-based oxide nanocomposites with enhanced photocatalytic activity. *Appl. Catal. B Environ.* **2016**, *188*, 1–12. [CrossRef]
76. Wang, L.; Liu, X.; Luo, J.; Duan, X.; Crittenden, J.; Liu, C.; Zhang, S.; Pei, Y.; Zeng, Y.; Duan, X. Self-Optimization of the Active Site of Molybdenum Disulfide by an Irreversible Phase Transition during Photocatalytic Hydrogen Evolution. *Angew. Chem. Int. Ed. Engl.* **2017**, *56*, 7610–7614. [CrossRef]
77. Che, H.; Che, G.; Dong, H.; Hu, W.; Hu, H.; Liu, C.; Li, C. Fabrication of Z-scheme $Bi_3O_4Cl$/g-$C_3N_4$ 2 D/2 D heterojunctions with enhanced interfacial charge separation and photocatalytic degradation various organic pollutants activity. *Appl. Surf. Sci.* **2018**, *455*, 705–716. [CrossRef]
78. Adepu, A.K.; Anumula, R.; Narayanan, V. Photocatalytic degradation of Rhodamine B over a novel mesoporous titanosilicate/g-C3N4 nanocomposite under direct sunlight irradiation. *Microporous Mesoporous Mater.* **2017**, *247*, 86–94. [CrossRef]
79. Cheng, H.; Huang, B.; Dai, Y. Engineering BiOX (X = Cl, Br, I) nanostructures for highly efficient photocatalytic applications. *Nanoscale* **2014**, *6*, 2009–2026. [CrossRef]
80. Collivignarelli, M.C.; Abbà, A.; Miino, M.C.; Damiani, S. Treatments for color removal from wastewater: State of the art. *J. Environ. Manag.* **2019**, *236*, 727–745. [CrossRef] [PubMed]
81. Chiu, Y.-H.; Chang, T.-F.M.; Chen, C.-Y.; Sone, M.; Hsu, Y.-J. Mechanistic Insights into Photodegradation of Organic Dyes Using Heterostructure Photocatalysts. *Catalysts* **2019**, *9*, 430. [CrossRef]
82. Descorme, C. Catalytic wastewater treatment: Oxidation and reduction processes. Recent studies on chlorophenols. *Catal. Today* **2017**, *297*, 324–334. [CrossRef]
83. Radke, E.; Braun, J.M.; Nachman, R.M.; Cooper, G.S. Phthalate exposure and neurodevelopment: A systematic review and meta-analysis of human epidemiological evidence. *Environ. Int.* **2020**, *137*, 105408. [CrossRef] [PubMed]
84. Zhang, B.; Jia, Y.; Wang, J.; Hu, X.; Zhao, Z.; Cheng, Y. Cysteine-assisted photoelectrochemical immunoassay for the carcinoembryonic antigen by using an ITO electrode modified with $C_3N_4$-BiOCl semiconductor and CuO nanoparticles as antibody labels. *Microchim. Acta* **2019**, *186*, 633. [CrossRef] [PubMed]
85. Wang, H.; Zhang, L.; Chen, Z.; Hu, J.; Li, S.; Wang, Z.; Liu, J.; Wang, X. Semiconductor heterojunction photocatalysts: Design, construction, and photocatalytic performances. *Chem. Soc. Rev.* **2014**, *43*, 5234–5244. [CrossRef]
86. Zhang, S.; Gu, P.; Ma, R.; Luo, C.; Wen, T.; Zhao, G.; Cheng, W.; Wang, X. Recent developments in fabrication and structure regulation of visible-light-driven g-$C_3N_4$-based photocatalysts towards water purification: A critical review. *Catal. Today* **2019**, *335*, 65–77. [CrossRef]
87. Bellamkonda, S.; Rao, G.R. Nanojunction-mediated visible light photocatalytic enhancement in heterostructured ternary BiOCl/CdS/g-$C_3N_4$ nanocomposites. *Catal. Today* **2019**, *321–322*, 18–25. [CrossRef]
88. Aghdam, S.M.; Haghighi, M.; Allahyari, S.; Yosefi, L. Precipitation dispersion of various ratios of BiOI/BiOCl nanocomposite over g-$C_3N_4$ for promoted visible light nanophotocatalyst used in removal of acid orange 7 from water. *J. Photochem. Photobiol. A Chem.* **2017**, *338*, 201–212. [CrossRef]
89. Bai, X.; Wang, L.; Wang, Y.; Yao, W.; Zhu, Y. Enhanced oxidation ability of g-$C_3N_4$ photocatalyst via C60 modification. *Appl. Catal. B Environ.* **2014**, *152–153*, 262–270. [CrossRef]
90. Li, H.; Li, J.; Ai, Z.; Jia, F.; Zhang, L. Oxygen Vacancy-Mediated Photocatalysis of BiOCl: Reactivity, Selectivity, and Perspectives. *Angew. Chem. Int. Ed.* **2018**, *57*, 122–138. [CrossRef]
91. Li, H.; Zhang, L. Oxygen vacancy induced selective silver deposition on the {001} facets of BiOCl single-crystalline nanosheets for enhanced Cr(vi) and sodium pentachlorophenate removal under visible light. *Nanoscale* **2014**, *6*, 7805–7810. [CrossRef]
92. Li, H.; Shang, J.; Zhu, H.; Yang, Z.; Ai, Z.; Zhang, L. Oxygen Vacancy Structure Associated Photocatalytic Water Oxidation of BiOCl. *ACS Catal.* **2016**, *6*, 8276–8285. [CrossRef]

93. Chen, Z.; Zhang, Q.; Luo, Y. Determining the Charge-Transfer Direction in a p-n Heterojunction BiOCl/g-$C_3N_4$ Photocatalyst by Ultrafast Spectroscopy. *ChemPhotoChem* **2017**, *1*, 350–354. [CrossRef]
94. Natarajan, T.S.; Thampi, K.R.; Tayade, R.J. Visible light driven redox-mediator-free dual semiconductor photocatalytic systems for pollutant degradation and the ambiguity in applying Z-scheme concept. *Appl. Catal. B Environ.* **2018**, *227*, 296–311. [CrossRef]
95. Guo, Q.; Li, H.; Zhang, Q.; Zhang, Y. Fabrication, characterization and mechanism of a novel Z-scheme $Ag_3PO_4$/NG/polyimide composite photocatalyst for microcystin-LR degradation. *Appl. Catal. B Environ.* **2018**, *229*, 192–203. [CrossRef]
96. Maeda, K. Z-Scheme Water Splitting Using Two Different Semiconductor Photocatalysts. *ACS Catal.* **2013**, *3*, 1486–1503. [CrossRef]
97. Abe, R.; Sayama, K.; Sugihara, H. Development of New Photocatalytic Water Splitting into $H_2$ and $O_2$ Using Two Different Semiconductor Photocatalysts and a Shuttle Redox Mediator $IO_3^-/I^-$. *J. Phys. Chem. B* **2005**, *36*, 16052–16061. [CrossRef]
98. Zhou, P.; Yu, J.; Jaroniec, M. All-Solid-State Z-Scheme Photocatalytic Systems. *Adv. Mater.* **2014**, *26*, 4920–4935. [CrossRef]
99. Sayama, K.; Mukasa, K.; Abe, R.; Abe, Y.; Arakawa, H. A new photocatalytic water splitting system under visible light irradiation mimicking a Z-scheme mechanism in photosynthesis. *J. Photochem. Photobiol. A Chem.* **2002**, *148*, 71–77. [CrossRef]
100. Sayama, K.; Mukasa, K.; Abe, R.; Abe, Y.; Arakawa, H. Stoichiometric water splitting into $H_2$ and $O_2$ using a mixture of two different photocatalysts and an $IO_3^-/I^-$ shuttle redox mediator under visible light irradiation. *Chem. Commun.* **2001**, 2416–2417. [CrossRef]
101. Tada, H.; Mitsui, T.; Kiyonaga, T.; Akita, T.; Tanaka, K. All-solid-state Z-scheme in CdS–Au–$TiO_2$ three-component nanojunction system. *Nat. Mater.* **2006**, *5*, 782–786. [CrossRef] [PubMed]
102. Zhao, W.; Wang, A.; Wang, Y.; Lv, C.; Zhu, W.; Dou, S.; Wang, Q.; Zhong, Q. Accessible fabrication and mechanism insight of heterostructured BiOCl/$Bi_2MoO_6$/g-$C_3N_4$ nanocomposites with efficient photosensitized activity. *J. Alloys Compd.* **2017**, *726*, 164–172. [CrossRef]
103. Shakeel, M.; Zhang, X.; Yasin, G.; Arif, M.; Abbas, Z.; Zaman, U.; Li, B. Fabrication of Amorphous BiOCl/$TiO_2$-$C_3N_4$ Heterostructure for Efficient Water Oxidation. *ChemistrySelect* **2019**, *4*, 8277–8282. [CrossRef]
104. Zhang, W.; Liang, Y. Facile Synthesis of Ternary g-$C_3N_4$@BiOCl/$Bi_{12}O_{17}Cl_2$ Composites with Excellent Visible Light Photocatalytic Activity for NO Removal. *Front. Chem.* **2019**, *7*, 231. [CrossRef] [PubMed]
105. Choi, J.U.; Jo, W.-K. g-$C_3N_4$/oxygen-deficient BiOCl nanocomposite assisted by distinguished properties of graphene quantum dots for the efficient photocatalytic removal of organic vapors. *Appl. Surf. Sci.* **2019**, *493*, 873–881. [CrossRef]
106. Hu, X.; Zhang, Y.; Wang, B.; Li, H.; Dong, W. Novel g-$C_3N_4$/$BiOCl_xI_{1-x}$ nanosheets with rich oxygen vacancies for enhanced photocatalytic degradation of organic contaminants under visible and simulated solar light. *Appl. Catal. B Environ.* **2019**, *256*, 117789. [CrossRef]
107. Feng, Y.; Du, Y.; Du, M.; Tian, X.; Jiang, N.; Liu, Y. Synthesis and enhanced visible light photocatalytic activity of g-$C_3N_4$/$BiOCl_xBr_{1-x}$ heterojunctions with adjustable energy band structure. *J. Phys. Chem. Solids* **2019**, *132*, 222–229. [CrossRef]
108. Kumar, A.; Kumar, A.; Sharma, G.; Al-Muhtaseb, A.H.; Naushad, M.; Ghfar, A.A.; Stadler, F.J. Quaternary magnetic BiOCl/g-$C_3N_4$/$Cu_2O$/$Fe_3O_4$ nano-junction for visible light and solar powered degradation of sulfamethoxazole from aqueous environment. *Chem. Eng. J.* **2018**, *334*, 462–478. [CrossRef]

catalysts

Review

# Olive Mill Wastewater Remediation: From Conventional Approaches to Photocatalytic Processes by Easily Recoverable Materials

Melissa G. Galloni [1,2], Elena Ferrara [1], Ermelinda Falletta [1,2,*] and Claudia L. Bianchi [1,2]

1 Department of Chemistry, Università degli Studi di Milano, Via Camillo Golgi 19, 20133 Milan, Italy
2 Consorzio Interuniversitario Nazionale per la Scienza e Tecnologia dei Materiali (INSTM), Via Giusti 9, 50121 Florence, Italy
* Correspondence: ermelinda.falletta@unimi.it; Tel.: +39-02-503114410

**Abstract:** Olive oil production in Mediterranean countries represents a crucial market, especially for Spain, Italy, and Greece. However, although this sector plays a significant role in the European economy, it also leads to dramatic environmental consequences. Waste generated from olive oil production processes can be divided into solid waste and olive mill wastewaters (OMWW). These latter are characterized by high levels of organic compounds (i.e., polyphenols) that have been efficiently removed because of their hazardous environmental effects. Over the years, in this regard, several strategies have been primarily investigated, but all of them are characterized by advantages and weaknesses, which need to be overcome. Moreover, in recent years, each country has developed national legislation to regulate this type of waste, in line with the EU legislation. In this scenario, the present review provides an insight into the different methods used for treating olive mill wastewaters paying particular attention to the recent advances related to the development of more efficient photocatalytic approaches. In this regard, the most advanced photocatalysts should also be easily recoverable and considered valid alternatives to the currently used conventional systems. In this context, the optimization of innovative systems is today's object of hard work by the research community due to the profound potential they can offer in real applications. This review provides an overview of OMWW treatment methods, highlighting advantages and disadvantages and discussing the still unresolved critical issues.

**Keywords:** olive oil production; olive mill; wastewater remediation; polyphenols; conventional photocatalysts; magnetic photocatalysts; floating devices; environmental remediation

**Citation:** Galloni, M.G.; Ferrara, E.; Falletta, E.; Bianchi, C.L. Olive Mill Wastewater Remediation: From Conventional Approaches to Photocatalytic Processes by Easily Recoverable Materials. *Catalysts* **2022**, *12*, 923. https://doi.org/10.3390/catal12080923

Academic Editors: Ioan Balint and Monica Pavel

Received: 31 July 2022
Accepted: 19 August 2022
Published: 21 August 2022



## 1. Introduction

Olive oil production is a fundamental sector for several European (EU) States, especially Spain, Italy, and Greece. In particular, Spain has the largest area of olive cultivation (estimated at *ca.* 2.47 million ha), followed by Italy (*ca.* 1.16) and Greece (about 0.81 million ha) [1,2]. However, olive oil production is responsible for several environmental concerns (soil contamination, underground seepage, water-body pollution, and odor emissions) due to poor waste management practices [3]. In this scenario, concerning olive mill wastewaters (OMWW), special attention must be paid to their high phenolic content, which is responsible for their antibacterial effect, phytotoxic effect, and dark colour.

Recently, phenols, fatty acids, and volatile acids have been recognized as potentially hazardous for environmental health: the former have pronounced antimicrobial and phytotoxic properties, whereas the latter show toxicity due to their long alkyl chain.

All these components make OMWW toxic to anaerobic bacteria, thus inhibiting conventional secondary and anaerobic treatments in municipal water plants. Furthermore, the high BOD (biological oxygen demand) and COD (chemical oxygen demand) levels, which cannot be reduced by anaerobic digestion, represent a further threat to receivers [2,4]. Moreover, land spreading and treatment in evaporation ponds could lead to problems related to groundwater pollution. The use of olive oil waste in agriculture may also affect the acidity, salinity, N immobilization, microbial response, leaching of nutrients, and concentration of lipids, organic acids, and phenolic compounds [5].

Alternative approaches based on physical treatments, such as dilution, evaporation, centrifugation, or sedimentation guarantee a high level of OMWW purification. However, they are expensive and energy-consuming, thus leading to an exponential increase in the processing cost. The olive oil industry, in its current status, composed of small and dispersed factories, cannot bear such high costs [6–13].

In recent years, advanced oxidation processes (AOPs), including photolysis, photo-oxidation, Fenton, and photo-Fenton reaction, have emerged as promising alternatives for simplicity and high organic removal efficiencies [14–20]. In particular, heterogeneous photocatalysis seems to be a successful technology in water decontamination due to its non-toxicity, low cost, and mineralization efficacy. However, due to the OMWW matrices' complexity, it is not easy to develop and successively optimize efficient photocatalytic systems that are so far characterized by common limitations (i.e., difficult recovery, poor stability, low reusability, fast deactivation).

Based on these premises, in the present work, for the first time, we illustrate the conventional methods commonly used to treat OMWW along with their related advantages and limitations. Then, a critical insight on alternative strategies for developing efficient photocatalytic systems based on recoverable catalysts is proposed. The latters can be used as alternatives to conventional photocatalysts. This topic is of fundamental importance for the research community as shown by the hard work currently been done for developing novel devices with high potential in real applications, acting as a bridge between environmental protection and circular economy.

## 2. An Insight into the EU Legislation

Olive oil is the desired product of the olives industry. Unfortunately, olive mill pomace and wastewater represent undesired by-products, requiring proper disposal treatments because of their complex composition (Figure 1).

The present work aims at discussing only the production and treatment of OMWW. OMWW composition is influenced by different factors, i.e., extraction methods, olives' type and origin, climate conditions, and cultivation/processing practices [21]. In general, it can be mainly summarized as follows (Figure 1): *ca.* 80–83 wt.% consists of water, *ca.* 15–18 wt.% relates to organic compounds (mainly polyphenols, phenols, and tannins), and the remaining 2 wt.% contains inorganic matter (i.e., potassium salts and phosphates). Specifically, phenols levels in OMWW range from 1 to 8 $g \cdot L^{-1}$, whereas micronutrients and mineral nutrients mainly consist of $K_2O$, and $P_2O_5$, which can be found in considerable amounts (2.4–10.8 or 0.3–1.5 $g \cdot L^{-1}$ intervals, respectively) [2]. Thus, it is critical to design efficient treatment methods, aligned to precise legislative constraints, whose general panorama is described below.

Concerning the processing of olive residues, the reform of standard agricultural policy related to olive oil does not provide specific provisions for their management [1]. It should be noted that a significant part of EU legislation acts according to Directives. These latter are legislative acts, setting objectives that all EU countries must reach and translate into their national legislation. This means that the member Countries have to adopt and impose complementary measures that should be compliant with the EU directives.

Following this scenario, an example is setting the emission limits and environmental quality standards. Of course, every Country can adopt laws and regulations that can be very different compared to others. Still, in the end, international norms are necessary for a common strategy to manage olive waste. In general, EU legislation governs each member state's framework of national legislation. Several EU laws regulate waste management, and the Waste Framework Directive, WFD (2008/98/EC), acts as core legislation, including hazardous waste and oil rules [22]. In addition, Landfill Directive 99/31/EC regulates landfill disposal [23]. In this case, the waste producer, such as the olive mill operator, is responsible for managing wastes up to their recovery and disposal [24].

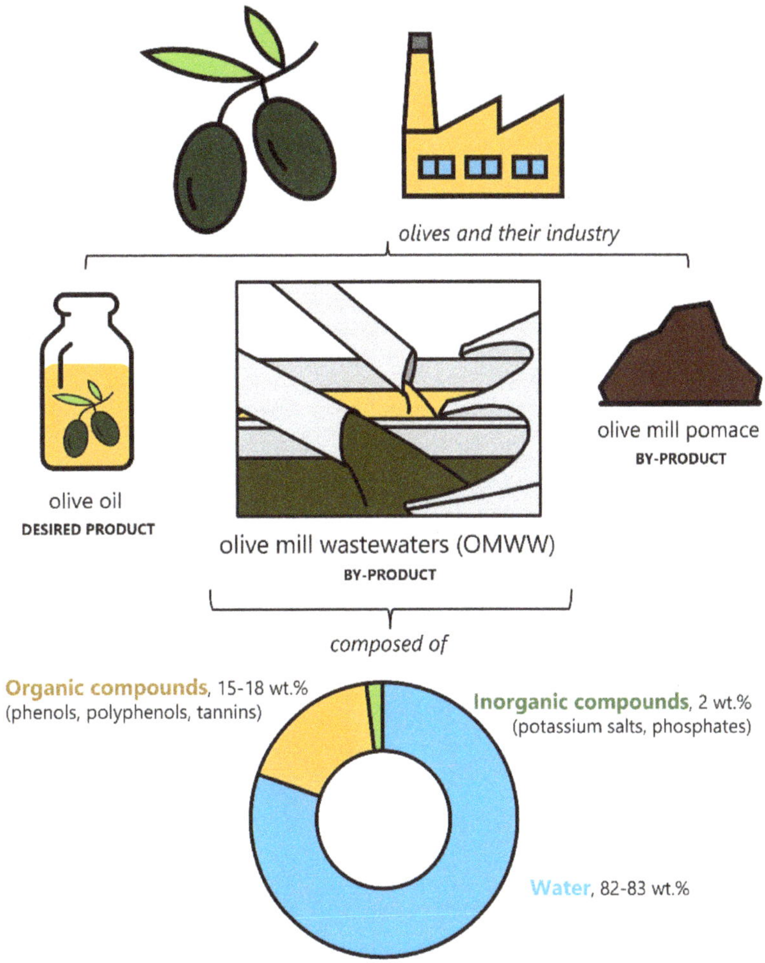


**Figure 1.** Scheme of products and by-products from the olive oil industry.

Here, the crucial point is to classify wastewaters as waste or by-products. If they are considered by-products, their further use as fertilizers with few restrictions is strongly recommended [25]. In this context, the EC Directive 2008/98 (point 22) clarifies the necessity to discriminate well between "waste" and "by-product", but unfortunately, considerable confusion is still present [22]. So, in many cases, law courts have to solve specific issues. To summarize, no EU legislation related to the management of OMWW exists today, and each EU country sets precise standard parameters.

## 3. Emerging Innovative Approaches for Olive Oil Production

Conventional techniques in olive oil extraction have not significantly changed in the last 25 years. Three main steps can be identified (Figure 2): crushing and malaxation, which mainly affect the oil quality and yield, and centrifugation [26,27].

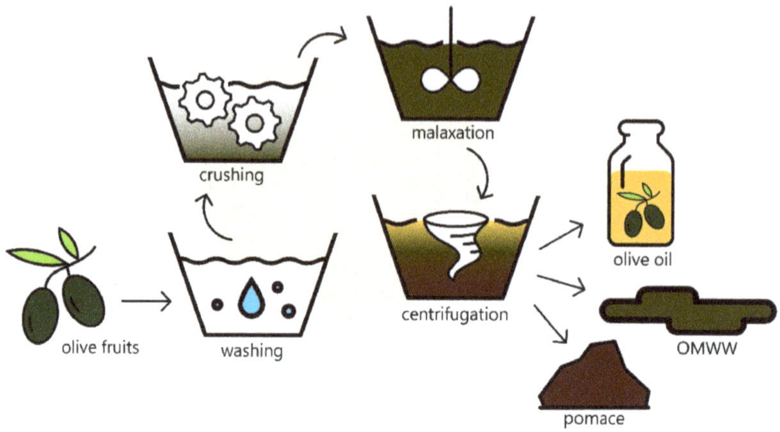


**Figure 2.** Scheme of olive oil production.

At first, stems, twigs, and leaves are separated from olive fruits [28]. These latter are then washed in a proper plant to remove dust, dirt, etc. In some plants, the washing water is recycled for the process after solid sedimentation or filtration, whereas in other cases, olives are directly processed without the washing step [29]. The next step involves malaxation: olives are ground up, mixed with/without their stones, and put in tanks, where the paste is divided into vegetation waters, pomace, and oil. Pomace, a brown-colored residue, is obtained by centrifugation and sedimentation after pressing olives [30,31]. Pomace mainly consists of skin pulp and pit fragments. Its separation is carried out using a horizontal decanter centrifuge and an olive oil press. The centrifuge step can be performed in two- or three-phases (Figure 3).

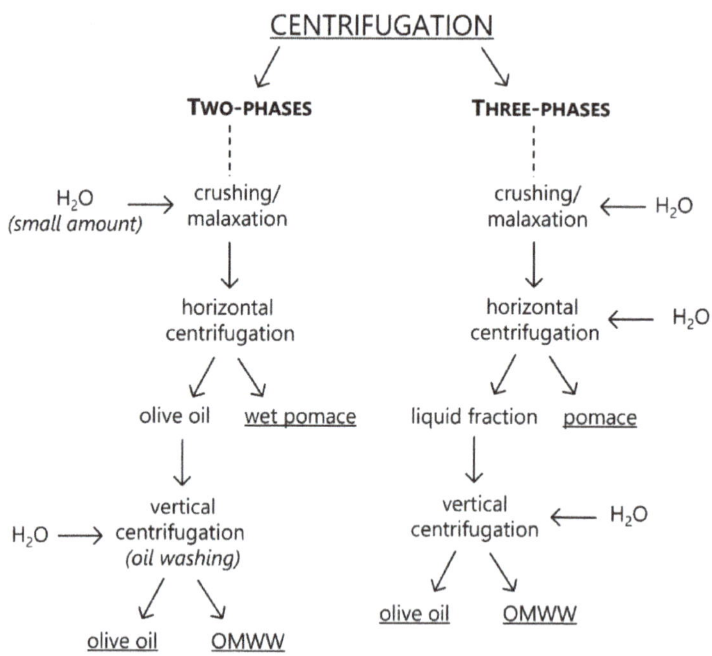


**Figure 3.** Scheme of two-phases and three-phases centrifugation strategies.

In the former case, wet pomace (also known as two-phase olive mill waste, TPOMW) and olive oil are obtained by horizontal centrifugation. Then, the obtained oil is centrifuged with water, producing olive oil and a small stream of OMWW [32,33]. In the latter, the olive paste is divided into pomace and a liquid fraction (olive oil *plus* OMWW), which is centrifuged with water to obtain high-quality olive oil and OMWW [32,33].

However, given the ever more urgent market demand, interesting novel methods characterized by minimal processing are currently the object of study. These approaches aim to obtain a final product with the same nutritional qualities in less time. In this context, numerous solutions, including the use of microwave, high-pressure processing, pulsed light, radio frequency, Ohmic heating, ultrasound, and pulsed electric field (PEF), have been investigated thanks to their advantages (enhanced extraction efficiency in reduced time with increased yield, and low energy consumption) [26,34–37].

Among them, ultrasound emerged as a powerful technology widely used in several extraction processes [37,38] and food processing methods (i.e., emulsification, filtration, crystallization, enzymes' and microorganisms' inactivation, thawing) [39,40]. Ultrasound can be applied to the olive paste to induce oil release from vacuoles in lower malaxation time. It has been demonstrated that high oil quality and yield are obtained [36,41–45].

Pulsed electric field (PEF) technology, used mainly in food science since 1960, consists of exposing food products (solid or liquid) to an electric field, inducing pore formation in cell membranes [46]. Recently, it has demonstrated its efficiency in reversible or irreversible permeabilization of cell membranes in different plants without causing significant temperature increase [34]. The possibility of maintaining low operating temperatures during the oil extraction process represents a valuable goal, as it allows the preservation of the product's organoleptic and nutritional characteristics.

An alternative to the two previous processes is microwave-assisted extraction (MAE), which represents a more efficient and successful strategy than the conventional ones because microwaves provide rapid heating and biological cell structure destruction. As a result, it leads to high-quality products with shallow energy requirements, inducing reduced environmental impact and financial costs [47].

Recently, emphasis has been placed on obtaining an increased Extra Virgin Olive Oil (EVOO) quality, preserving its sensory characteristic and favorable health properties. The quality of the EVOO strongly depends on the presence of phenolic and volatile compounds [43,44]. So, the development of emerging technologies to increase the oil yield while protecting and improving the bioactive oil compounds and quality is of fundamental importance.

Table 1 summarizes some interesting studies related to innovative technologies applied to olive oil extraction, including the maximum extraction yield obtained (i.e., the percentage value given by the ratio between the weights of the extracted oil and olives).

**Table 1.** Emerging extraction methods for olive oil production. Adapted from Ref. [48].

| Olives' Variety | Used Technology [a] | Investigated Parameters | Dependent Variables | Maximum Extraction Yield (%) [b] | Ref. |
|---|---|---|---|---|---|
| Edremit | HPU | Ultrasound time, ultrasound temperature, malaxation time | Oil yield, acidity, peroxide value, and antioxidant properties | 9 | [26] |
| Coratina | HPU | Ultrasound application step (After crushing/ before crushing) | Olive paste temperature, energy balance, oil yield, quality indices of oil, minor compounds | 16 | [36] |
| Picual | HPU | Direct/indirect application of ultrasound | Olive paste temperature | 16 | [49] |

**Table 1.** *Cont.*

| Olives' Variety | Used Technology [a] | Investigated Parameters | Dependent Variables | Maximum Extraction Yield (%) [b] | Ref. |
|---|---|---|---|---|---|
| Picual | HPU | Continuous ultrasound application before centrifugation | Oil yield, quality indices, volatile and minor compounds, fatty acid composition | 53 | [50] |
| Edremit, Gemlik, Uslu | HPU | Ultrasound and malaxation time | Oil yield, UV absorbance values acidity, peroxide value, total phenolic content | 68 | [51] |
| Picual | HPU | Olive paste flow, HPU intensity, fruit temperature, olive moisture, and fat content | Olive paste temperature | 17 | [52] |
| Ogliarola Barese | HPU, MW | Thermal effect of US and MW | Malaxation time, oil yield, quality characteristics, and energy efficiency | 17 | [45] |
| Arbequina | PEF | PEF application | Oil yield, acidity, quality characteristics, total phenols, sensory properties | n.d. [c] | [35] |
| Chemlal | MW | Extraction time, acetic acid content in hexane, irradiation power | Oil yield, total phenols, quality parameters | 6 | [53] |
| Peranzana | MW | Malaxation time and MW | Energy consumption, oil yield, structure modifications of olive pastes | n.d. [c] | [54] |
| Coratina | HPU | Sonication time | Oil yield, oil quality indices, phenolic composition | n.d. [c] | [55] |

[a] HPU: high-power ultrasound; PEF: pulsed electric field; MW: microwave; [b] expressed as percentage given by the ratio between the weights of the extracted oil and the olives; [c] not defined.

## 4. Olive Mill Wastewater Treatment

As reported above, OMWW is the waste of olive oil production characterized by high organic content and phytotoxic features mainly due to the presence of phenols, which are responsible for the olive oil's antimicrobial and antioxidant qualities. This makes waste biodegradation difficult in conventional treatment facilities (e.g., anaerobic digestion processes) that generally use microorganisms for waste biodegradation.

According to these premises, the OMWW treatment has faced several traditional approaches, which can be categorized as: disposal, physicochemical, biological, and advanced oxidation methods. Figure 4 schematizes their potentialities and weakness.

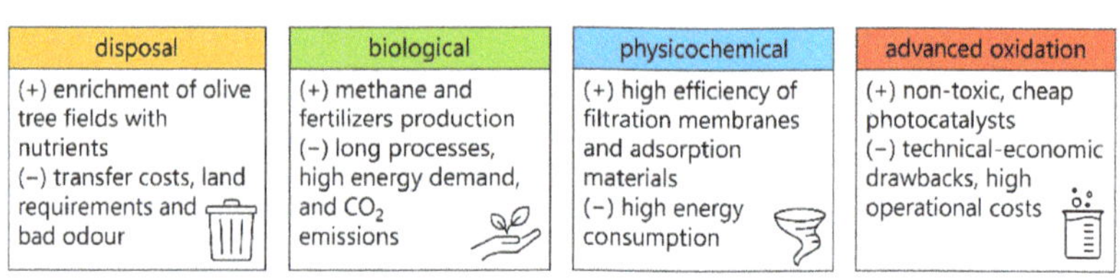


**Figure 4.** Main advantages/disadvantages of the standard technologies used for OMWW treatment.

As can be observed, all the mentioned technologies have specific advantages, but also cost problems. Therefore, some alternatives are currently the object of hard work by the research community. Among all the possibilities, one emerging and valid strategy is the OMWW steam reforming (OMWWSR), which permits valorization of wastes, producing green $H_2$, following the circular economy perspective [32]. It is a promising strategy in the view of the future projections of the $H_2$ market demand recently published by the International Energy Agency [56–58]. This method is described as having a high potential because of the environmental attractiveness of $H_2$, which is able to reduce $CO_2$ emissions in the atmosphere [56]. In this way, OMWWSR could contribute to air pollution reduction and, at the same time, valorize the waste from the olive oil industry [32]. However, this process still deserves to be properly studied and optimized because of some drawbacks affecting the catalyst formulation (e.g., low stability, deactivation, coke poisoning) [32].

The following paragraphs provide an accurate description of each traditional technology.

### 4.1. *Disposal Methods*

They mainly consist of treatment with calcium oxide (neutralization and coagulation) followed by the disposal of waterproof lagoons [59]. Unfortunately, they are affected by some disadvantages, such as foul odors, mosquitoes, and transfer costs because they require land far from residential areas. As an alternative, OMWW can be carried to fields of olive trees and then spread, thus enriching the soil with nutritive compounds [60].

### 4.2. *Biological Methods*

Bioremediation is a technology that exploits the metabolic potential of microorganisms to purify contaminated sites. It can be performed in a non-sterile and open area containing numerous organisms. Among these, bacteria have a central role in the process thanks to their ability to degrade pollutants. In addition, fungi and other components (e.g., grazing protozoa) can also affect the process [61]. All these species require nutrients (i.e., carbon, nitrogen, phosphates, metal traces) to survive, so they break down organic compounds to attain them. Bioremediation can occur under aerobic or anaerobic conditions [62]. In the former case, the survival of microorganisms is mainly due to the consumption of atmospheric oxygen. In contrast, in the latter, microorganisms gain food by breaking down chemical compounds in the soil [27].

In recent decades, OMWW has been used in this way, acting as substrate for microorganisms' growth by providing nutritive substances. Some yeast species (i.e., *Candida tropicalis, Yarrowia lipolytica*) together with bacteria of the *Azotobacter vinelandii, Pseudomonas, Sphingomonas, Ralstonia* species have been proven to be helpful in the OMWW aerobic biodegradation and detoxification [63–65]. By way of example, the activity of a free-living $N_2$-fixing bacterium, *Azotobacter vinelandii,* was investigated. In particular, OMWW was initially treated with calcium hydroxide to achieve the pH value of *ca.* 8–10 (stage I). Successively, it was mixed in a bioreactor in the presence of *Azotobacter vinelandii* (stage II). The process carried out according to the procedure reported by Arvanitoyannis et al. [27] resulted in an increasing level of nitrogen and its ammonium form throughout the whole remediation period. On the other hand, regarding phenols and sugar degradation, ca. 66–99% and 100% of phenols'abatement was observed after 3 and 7 days, respectively, whereas sugars were wholly degraded in only 3 days. Low phytotoxic features characterize the final product so it can be exploited as fertilizer [66].

In general, to reduce the high content of phenolic compounds in OMWW, water dilution represents a suitable strategy for a successful aerobic treatment. In fact, phenols are responsible for inhibition of microorganism growth [67,68]. Alternatively, OMWW could be mixed with additional waste and digested with the help of a solid substrate (i.e., straw, sesame bark, olive leaves, vineyard leaves, wood chips, animal manure) [69–71]. Then, when the phenol content of waste decreases, usually after 6–7 months, the final product can be exploited as fertilizer, giving profit [70,71]. Additionally, the composting stage could be coupled with physicochemical processes [72–74]. This last method requires high energy demand and consequent high $CO_2$ emissions. However, the energy demand can be reduced thanks to simultaneous methane production [75].

### 4.3. Physicochemical Treatments

Among physicochemical treatments, dilution, evaporation, sedimentation, filtration, and centrifugation are commonly used to treat OMWW.

OMWW dilution is usually employed before biological treatments with the final aim to reduce its toxicity to microorganisms. On the other hand, evaporation and sedimentation result in a concentrated OMWW (*ca.* 70–75% more concentrated) thanks to both phase separation/dehydration and organic matter degradation [6,7]. In this context, solar distillation applied to OMWW can remove 80% COD in the distillate in 9 days, maintaining 25% water content [8].

Other strategies have also been investigated, mainly consisting of irreversible thermal treatments. This is the case with combustion and pyrolysis that require a reduced volume of waste and provide energy recovery. Still, unfortunately, they need expensive facilities, emit toxic substances into the atmosphere, and require an OMWW pre-concentration step [9,10].

Centrifugation and filtration increase the effluent pH and conductivity, removing the organic matter using phase separation and exclusion. Ordinarily, combining physical processes, coupled with coagulation/flocculation or adsorption techniques, gives rise to more efficient removal of organic matter. For example, it was found that when the sedimentation is followed by centrifugation and filtration, 21% and 15% decrease in COD and BOD, respectively, was observed, with the further 16% reduction in BOD due to the final filtration [11]. OMWW adsorption on activated clay causes an additional 71% COD reduction. However, a particular focus has to be put on the adsorption/desorption equilibrium since organic and phenolic features start to desorb after a precise contact time. The combination of treatment stages, i.e., settling, centrifugation, filtration, and adsorption on activated carbon, induce a maximum of 94% phenol abatement and 83% organic matter removal [12].

Regarding filtration, it is fundamental to point out that, besides the high efficiency of membranes, these processes require high operative pressures and energy consumption. However, proper membranes can be exploited to recover valuable by-products, such as phenols, which are mainly required for the pharmaceutical and chemical industry [13].

Lime treatment has been selected as a pre-treatment step for reducing OMWW polluting effect due to its inexpensiveness [76–79].

In this context, coagulation-flocculation is a very similar technology to lime treatment. Different coagulants (i.e., ferric chloride, polyelectrolytes) can be exploited [80]. On the other hand, electro-coagulation mainly consists of the suspension and precipitation of charged particles in the waste thanks to an applied voltage. Since this process is characterized by low cost and energy consumption, it is not so efficient in removing organic waste species.

### 4.4. Advanced Oxidation Methods

The addition of strong oxidizing agents can influence the efficiency of wastewater treatment, mainly in terms of the breakdown of recalcitrant and toxic compounds. In this context, high mineralization levels can occur depending on the oxidizing power of the agent employed and contact time. In recent decades, the scientific community has addressed the efforts to exploit advanced oxidation processes (AOPs) to treat industrial effluents and OMWW [14,15]. In general, AOPs combine ozone ($O_3$), light irradiation (UV, solar, visible), hydrogen peroxide ($H_2O_2$), and/or catalysts to produce unstable radical species able to degrade both organic and inorganic pollutants.

In electrolysis, the oxidation of the content of organic species directly occurs on the anode or indirectly by oxidizing agents present in the solution [81]. Over the years, several materials for anodes' production have been studied (i.e., Pt/Ir, Ti/$IrO_2$, Pt/Ti, and boron-doped diamond) [81–84]. However, this is a high-energy consuming approach. In contrast,

Fenton oxidation is based on the addition of Fenton's reagent ($H_2O_2$ and Fe(II)) into the waste [16]. In this case, the oxidation process is due to a cascade of different reactions in the solution. Although it is low energy consumption, $H_2O_2$ makes this technology quite expensive. The photo-Fenton method is very similar to the Fenton one, but the UV radiation accelerates $Fe^{2+}$ regeneration, enhancing, as a consequence, the process efficiency. However, the necessity to employ UV radiation causes high energy consumption [85]. Supercritical water oxidation consists of waste oxidation in the presence or absence of catalyst above the water critical temperature and at high pressures [86–88]. It is a very efficient technology for organic content reduction, but the energy consumption is high due to the high temperatures and pressures required. Finally, ozonation employs $O_3$ as oxidant species for waste oxidation. It is not so efficient in the organic content reduction, but that of phenols is high. Unfortunately, using $O_3$ increases the process costs [89–91].

#### 4.4.1. Photocatalytic Treatments

Photocatalysis can be described as an advanced oxidation process able to fully mineralize the contamination in liquid as well as the gas phase under room pressure and temperature [92]. Its efficiency is mainly due to the capability to generate powerful oxidizing agents [14,15,93,94]. In this way, the chemical transformation rate is enhanced by the chosen photocatalyst under light irradiation [95]. Following these perspectives, photocatalysis has found a successful application in the water decontamination field [96], providing promising results in the removal of a large variety of contaminants (e.g., aromatics, pesticides, drugs, oils) [97].

In this context, photocatalytic treatments can be applied in the field of OMWW degradation using both homogeneous and heterogeneous photocatalysts in the presence of UV, visible and solar light irradiation. In this class of treatments, photo-Fenton and solar-Fenton processes are also included [18,98,99].

In the following paragraphs, deeper insights into the current approaches used in the literature are reported with the aim of fully describing the scenarios related to these technologies.

##### UV Photocatalysis

Data summarized in Table 2 show how UV photocatalysis finds application in the OMWW treatment in the presence of both homogeneous and heterogeneous catalysts.

**Table 2.** State of the art of UV photocatalysis used to treat OMWW. Adapted from Reference [100].

| OMWW Origin | Type of Process Treatment and Scale | Obtained Results | Ref. |
|---|---|---|---|
| Jordan | (i) $O_3$/UV or (ii) UV/$O_3$, followed by (iii) biodegradation—laboratory scale | COD removal efficiencies up to (i) 91% by UV/$O_3$ followed by biodegradation | [101] |
| Greece | Photocatalytic treatment with $TiO_2$ (Degussa P25)—laboratory scale | 200 mg·L$^{-1}$ COD residual and complete total phenol removal | [98] |
| Spain | pH-temperature flocculation + ferromagnetic core $TiO_2$ + UV photocatalysis—laboratory and pilot scale | 58.3% COD and 27.5% total phenols removal efficiencies; overall COD removal efficiency up to 91% | [100,102] |
| Portugal | nano-$TiO_2$ immobilized in nonwoven paper—laboratory scale | 90.8 ± 2.7% removal of the phenolic content | [103] |
| Italy | UV/$TiO_2$—laboratory and pilot scale | COD reduction around 50% upon 1.5 g·L$^{-1}$ nanocatalyst dosage | [104,105] |

Since 1972, titanium dioxide ($TiO_2$)-based photocatalysts have been investigated [106] and then widely used for their effective semiconductor features, enabling the removal of various pollutants in environmental remediation [107–109]. Interesting properties characterize these systems, like chemical stability, long-term stability, remarkable oxidation ability, and low-cost [110–112]. Heterojunction photocatalysts based on $TiO_2$ have been studied mainly for the mineralization of targeted pollutants into harmless products, thanks to the generation of electron-hole ($e^-/h^+$) pairs if the semiconductor is under UV radiation [97]. In this frame, 2.80 V oxidizing power was produced by hydroxyl radicals produced during the photocatalytic step [96]. Besides the high chemical and physical stability of $TiO_2$, this material tends to go through phase transformation from anatase to rutile [113]. This induces a detrimental effect on the resulting $TiO_2$-materials because the rutile-phase has a lower surface area, negatively impacting the photocatalytic behaviour because of the ($e^-/h^+$) pairs' recombination [114].

In this regard, Chatzisymeon et al. explored the photocatalytic treatment of a three-phase OMWW remediation approach using $TiO_2$ in a laboratory-scale photoreactor. By properly optimizing the contact time, they observed the enhancement of COD removal. The product was a non-toxic effluent with 200 $mg{\cdot}L^{-1}$ COD organic content [98].

In this context, the high surface/volume ratio of $TiO_2$ nanoparticles, the possibility to dope them to increase the activation under solar irradiation, and the resistance to photo-corrosion are advantages related to the use of $TiO_2$-based photocatalysts.

This hitch can be minimized with the introduction of a second metal oxide component (e.g., $MnO_2$, NiO, $La_2O_3$, $SiO_2$, $SnO_2$, ZnO, $ZrO_2$), which has been recognized to induce significant degradation under UV irradiation [115–119], generating oxygen vacancies by the substitution of di- or tri-valent atoms by tetravalent atoms and providing particle-particle interaction [120]. In this context, very promising results have been obtained in terms of improved chemical stability and photocatalytic activities of the obtained materials, as demonstrated by many researchers in the last decades [121–123] and recently by Yaacob et al. for $ZrO_2$-$TiO_2$ materials [124].

However, $TiO_2$ has been recently recognized as a carcinogenic substance [125], so an unavoidable challenge is the development of alternative systems able to maintain the same or better photocatalytic activity. In this scenario, among all the potential candidates, one could be zinc oxide (ZnO), which is able to absorb a wide fraction of the solar spectrum and more than $TiO_2$ [126]. Many researchers have demonstrated its efficiency in the photodegradation of organic pollutants in water matrixes [127]. Additional features describe ZnO more than $TiO_2$ [128]; by way of example, it can be used in acidic or alkaline environments through proper treatment [129,130]. Moreover, the optimum pH for the ZnO process is *ca.* 7, whereas that of $TiO_2$ lies at acidic values, implying lower operational costs and higher efficiency than $TiO_2$ in the advanced oxidation of pulp mill bleaching wastewater [131], phenol and 2-phenyl phenol photooxidations [132,133]. In addition, it is highly photosensitive, stable, and possesses a bandgap of *ca.* 3.2 eV [134]. However, besides the numerous studies on using this material in this field, efforts to overcome drawbacks are necessary.

Visible/Solar Photocatalysis

As discussed so far, each step of the industrial sector for olive oil production implies high operational costs. In this context, any improvements introduced to reduce treatment costs must be carefully considered. Among these, for photocatalytic remediation, solar energy has to be properly developed, especially in the Mediterranean countries, with the final aim of cost-effectiveness.

Visible/solar photocatalytic strategies employ adequately designed heterogeneous and homogeneous photocatalysis, photo-Fenton, and solar-Fenton reagents. Some examples are reported in Table 3.

Gernjak et al. investigated OMWW from Portugal and Spain by solar photocatalysis [105]. In more detail, two solar reactors i were employed at pilot scale: (i) a conventional compound parabolic collector type (CPC); (ii) an open non-concentrating falling film reactor (FFR). Different solar photocatalytic systems were tested, but the photocatalyst with the higher amount of Fe (10 mM) showed the most increased activity.

**Table 3.** State of the art in visible/solar photocatalytic processes for OMWW treatment. Adapted with permission from Reference [135].

| OMWW Origin | Type of Process Treatment and Scale | Obtained Results | Ref. |
|---|---|---|---|
| Spain and Portugal | (i) Solar photocatalysis with $TiO_2$ or added peroxydisulphate, or (ii) solar photo-Fenton—pilot plant | (i) Solar photocatalytic systems did not present sufficientefficacy (ii) 85% COD and up to 100% phenols concentration removal | [105] |
| Italy | (i) Centrifugation + solar photolysis, or (ii) centrifugation + solar modified photo Fenton—laboratory scale | (ii) COD and phenolics removal efficiencies up to 29.3% and 63.6% | [136] |
| Italy | Fenton preceded by coagulation—laboratory scale | 85% COD removal (2 h) | [137] |
| Portugal | Biological (fungi *Pleurotus sajor caju*) and photo-Fenton oxidation—laboratory scale | COD removal efficiency up to 76% and total phenols up to 92% | [99] |
| Cyprus | Coagulation–flocculation, extraction of phenolics and post-oxidation by photo Fenton—laboratory scale | COD removal about 73 ± 2.3% and total phenols of 87 ± 3.1% | [18] |
| Turkey | Sequential adsorption, biological and photo-Fenton treatment—laboratory scale | 99% phenols reduction and 90% total organic content | [138] |
| Spain | $UV/H_2O_2$—laboratory scale | COD removal of 40–48% (30 min) | [139] |

Ruzmanova et al. studied the photocatalytic treatment of a three-phase OMWW photodegradation process using reusable N-doped $TiO_2$ sol-gel compounds, demonstrating the higher activities of doped-catalysts compared to the non-doped ones, reaching a COD removal more elevated than 60% [140]. Additionally, N-doped materials maintain high efficiency when used for several cycles.

In addition, the role of photochemistry in the Fenton-like process is gaining attention thanks to ultraviolet and/or visible light to reduce the catalyst loading, enhancing the catalytic behaviour. In particular, Gernjak et al. investigated OMWW treatment processes by solar-photo Fenton approach on a pilot-plant scale, successfully removing up to 85% COD and 100% phenols [105].

Andreozzi et al. proposed an OMWW treatment based on a three-phase method exploiting (i) centrifugation followed by solar photolysis, (ii) centrifugation and solar photo-Fenton, and (iii) centrifugation coupled with solar photo-Fenton and ozonation. In this context, the ferric catalyst is responsible for COD and phenol removal (up to *ca.* 30% and 64%, respectively) [136].

Rizzo et al. investigated OMWW treatment by photo-Fenton, preceded by coagulation. In this case, the maximum efficiency of organic matter removal was *ca.* 95% in 1 h [137].

Justino et al. studied the combination of fungi *Pleurotus sajor caju* and photo-Fenton oxidation [99]. The treatment by fungi confirmed the reduction of OMWW toxicity towards *Daphnia longispina* and resulted in 72.9% total phenolic compounds removal and 77% COD reduction. When the treatment is preceded by photo-Fenton oxidation, the biological treatment with fungi is more efficient.

Papaphilippou et al. proposed a treatment process for OMWW by coupling coagulation–flocculation and Fenton oxidation. Following the photo-Fenton oxidation, COD and phenol removals were approximately 73% and 87%, respectively [18].

Finally, Aytar et al. reached 99% phenol and 90% total organic content reduction using adsorption, biological (*T. versicolor*), and photo-Fenton treatment in sequence [138].

Considering the depicted scenarios, it emerges that a proper comparison among the performances of the studied technologies to treat OMWW is not a trivial task. Indeed, the numerous variables in play (i.e., OMWW origin, process type and operative conditions, used scale) do not allow identification of a method that guarantees the best results in terms of OMWW removal. Only a rough evaluation in terms of COD removal can be done, but in this case, all the advantages and/or drawbacks of each strategy must be considered. In general, looking at the COD removal values reported in Table 3, interesting results were obtained when working on a laboratory scale and in pilot plants, suggesting promising avenues that deserve to be investigated.

## 5. From Conventional to Easily Recoverable Magnetic Photocatalysts

As described in the previous sections, many approaches have been investigated for OMWW treatment [124,141–146]. Still, most of them suffer from not trivial and not negligible drawbacks (i.e., expensive maintenance, lateness in the separation time, high retention time).

In this regard, technologies based on photocatalysis can be advantageous for their environmental friendliness and high oxidation efficiency [147–149]. To develop even more efficient photocatalytic systems for real applications, research continuously moves the efforts toward exploring different materials.

Conventional nano-or micro-powder photocatalysts are developed for continuous, safe, and efficient photocatalytic reactions. Still, at the same time, their use is limited by the difficult separation and recovery from the reaction mixture for their sustainable reuse [150,151]. The recovery cost could invalidate the technology from an economic viewpoint [152]. To overcome this issue, the introduction of magnetic features in photocatalytic systems seems to be one of the best solutions, giving the possibility to maintain the catalytic performances of samples while making their separation from the reaction a more accessible medium.

Several approaches have been recently explored to develop advanced magnetic photocatalytic materials for wastewater remediation. However, unfortunately, few studies have mainly focused on applying these materials in the treatment of OMWW.

For this purpose, different magnetic nanoparticles (i.e., $\gamma$-$Fe_2O_3$, $Fe_3O_4$, $MFe_2O_4$, where M = Mg, Ni, Zn, Cu, Co) have been introduced in photocatalysts, giving rise to composite materials with magnetic features [153–156]. In this context, electron and hole migration between the magnetic and semiconductor components results in the separation of the photo-induced charge carriers, enhancing the light absorption ability [153–156]. This class of innovative materials has been studied regarding several pollutants in wastewater decontamination. Shen et al. prepared $Fe_3O_4$@$TiO_2$@Ag-Au microspheres with promising magnetic and photocatalytic properties [157]. Singh et al. immobilized BiOI/$Fe_3O_4$ photocatalyst on graphene oxide to degrade 2, 4-dinitrophenol [158]. Furthermore, the potentialities of other magnetic composite photocatalysts have been explored, such as $Cu_2V_2O_7$/$CoFe_2O_4$/g-$C_3N_4$ [159], $MnFe_2O_4$/$SnO_2$ [160], $MoO_3$/$CoFe_2O_4$ [161]. As already mentioned by Ma et al., the research efforts in this field have resulted in the development of several simple and magnetic photocatalytic materials, such as magnetic bismuth-based photocatalysts [162]. In addition, Ruzmanova et al. developed magnetic core $TiO_2$/$SiO_2$/$Fe_3O_4$ nanoparticles to degrade organic compounds in OMWW. 1.5 $g{\cdot}L^{-1}$ of catalyst dosage optimized the photodegradation process, providing high efficiency and an easy catalyst recovery [140]. Successively, Vaiano et al., using ferromagnetic N-$TiO_2$/$SiO_2$/$Fe_3O_4$ nanoparticles, achieved 64% phenol removal and 55% TOC reduction after an irradiation time of 270 min, as well as good stability of the photocatalytic materials after four operation/regeneration cycles [163]. Hesas et al. explored a magnetically separable $Fe_3O_4$ on modernite zeolite to purify OMWW from Kermanshah. They identified the key parameters influencing COD and BOD removal: pH (optimized at the value of 7.8) and turbidity of the treated solution. In addition, in this case, the regenerated $Fe_3O_4$/mordenite zeolite could be reused for five consecutive cycles [164].

In addition, the research community is currently working hard on novel alternatives.

## 6. Perspectives

Considering the high impact of OMWW treatment on the environment and human health, all the sustainability and circular economy principles should be adequately assessed. In this context, perspectives related to the development of efficient, sustainable alternatives to nano- or micro-sized photocatalysts to treat OMWW (Figure 5) can be mainly divided into two categories: (i) eco-friendly materials (mainly characterized by magnetic features) already investigated in the treatment of several "model pollutants"; and (ii) other emerging eco-friendly materials (floating devices, membranes).

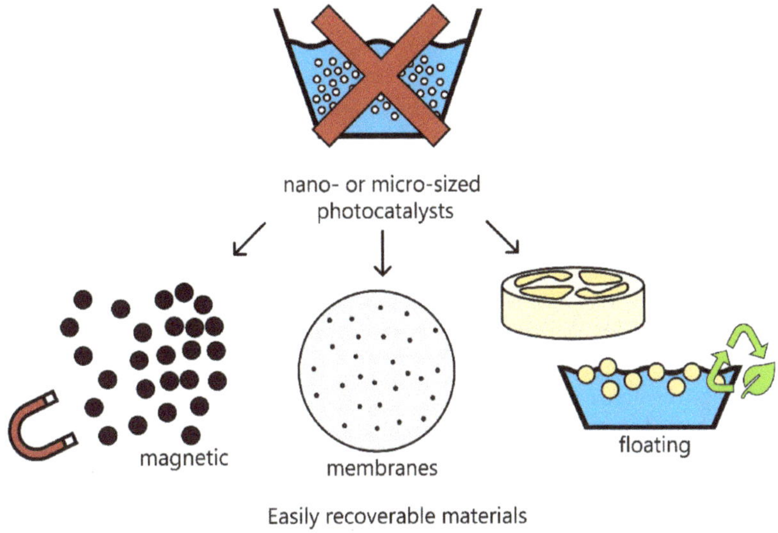


**Figure 5.** Proposed eco-friendly alternatives to nano- or micro-sized photocatalysts to treat OMWW.

### 6.1. Eco-Friendly Materials Used to Treat "Model Pollutants"

Several materials have already been investigated for the degradation of "model pollutants". They show promise for treating OMWW efficiently, and the scientific community could draw inspiration for appropriate evaluations. By way of example, magnetic bismuth-based photocatalysts have been largely used in the wastewater remediation field, and they could also find successful application in OMMW treatment, on which only preliminary studies have been reported.

In general, magnetic bismuth-based materials can be classified as magnetic bismuth-based oxyacid salt, magnetic oxyhalides, magnetic sulfides, and magnetic oxides.

Bismuth-based oxyacid salts (commonly labeled as $Bi_aAO_b$) have gained attention for their excellent visible-light absorption, band potential, and interesting chemical stability [165]. Their specific crystal phase confers good electron transport ability [166]. The introduction of proper magnetic components makes them easily recoverable and reusable for real applications.

In more detail, bismuth ferrite materials ($BiFeO_3$) are characterized by ferroelectricity and ferromagnetic features [167]. They have been explored as nanofibers [168], nanoparticles [169–171], nanosheets [172], nanotubes [173], microspheres [174], and nanorods [175], exploiting their magnetic properties and the 2.2 eV bandgap. Li et al. [173] compared the photocatalytic behaviour of $BiFeO_3$ in the form of nanoparticles, nanofibers, and hollow nanotubes, discovering the superior photoactivity of the latter due to the ultra-thin wall thickness and unique material structure. $BiFeO_3$ nanosheets of 140–230 nm side length and 30 nm thickness were synthesized by Zhu et al. [172] by hydrothermal procedures, demonstrating their high capability to degrade 89% rhodamine B (RhB) under 180 min of visible light irradiation. Bharathkumar et al. [176] prepared $BiFeO_3$ mat and mesh nanostructure materials by an electrospinning method, discovering that the photocatalytic degradation of the mesh sample was greater than that of the mat sample, probably due to the decrease of band gap energy. However, a limitation of the photocatalytic activity of $BiFeO_3$ is related to the fast photogenerated electron-hole recombination. In this context, some studies pointed out that metal deposition and doping have a positive effect, reducing the charge recombination and improving their resulting photocatalytic performance [177].

Other bismuth-based oxyacid salts with narrow band gaps exist, such as $BiVO_4$ (2.26–2.51 eV), $Bi_2WO_6$ (2.56–2.92 eV), $Bi_2MoO_6$ (2.49–2.66 eV), and $Bi_2O_2CO_3$ (2.8–3.4 eV), which can be combined with magnetic components to obtain interesting and advanced materials with enhanced photocatalytic activity [178,179]. By way of example, Cam et al. introduced $MnFe_2O_4$ on $BiVO_4$, obtaining an innovative material with good photocatalytic activity and magnetic recovery [180]. Sakhare et al. [181] produced $BiVO_4/NiFe_2O_4$ composites able to degrade 98% methylene blue in 240 min of collected sunlight illumination and to maintain excellent stability even after four cycles. Bastami et al. [182] prepared magnetic $Fe_3O_4/Bi_2WO_6$ nanohybrids to degrade ibuprofen under solar light. Xiu et al. [183] developed 3D magnetic $Fe_3O_4/Ag/Bi_2MoO_6$ spheres, obtaining an advanced photocatalytic-Fenton coupling system, which exhibited excellent photocatalytic behaviors in the Aatrex degradation.

Bismuth oxyhalides (BiOX, X = Br, Cl, I) represent another family of bismuth-based materials, which have recently attracted scientific research due to their band gap, high stability, and non-toxicity [184,185]. They exhibit a tetragonal matlockite structure interlaced with $[Bi_2O_2]^{2+}$ flat plates and double halogen atomic layers, which reduce the electron-hole pairs' recombination, producing good photocatalytic behaviour [186,187]. In this context, the combination of BiOX and magnetic components represents an interesting perspective to obtain easily recoverable photocatalytic compounds on which many researchers are working. Briefly, Cao et al. [188] investigated the performances of $BiOBr/Fe_3O_4$ composites, prepared by solvothermal method, under visible light irradiation to degrade glyphosate. Li et al. [189] produced $BiOBr/NiFe_2O_4$ materials of different mass ratios according to a conventional hydrothermal approach, and their photocatalytic performances were explored in the photodegradation of methylene blue and phenol. The authors additionally synthesized BiOBr nanosheets decorated with $NiFe_2O_4$ nanoparticles and tested the samples in the rhodamine-B photodegradation [190], observing that the $BiOBr/NiFe_2O_410$ (having 10 wt.% $NiFe_2O_4$) composite was able to degrade rhodamine-B more efficiently than the pure BiOBr and $NiFe_2O_4$ (99.8% rhodamine-B degradation after 30 min radiation). Sin et al. [191] prepared N-$BiOBr/NiFe_2O_4$ composites by a hydrothermal strategy, demonstrating the enhanced photocatalytic behaviour towards phenol and Cr(VI) removal.

Moreover, systems based on BiOCl and BiOI were additionally developed, and their photocatalytic performances have been properly investigated. In particular, Ma et al. [192] prepared magnetic $BiOCl/ZnFe_2O_4$ samples, showing their high photocatalytic activity towards penicillin-G degradation (99% penicillin-G degradation within 180 min under visible-light irradiation). Zhou et al. [193] studied ternary magnetic $Ag_2WO_4/BiOI/CoFe_2O_4$ hybrid compounds, evaluating their photocatalytic activity towards toxic elemental mercury Hg(0) removal. In addition, $BiOI/CoFe_2O_4$ composites modified with $AgIO_3$ [194] and $Ag_2CO_3$ [195] were found to be highly efficient in the photocatalytic reduction of Hg(0).

Finally, magnetic sulfides and oxides deserve to be also mentioned. The former (labeled as $Bi_2S_3$) is described by the 1.3 eV energy bandgap and complete visible light region response [196]. They can be combined with materials with magnetic features to promote charge separation and guarantee good recyclability. For example, Li et al. explored the potentialities of $Fe_3O_4/Bi_2S_3/BiOBr$ samples in the photodegradation of diclofenac and ibuprofen, observing *ca.* 94 and 97% conversion of the studied pollutants, respectively, after 40 and 30 min under visible light irradiation [197]. On the other hand, Zhu et al. tested $Fe_3O_4/Bi_2S_3$ microspheres towards Congo red removal, discovering good stability for continuous tests. The latter (commonly named $Bi_2O_3$) is an attractive material possessing high redox reversibility, bandgap spanning from 2.6 to 2.8 eV, and good electrochemical stability [198]. Several researchers combined it with magnetic compounds to obtain final easily recoverable materials. In particular, Abbasi et al. prepared 3D flower-like $Fe_3O_4@Bi_2O_3/g\text{-}C_3N_4$ nanocomposites, successively evaluating their photocatalytic activity towards indigo carmine degradation [199]. In this case, introducing the conductive

C layer in the nanocomposite sample could improve the photocatalytic behaviour. In addition, Gao et al. first obtained a C/$Fe_3O_4$ composite and then a double conductive C/$Fe_3O_4$/$Bi_2O_3$ photocatalyst. In this case, electron-hole pairs' recombination and the reverse electron transfer to $Bi_2O_3$ can be prevented [200].

### 6.2. Other Emerging Eco-Friendly Materials (Floating Devices, Membranes)

Due to their floating properties and good visible light utilization, floating photocatalysts could be considered an excellent choice to gradually substitute conventional photocatalysts [201]. In fact, since 1993, floating $TiO_2$-based materials have been studied [202]. In general, a floating device exploits a lightweight material to float on the water surface, and the photocatalytic performances are maximized thanks to its exposed large surface [203,204]. At the same time, due to its peculiar structure, it minimizes photocatalyst loss, avoiding the long-term contact between photocatalyst and pollutants, which can decrease photocatalytic activity. In the last decades, various supports (i.e., perlite, vermiculite, glass, cork, graphite, polymer) have been investigated as candidates for developing efficient floating photocatalysts [201].

Among them, by way of example, some of the authors studied the performance of aerogel water-floating based materials prepared by poly (vinyl alcohol) and polyvinylidene fluoride as a polymer platform and loaded with different semiconductors, such as g-$C_3N_4$, $MoO_3$, $Bi_2O_3$, $Fe_2O_3$ or $WO_3$, obtaining interesting results towards the reduction of Cr(VI) under visible light [204]. Moreover, Wang et al. [205] recently investigated the use of advanced spongy foam photocatalysts composed of BiOX compounds deposited onto polyurethane foams to degrade targeted pollutants, such as methyl orange, phenol, and chlortetracycline. These systems showed a high potential because they can conjugate high stability, excellent adaptability, and easy recovery, with high photocatalytic performances and good reusability.

In the present panorama, the possibility of using supports characterized by eco-friendly features (i.e., low-cost, non-toxicity, bioavailability) is a priority for further evaluation, and will require strenuous investigation efforts. Some researchers have already considered luffa cylindrica, alginate sphere, or light expanded clay aggregate (LECA), but their potentialities are still the object of study today. Following this perspective, Chawla et al. immobilized $MoSe_2$/$BiVO_4$ on luffa cylindrica, and then they tested it in phenol degradation, observing up to 97% removal within 2 h of visible light irradiation [206]. Huang et al. recently investigated the possibility of combining alginate spheres with magnetic components, finding exciting results. In this case, the excellent floating performance, together with the availability of reaction sites offered by the material, resulted in the degradation of the selected pollutants (e.g., methyl orange) [207].

Finally, the use of membranes deserves also to be cited. This technology has been investigated in the OMWW treatments for several advantages (simplicity, modulability, easy maintenance, high separation efficiency, small footprint, and easy scale-up) [208]. Several membrane types have been developed and produced, from the polymeric-based ones [209,210] to the inorganic-based ones [211]. All of them have shown excellent performance in the separation of targeted pollutants. However, membrane technology is characterized by some drawbacks. By way of examples, they may be limited by the high concentration of suspended solids present in the OMWW to be treated, and they suffer from foulant deposition due to contaminants separated from the feed. Thus, further treatments are usually required. In this context, Dzinun et al. [212–214] tried to develop a photocatalytic membrane to overcome the membrane fouling and use it as support for photocatalysts. In this case, the photocatalyst addition should minimize the fouling rate. Unfortunately, photocatalytic membranes are also affected by some drawbacks. For example, prolonged exposure to irradiation may ruin their structure, causing damage to the active surface area, which strongly impacts the photocatalytic efficiency [215]. In this context, many ideas are currently put into action by several researchers, as recently reported by Salim et al. [216,217].

All these interesting and promising results obtained in the decontamination of targeted pollutants present in wastewater can be a starting point to investigate more in detail what happens in the case of such complex matrices as OMWW.

## 7. Conclusions

This review provides a critical insight into the current status and the consequent advances related to OMWW treatments, underlying their potentialities and drawbacks. A particular focus on developing innovative eco-friendly photocatalysts, which could become valid alternatives to conventional systems, if properly optimized, is provided.

Nowadays, the OMWW sector plays a fundamental role in the European economy, but at the same time, it also leads to dramatic consequences on the environment and human health. In this context, the current challenge involves optimizing well-known and conventional technologies. Still, the most captivating challenge is the development of innovative advanced strategies, such as those based on photocatalysis. These latter offer many advantages (i.e., high efficiency, low cost) but require the use of novel materials to overcome the common issues related to using slurry reactors and difficult photocatalyst recovery.

In this scenario, the potential use of easily recoverable magnetic compounds as well as floating- and membrane-based devices points to new horizons for sustainability, alternative to conventional $TiO_2$-based systems. The application of these advanced systems still needs hard work by the research world. Their future success in real applications will create a bridge between environmental protection and a circular economy.

**Author Contributions:** Conceptualization, E.F. (Ermelinda Falletta) and C.L.B.; methodology, E.F. (Ermelinda Falletta); visualization, M.G.G., E.F. (Elena Ferrara) and E.F. (Ermelinda Falletta); literature collection and analysis, M.G.G., E.F. (Elena Ferrara) and E.F. (Ermelinda Falletta); Content design, E.F. (Ermelinda Falletta); Writing—original draft preparation, M.G.G., E.F. (Elena Ferrara) and E.F. (Ermelinda Falletta); writing—review and editing, E.F. (Ermelinda Falletta) and C.L.B.; supervision, C.L.B.; project administration, E.F. (Ermelinda Falletta) and C.L.B.; funding acquisition, C.L.B. All authors have read and agreed to the published version of the manuscript.

**Funding:** Velux Stiftung Foundation is gratefully acknowledged for its financial support through project 1381, "SUNFLOAT—Water decontamination by sunlight-driven floating photocatalytic systems".

**Data Availability Statement:** The data that support the plots within this paper are available from the corresponding author on reasonable request.

**Acknowledgments:** This work was supported by the Department of Chemistry, Università degli Studi di Milano, Italy (Piano Sostegno alla Ricerca, PSR, grant 2021).

**Conflicts of Interest:** The authors declare no conflict of interest.

## References

1. Ducom, G.; Gautier, M.; Pietraccini, M.; Tagutchou, J.P.; Lebouil, D.; Gourdon, M. Comparative analyses of three olive mill solid residues from different countries and processes for energy recovery by gasification. *Renew. Energy* **2020**, *145*, 180–189. [CrossRef]
2. Inglezakis, V.J.; Moreno, J.L.; Doula, M. Olive oil waste management EU Legislation: Current situation and policy recommendations. *Int. J. Chem. Environ. Eng. Syst.* **2012**, *3*, 65–77.
3. Mercé Sole, M.; Pons, L.; Conde, M.; Gaidau, C.; Baccardit, A. Characterization of Wet Olive Pomace Waste as Bio Based Resource for Leather Tanning. *Materials* **2021**, *14*, 5790. [CrossRef] [PubMed]
4. Kapellakis, I.E.; Tsagarakis, K.P.; Crowther, J.C. Olive oil history, production and by-product management. *Rev. Environ. Sci. Bio/Technol.* **2008**, *7*, 1–26. [CrossRef]
5. Azaizeh, H.; Abu Tayeh, H.N.; Gerchman, Y. Chapter 2: Valorisation of olive oil indus-try solid waste and production of ethanol and high value-added biomolecules. In *Biovalorisation of Wastes to Renewable Chemicals and Biofuels*; Elsevier: Amsterdam, The Netherlands, 2019; pp. 27–40. ISBN 9780128179512.
6. Paredes, C.; Cegarra, J.; Roig, A.; Sanchez-Monedero, M.A.; Bernal, M.P. Characterisation of olive mill wastewater (alpechin) and its sludge for agricultural purposes. *Bioresour. Technol.* **1999**, *67*, 111–115. [CrossRef]
7. Paraskeva, P.; Diamadopoulos, E. Technologies for olive mill wastewater (OMW) treatment: A review. *J. Chem. Technol. Biotechnol.* **2006**, *81*, 1475–1485. [CrossRef]
8. Potoglou, D.; Kouzeli-Katsiri, A.; Haralambopoulos, D. Solar distillation of olive mill wastewater. *Renew. Energy* **2004**, *29*, 569–579. [CrossRef]

9. Niaounakis, M.; Halvadakis, C.P. *Olive-Mill Waste Management—Literature Review and Patent Survey*; Typothito-George Dardanos: Athens, Greece, 2004.
10. Caputo, A.C.; Scacchia, F.; Pelagagge, P.M. Disposal of byproducts in olive oil industry: Waste-to-energy solutions. *Appl. Therm. Eng.* **2003**, *23*, 197–214. [CrossRef]
11. Al-Malah, K.; Azzam, M.O.J.; Abulail, N.I. Olive mills effluent (OME) wastewater post-treatment using activated clay. *Sep. Purif. Technol.* **2000**, *20*, 225–234. [CrossRef]
12. Azzam, M.O.J.; Al-Malah, K.I.; Abu-Lail, N.I. Dynamic posttreatment response of olive mill effluent wastewater using activated carbon. *J. Environ. Sci. Health A* **2004**, *39*, 269–280. [CrossRef]
13. Arvaniti, E.C.; Zagklis, D.P.; Papadakis, V.G.; Paraskeva, C.A. High-added value materials production from OMW: A technical and economical optimization. *Int. J. Chem. Eng.* **2012**, *2012*, 607219. [CrossRef]
14. Djellabi, R.; Giannantonio, R.; Falletta, E.; Bianchi, C.L. SWOT analysis of photocatalytic materials towards large scale environmental remediation. *Curr. Opin. Chem. Eng.* **2021**, *33*, 100696. [CrossRef]
15. Galloni, M.G.; Cerrato, G.; Giordana, A.; Falletta, E.; Bianchi, C.L. Sustainable Solar Light Photodegradation of Diclofenac by Nano- and Micro-Sized $SrTiO_3$. *Catalysts* **2022**, *12*, 804. [CrossRef]
16. Lucas, M.S.; Peres, J.A. Removal of COD from olive mill wastewater by Fenton's reagent: Kinetic study. *J. Hazard. Mater.* **2009**, *168*, 1253–1259. [CrossRef] [PubMed]
17. Stoller, M.; Bravi, M. Critical flux analyses on differently pretreated olive vegetation wastewater streams: Some case studies. *Desalination* **2010**, *250*, 578–582. [CrossRef]
18. Papaphilippou, P.C.; Yiannapas, C.; Politi, M.; Daskalaki, V.M.; Michael, C.; Kalogerakis, N.; Mantzavinos, D.; Fatta-Kassinos, D. Sequential coagulation–flocculation, solvent extraction and photo-Fenton oxidation for the valorization and treatment of olive mill effluent. *Chem. Eng. J.* **2013**, *224*, 82–88. [CrossRef]
19. Aziz, K.H.H.; Omer, K.M.; Mahyar, A.; Miessner, H.; Mueller, S.; Moeller, D. Applica-tion of Photocatalytic Falling Film Reactor to Elucidate the Degradation Pathways of Pharmaceutical Diclofenac and Ibuprofen in Aqueous Solutions. *Coatings* **2019**, *9*, 465. [CrossRef]
20. Aziz, K.H.H. Application of different advanced oxidation processes for the removal of chloroacetic acids using a planar falling film reactor. *Chemosphere* **2019**, *228*, 377–383. [CrossRef]
21. Podgornik, M.; Bucar-Miklavcic, M.; Levart, A.; Salobir, J.; Rezar, V.; Butinar, B. Chemical Characteristics of Two-Phase Olive-Mill Waste and Evaluation of Their Direct Soil Application in Humid Mediterranean Regions. *Agronomy* **2022**, *12*, 1621. [CrossRef]
22. Directive 2008/98/EC of the European Parliament and of the Council of 19 November 2008 on Waste and Repealing Certain Directives. Available online: https://eur-lex.europa.eu/legal-content/EN/TXT/?uri=celex%3A32008L0098 (accessed on 15 July 2022).
23. Council Directive 1999/31/EC on the Landfill of Waste. Available online: https://eur-lex.europa.eu/legal-content/EN/TXT/PDF/?uri=CELEX:01999L0031-20180704&from=FI (accessed on 10 July 2022).
24. Taccogna, G. *The Legal Regime of Olive Pomace Deriving from Olive Oil Extraction at Olive Mills, Waste, By-Products and Biomass*; On behalf of ARE S.p.A. Agenzia regionale per l'energia della Liguria, member of the community project: "MORE: Market of Olive Residues for Energy"; Department of Public and Procedural Law, University of Genoa: Genoa, Italy, 2010.
25. EC-DG. Survey of Wastes Spread on Land, Final Report, European Commission, Directorate-General for Environment. 2001. Available online: https://ec.europa.eu/environment/pdf/waste/compost/econanalysis_finalreport.pdf (accessed on 30 June 2022).
26. Aydar, A.Y.; Bagdatlioglu, N.; Köseoglu, O. Effect of ultrasound on olive oil extraction and optimization of ultrasound-assisted extraction of extra virgin olive oil by response surface methodology (RSM). *Grasas Aceites* **2017**, *68*, e189. [CrossRef]
27. Arvanitoyannis, I.S.; Kassaveti, A.; Stefanatos, S. Olive Oil Waste Treatment: A Comparative and Critical Presentation of Methods, Advantages & Disadvantages. *Crit. Rev. Food Sci. Nutr.* **2007**, *47*, 187–229. [PubMed]
28. Facts and Definitions: The Olive Oil Source. Available online: http://www.oliveoilsource.com/millandpressfacts.htm (accessed on 25 June 2022).
29. Olive Oil Waste Treatment. Available online: http://www.ucm.es/info/improliv/allgem.htm (accessed on 25 June 2022).
30. Dakhli, R.D. Agronomic Application of Olive Mill Waste Water: Short-Term Effect on Soil Chemical Properties and Barley Performance Under Semiarid Mediterranean Conditions. *EQA-Int. J. Environ. Qual.* **2018**, *27*, 1–17.
31. Fedeli, E.; Camurati, F. Valorisation des margines et des grignons´epuises par recuperation de quelques composants. In Proceedings of the Seminaire International sur la Valorisation des Sous-Produits de L'olivier PNUD/FAO/COI, Monastir, Tunisia, 15–17 December 1981.
32. Rocha, C.; Soria, M.A.; Madeira, L.M. Olive Mill Wastewater Valorization through Steam Reforming Using Multifunctional Reactors: Challenges of the Process Intensification. *Energies* **2022**, *15*, 920. [CrossRef]
33. Production of Olive Oil. Available online: http://www.oliveoilsource.com/millandpressfacts3.htm (accessed on 26 June 2022).
34. Abenoza, M.; Benito, M.; Saldaña, G.; Alvarez, I.; Raso, J.; Sanchez-Gimeno, A.C. Effects of pulsed electric field on yield extrac-tion and quality of olive oil. *Food Bioprocess Technol.* **2013**, *6*, 1367–1373. [CrossRef]
35. Puértolas, E.; Martínez de Marañón, I. Olive oil pilot-production assisted by pulsed electric field: Impact on extraction yield, chemical parameters and sensory properties. *Food Chem.* **2015**, *167*, 497–502. [CrossRef] [PubMed]
36. Clodoveo, M.L.; Durante, V.; La Notte, D. Working towards the development of innovative ultrasound equipment for the extraction of virgin olive oil. *Ultrason. Sonochem.* **2013**, *20*, 1261–1270. [CrossRef]

37. Aydar, A.Y. *Utilization of Response Surface Methodology in Optimization of Extraction of Plant Materials*; Intech Open: London, UK, 2018; pp. 157–169.
38. Meroni, D.; Djellabi, R.; Ashokkumar, M.; Bianchi, C.L.; Boffito, D.C. Sonoprocessing: From Concepts to Large-Scale Reactors. *Chem. Rev.* **2022**, *122*, 3219–3258. [CrossRef] [PubMed]
39. Bermúdez-Aguirre, D.; Mobbs, T.; Barbosa-cánovas, G.V. *Ultrasound Technologies for Food and Bioprocessing*; Epub ahead of print; Springer: Berlin/Heidelberg, Germany, 2011. [CrossRef]
40. Chemat, F.; Zill-E-Huma, R.; Khan, M.K. Applications of ultrasound in food technology: Processing, preservation and extraction. *Ultrason. Sonochem.* **2011**, *18*, 813–835. [CrossRef] [PubMed]
41. Veneziani, G.; Sordini, B.; Taticchi, A.; Esposto, S.; Selvaggini, R.; Urbani, S.; Di Maio, I.; Servili, M. Improvement of Olive Oil Mechanical Extraction: New Technologies, Process Efficiency, and Extra Virgin Olive Oil Quality. In *Products from Olive Tree Dimitrios Boskou and Maria Lisa Clodoveo*; Intech Open: London, UK, 2016; Available online: https://www.intechopen.com/books/products-from-olive-tree/improvement-of-olive-oil-mechanicalextraction-new-technologies-processefficiency-and-extra-virgin (accessed on 20 June 2022). [CrossRef]
42. Clodoveo, M.L. An overview of emerging techniques in virgin olive oil extraction process: Strategies in the development of innovative plants. *J. Agric. Eng.* **2013**, *44*, 49–59. [CrossRef]
43. Clodoveo, M.L. New advances in the development of innovative virgin olive oil extraction plants: Looking back to see the future. *Food Res. Int.* **2013**, *54*, 726–729. [CrossRef]
44. Clodoveo, M.L.; Hachicha Hbaieb, R. Beyond the traditional virgin olive oil extraction systems: Searching innovative and sustainable plant engineering solutions. *Food Res. Int.* **2013**, *54*, 1926–1933. [CrossRef]
45. Clodoveo, M.L.; Camposeo, S.; Amirante, R.; Dugo, G.; Cicero, N.; Boskou, D. Research and Innovative Ap-proaches to Obtain Virgin Olive Oils with a Higher Level of Bioactive Constituents. *Olive Olive Oil Bioact. Const.* **2015**, *7*, 179–215. [CrossRef]
46. Sun, D.W. *Emerging Technologies for Food Processing*, 2nd ed.; Elsevier Inc.: Dublin, Ireland, 2014.
47. Çavdar, H.K.; Yanık, D.K.; Gök, U.; Gogus, F. Optimisation of microwave-assisted extraction of pomegranate (*Punica granatum* L.) seed oil and evaluation of Its physicochemical and bioactive properties. *Food Technol. Biotechnol.* **2017**, *55*, 86–94. [PubMed]
48. Aydar, A.Y. Chapter: Emerging extraction technologies in olive oil production. In *Technological Innovation in the Olive Oil Production Chain*; Intech Open: London, UK, 2018. [CrossRef]
49. Jiménez, A.; Beltrán, G.; Uceda, M. High-power ultra-sound in olive paste pretreatment. Effect on process yield and virgin olive oil characteristics. *Ultrason. Sonochem.* **2007**, *14*, 725–731. [CrossRef] [PubMed]
50. Bejaoui, M.A.; Beltrán, G.; Sánchez-Ortiz, A.; Sanchez, S.; Jimenez, A. Continuous high power ultrasound treatment before malaxation, a laboratory scale approach: Effect on virgin olive oil quality criteria and yield. *Eur. J. Lipid Sci. Technol.* **2016**, *118*, 332–336. [CrossRef]
51. Aydar, A.Y. Physicochemical characteristics of extra virgin olive oils obtained by ultrasound assisted extraction from different olive cultivars. *Int. J. Sci. Technol. Res.* **2018**, *4*, 1–10.
52. Bejaoui, M.A.; Beltran, G.; Aguilera, M.P.; Jimenez, A. Continuous conditioning of olive paste by high power ultrasounds: Response surface methodology to predict temperature and its effect on oil yield and virgin olive oil characteristics. *LWT-Food Sci. Technol.* **2016**, *69*, 175–184. [CrossRef]
53. Kadi, H.; Moussaoui, R.; Djadoun, S.; Sharrock, P. Microwave assisted extraction of olive oil pomace by acidic hexane. *Iran. J. Chem. Chem. Eng.* **2016**, *35*, 73–79.
54. Leone, A.; Tamborrino, A.; Zagaria, R.; Sabella, E.; Romaniello, R. Plant innovation in the olive oil extraction process: A comparison of efficiency and energy consumption between microwave treatment and traditional malaxation of olive pastes. *J. Food Eng.* **2015**, *146*, 44–52. [CrossRef]
55. Clodoveo, M.L.; Paduano, A.; Di Palmo, T.; Crupi, P.; Corbo, F.; Pesce, V.; Distaso, E.; Tamburrano, P.; Amirante, R. Engineering design and prototype development of a full scale ultrasound system for virgin olive oil by means of numerical and experimental analysis. *Ultrason. Sonochem.* **2017**, *37*, 169–181. [CrossRef]
56. Leal, A.L.; Soria, M.A.; Madeira, L.M. Autothermal reforming of impure glycerol for $H_2$ production: Thermodynamic study including in situ $CO_2$ and/or $H_2$ separation. *Int. J. Hydrogen Energy* **2016**, *41*, 2607–2620. [CrossRef]
57. Agency, I.E. Hydrogen. 2021. Available online: https://www.iea.org/reports/hydrogen (accessed on 12 August 2022).
58. Agency, I.E. Global Hydrogen Demand by Sector in the Net Zero Scenario, 2020–2030. 2021. Available online: https://www.iea.org/data-and-statistics/charts/global-hydrogen-demand-by-sector-in-the-net-zero-scenario-2020-2030 (accessed on 12 August 2022).
59. Kapellakis, I.E.; Tsagarakis, K.P.; Avramaki, C.; Angelakis, A.N. Olive mill wastewater management in river basins: A case study in Greece. *Agric. Water Manag.* **2006**, *82*, 354–370. [CrossRef]
60. Chartzoulakis, K.; Psarras, G.; Moutsopoulou, M.; Stefanoudaki, E. Application of olive mill wastewater to a Cretan olive orchard: Effects on soil properties, plant performance and the environment. *Agric. Ecosyst. Environ.* **2010**, *138*, 293–298. [CrossRef]
61. Watanabe, K. Microorganisms relevant to bioremediation. *Curr. Opin. Biotechnol.* **2001**, *12*, 237–241. [CrossRef]
62. Enhanced Bioremediation. Available online: http://www.cpeo.org/techtree/ttdescript/ensolmx.htm (accessed on 15 June 2022).
63. Ehaliotis, C.; Papadopoulou, K.; Kotsou, M.; Mari, I.; Balis, C. Adaptation and population dynamics of Azotobacter vinelandii during aerobic biological treatment of olive-mill wastewater. *FEMS Microbiol. Ecol.* **1999**, *30*, 301–311. [CrossRef]

64. Di Gioia, D.; Barberio, C.; Spagnesi, S.; Marchetti, L.; Fava, F. Characterization of four olive-mill-wastewater indigenous bacterial strains capable of aerobically degrading hydroxylated and methoxylated monocyclic aromatic compounds. *Arch. Microbiol.* **2002**, *178*, 208–217. [CrossRef] [PubMed]
65. Di Gioia, D.; Bertin, L.; Fava, F.; Marchetti, L. Biodegradation of hydroxylated and methoxylated benzoic, phenylacetic and phenylpropenoic acids present in olive mill wastewaters by two bacterial strains. *Res. Microbiol.* **2001**, *152*, 83–93. [CrossRef]
66. Piperidou, C.I.; Chaidou, C.I.; Stalikas, C.D.; Soulti, K.; Pilidis, G.A.; Balis, C. Bioremediation of olive oil mill wastewater: Chemical alternations induced by Azotobacter vinelandii. *J. Agric. Food Chem.* **2000**, *48*, 1941–1948. [CrossRef]
67. Yesilada, O.; Sik, S.; Sam, M. Biodegradation of olive oil mill wastewater by Coriolus versicolor and Funalia trogii: Effects of agitation, initial COD concentration, inoculum size and immobilization. *World J. Microbiol. Biotechnol.* **1997**, *14*, 37–42. [CrossRef]
68. Zagklis, D.P.; Arvaniti, E.C.; Papadakis, V.G.; Paraskeva, C.A. Sustainability analysis and benchmarking of olive mill wastewater treatment methods. *J. Chem. Technol. Biotechnol.* **2013**, *88*, 742–750. [CrossRef]
69. Tomati, U.; Galli, E.; Fiorelli, F.; Pasetti, L. Fertilizers from composting of olive-mill wastewaters. *Int. Biodeterior. Biodegrad.* **1996**, *38*, 155–162. [CrossRef]
70. Hachicha, S.; Cegarra, J.; Sellami, F.; Hachicha, R.; Drira, N.; Medhioub, K.; Ammar, E. Elimination of polyphenols toxicity from olive mill wastewater sludge by its co-composting with sesame bark. *J. Hazard. Mater.* **2009**, *161*, 1131–1139. [CrossRef] [PubMed]
71. New Technologies for Husks and Waste Waters Recycling. Available online: https://webgate.ec.europa.eu/life/publicWebsite/project/details/1782 (accessed on 5 June 2022).
72. Gavala, H.N.; Skiadas, I.V.; Bozinis, N.A.; Lyberatos, G. Anaerobic codigestion of agricultural industries' wastewaters. *Water Sci. Technol.* **1996**, *34*, 67–75. [CrossRef]
73. Zouari, N.; Ellouz, R. Toxic effect of coloured olive compounds on the anaerobic digestion of olive oil mill effluent in UASB-like reactors. *J. Chem. Technol. Biotechnol.* **1996**, *66*, 414–420. [CrossRef]
74. Stamatelatou, K.; Kopsahelis, A.; Blika, P.S.; Paraskeva, C.A.; Lyberatos, G. Anaerobic digestion of olive mill wastewater in a periodic anaerobic baffled reactor (PABR) followed by further effluent purification via membrane separation technologies. *J. Chem. Technol. Biotechnol.* **2009**, *84*, 909–917. [CrossRef]
75. Boari, G.; Brunetti, A.; Passino, R.; Rozzi, A. Anaerobic digestion of olive oil mill wastewaters. *Agric. Wastes* **1984**, *10*, 161–175. [CrossRef]
76. Aktas, E.S.; Imre, S.; Ersoy, L. Characterization and lime treatment of olive mill wastewater. *Water Res.* **2001**, *35*, 2336–2340. [CrossRef]
77. Boukhoubza, F.; Jail, A.; Korchi, F.; Idrissi, L.L.; Hannache, H.; Duarte, J.C.; Hassani, L.; Nejmeddine, A. Application of lime and calcium hypochlorite in the dephenolisation and discolouration of olive mill wastewater. *J. Environ. Manag.* **2009**, *91*, 124–132. [CrossRef]
78. Saglık, S.; Ersoy, L.; Imre, S. Oil recovery from lime-treated wastewater of olive mills. *Eur. J. Lipid Sci. Technol.* **2002**, *104*, 212–215. [CrossRef]
79. Ugurlu, M.; Kula, I. Decolourization and removal of some organic compounds from olive mill wastewater by advanced oxidation processes and lime treatment. *Environ. Sci. Pollut. Res.* **2007**, *14*, 319–325. [CrossRef]
80. Inan, H.; Dimoglo, A.; Simsek, H.; Karpuzcu, M. Olive oil mill wastewater treatment by means of electro-coagulation. *Sep. Purif. Technol.* **2004**, *36*, 23–31. [CrossRef]
81. Israilides, C.J.; Vlyssides, A.G.; Mourafeti, V.N.; Karvouni, G. Olive oil wastewater treatment with the use of an electrolysis system. *Bioresour. Technol.* **1997**, *61*, 163–170. [CrossRef]
82. Giannis, A.; Kalaitzakis, M.; Diamadopoulos, E. Electrochemical treatment of olive mill wastewater. *J. Chem. Technol. Biotechnol.* **2007**, *82*, 663–671. [CrossRef]
83. Papastefanakis, N.; Mantzavinos, D.; Katsaounis, A. DSA electrochemical treatment of olive mill wastewater on $Ti/RuO_2$ anode. *J. Appl. Electrochem.* **2010**, *40*, 729–737. [CrossRef]
84. Chatzisymeon, E.; Dimou, A.; Mantzavinos, D.; Katsaounis, A. Electrochemical oxidation of model compounds and olive mill wastewater over DSA electrodes: 1. The case of $Ti/IrO_2$ anode. *J. Hazard. Mater.* **2009**, *167*, 268–274. [CrossRef]
85. Ahmed, B.; Limem, E.; Abdel-Wahab, A.; Nasr, B. Photo-Fenton treatment of actual agro-industrial wastewaters. *Ind. Eng. Chem. Res.* **2011**, *50*, 6673–6680. [CrossRef]
86. Aki, S.N.V.K.; Abraham, M.A. An economic evaluation of catalytic supercritical water oxidation: Comparison with alternative waste treatment technologies. *Environ. Prog.* **1998**, *17*, 246–255. [CrossRef]
87. Rivas, F.J.; Gimeno, O.; Portela, J.R.; de la Ossa, E.M.; Beltran, F.J. Supercritical water oxidation of olive oil mill wastewater. *Ind. Eng. Chem. Res.* **2001**, *17*, 246–255. [CrossRef]
88. Chatzisymeon, E.; Diamadopoulos, E.; Mantzavinos, D. Effect of key operating parameters on the non-catalytic wet oxidation of olive mill wastewaters. *Water Sci. Technol.* **2009**, *59*, 2509–2518. [CrossRef]
89. Weichgrebe, D.; Vogelpohl, A. A comparative study of wastewater treatment by chemical wet oxidation. *Chem. Eng. Process. Process. Intensif.* **1994**, *33*, 199–203. [CrossRef]
90. Katsoyiannis, I.A.; Canonica, S.; von Gunten, U. Efficiency and energy requirements for the transformation of organic micropollutants by ozone, $O_3/H_2O_2$ and $UV/H_2O_2$. *Water Res.* **2011**, *45*, 3811–3822. [CrossRef] [PubMed]
91. Karageorgos, P.; Coz, A.; Charalabaki, M.; Kalogerakis, N.; Xekoukoulotakis, N.P.; Mantzavinos, D. Ozonation of weathered olive mill wastewaters. *J. Chem. Technol. Biotechnol.* **2006**, *81*, 1570–1576. [CrossRef]

92. Mousavi, M.; Habibi-Yangjeh, A.; Pouran, S.R. Review on magnetically separable graphitic carbon nitride-based nanocomposites as promising visible-light-driven photocatalysts. *J. Mater. Sci. Mater. Electron.* **2018**, *29*, 1719–1747. [CrossRef]
93. Shekofteh-Gohari, M.; Habibi-Yangjeh, A.; Abitorabi, M.; Rouhi, A. Magnetically separable nanocomposites based on ZnO and their applications in photocatalytic processes: A review. *Crit. Rev. Environ. Sci. Technol.* **2018**, *48*, 806–857. [CrossRef]
94. Djellabi, R.; Noureen, L.; Dao, V.D.; Meroni, D.; Falletta, E.; Dionysiou, D.D.; Bianchi, C.L. Recent advances and challenges of emerging solar-driven steam and the contribution of photocatalytic effect. *Chem. Eng. J.* **2022**, *431*, 134024. [CrossRef]
95. Pirhashemi, M.; Habibi-Yangjeh, A.; Rahim Pouran, S. Review on the criteria anticipated for the fabrication of highly efficient ZnO-based visible-light-driven photo-catalysts. *J. Ind. Eng. Chem.* **2018**, *62*, 1–25. [CrossRef]
96. Ahmad, R.; Ahmad, Z.; Khan, A.U.; Mastoi, N.R.; Aslam, M.; Kim, J. Photocatalytic systems as an advanced environmental remediation: Recent developments, limitations and new avenues for applications. *J. Environ. Chem. Eng.* **2016**, *4*, 4143–4164. [CrossRef]
97. Tolosana-Moranchel, A.; Ovejero, D.; Barco, B.; Bahamonde, A.; Díaz, E.; Faraldos, M. An approach on the comparative behavior of chloro/nitro substituted phenols photocatalytic degradation in water. *J. Environ. Chem. Eng.* **2019**, *7*, 103051. [CrossRef]
98. Chatzisymeon, E.; Xekoukoulotakis, N.P.; Mantzavinos, D. Determination of key operating conditions for the photocatalytic treatment of olive mill wastewaters. *Catal. Today* **2009**, *144*, 143–148. [CrossRef]
99. Justino, C.I.; Duarte, K.; Loureiro, F.; Pereira, R.; Antunes, S.C.; Marques, S.M.; Gonçalves, F.; Rocha-Santos, T.A.P.; Freitas, A.C. Toxicity and organic content characterization of olive oil mill wastewater undergoing a sequential treatment with fungi and photo-Fenton oxidation. *J. Hazard. Mater.* **2009**, *172*, 1560–1572. [CrossRef]
100. Ochando-Pulido, J.M.; Hodaifa, G.; Víctor-Ortega, M.D.; Martínez-Férez, A. A novel photocatalyst with ferromagnetic core used for the treatment of olive oil mill effluents from two-phase production process. *Sci. World J.* **2013**, *2013*, 196470. [CrossRef] [PubMed]
101. Lafi, W.K.; Shannak, B.; Al-Shannag, M.; Al-Anber, Z.; Al-Hasan, M. Treatment of olive mill wastewater by combined advanced oxidation and biodegradation. *Sep. Purif. Technol.* **2009**, *70*, 141–146. [CrossRef]
102. Ochando-Pulido, J.M.; Stoller, M. Kinetics and boundary flux optimization of integrated photocatalysis and ultrafiltration process for two-phase vegetation and olive washing wastewaters treatment. *Chem. Eng. J.* **2015**, *279*, 387–395. [CrossRef]
103. Costa, J.C.; Alves, M.M. Posttreatment of olive mill wastewater by immobilized $TiO_2$ photocatalysis. *Photochem. Photobiol.* **2013**, *89*, 545–551. [CrossRef]
104. Ruzmanova, Y.; Stoller, M.; Chianese, A. Photocatalytic treatment of olive mill wastewater by magnetic core titanium dioxide nanoparticles. *Chem. Eng. Trans.* **2013**, *32*, 2269–2274.
105. Gernjak, W.; Maldonado, M.I.; Malato, S.; Cáceres, J.; Krutzler, T.; Glaser, A.; Bauer, R. Pilot-plant treatment of olive mill wastewater (OMW) by solar $TiO_2$ photocatalysis and solar photo-Fenton. *Sol. Energy* **2004**, *77*, 567–572. [CrossRef]
106. Fujishima, A.; Honda, K. Electrochemical photolysis of water at a semiconductor electrode. *Nature* **1972**, *238*, 37–38. [CrossRef] [PubMed]
107. Xing, Z.; Zhang, J.; Cui, J.; Yin, J.; Zhao, T.; Kuang, J.; Xiu, Z.; Wan, N.; Zhou, W. Recent advances in floating $TiO_2$-based photocatalysts for environmental application. *Appl. Catal. B Environ.* **2018**, *225*, 452–467. [CrossRef]
108. Zhao, Y.; Wang, Y.; Xiao, G.; Su, H. Fabrication of biomaterial/$TiO_2$ composite photocatalysts for the selective removal of trace environmental pollutants. *Chin. J. Chem. Eng.* **2019**, *27*, 1416–1428. [CrossRef]
109. Tomar, L.J.; Chakrabarty, B.S. Synthesis, structural and optical properties of $TiO_2$-$ZrO_2$ nanocomposite by hydrothermal method. *Adv. Mater. Lett.* **2013**, *4*, 64–67. [CrossRef]
110. Habibi-Yangjeh, A.; Feizpoor, S.; Seifzadeh, D.; Ghosh, S. Improving visible-light-induced photocatalytic ability of $TiO_2$ through coupling with $Bi_3O_4Cl$ and carbon dot nanoparticles. *Sep. Purif. Technol.* **2020**, *238*, 116404. [CrossRef]
111. Feizpoor, S.; Habibi-Yangjeh, A.; Yubuta, K. Integration of carbon dots and polyaniline with $TiO_2$ nanoparticles: Substantially enhanced photocatalytic activity to removal various pollutants under visible light. *J. Photochem. Photobiol. A Chem.* **2018**, *367*, 94–104. [CrossRef]
112. Divya, K.S.; Madhu, A.K.; Umadevi, T.U.; Suprabha, T.; Nair, P.R.; Suresh, M. Improving the photocatalytic performance of $TiO_2$ via hybridizing with graphene. *J. Semicond.* **2017**, *38*, 063002. [CrossRef]
113. Kubiak, A.; Siwinska-Ciesielczyk, K.; Jesionowski, T. Titania-based hybrid materials with ZnO, $ZrO_2$ and $MoS_2$: A review. *Materials* **2018**, *11*, 2295. [CrossRef]
114. Shao, G.N.; Imran, S.M.; Jeon, S.J.; Engole, M.; Abbas, N.; Salman Haider, M.; Kang, S.J.; Kim, H.T. Sol–gel synthesis of photoactive zirconia–titania from metal salts and investigation of their photocatalytic properties in the photodegradation of methylene blue. *Powder Technol.* **2014**, *258*, 99–109. [CrossRef]
115. Chen, Q.; Wei, W.; Tang, J.; Lin, J.; Li, S.; Zhu, M. Dopamine-assisted preparation of $Fe_3O_4$@$MnO_2$ yolk@shell microspheres for improved pseudocapacitive performance. *Electrochim. Acta* **2019**, *317*, 628–637. [CrossRef]
116. Karafas, E.S.; Romanias, M.N.; Stefanopoulos, V.; Binas, V.; Zachopoulos, A.; Kiriakidis, G.; Papagiannakopoulos, P. Effect of metal doped and co-doped $TiO_2$ photocatalysts oriented to degrade indoor/outdoor pollutants for air quality improvement. A kinetic and product study using acetaldehyde as probe molecule. *J. Photochem. Photobiol. A Chem.* **2019**, *371*, 255–263. [CrossRef]
117. Mokoena, T.P.; Tshabalala, Z.P.; Hillie, K.T.; Swart, H.C.; Motaung, D.E. The blue luminescence of p-type NiO nanostructured material induced by defects: $H_2S$ gas sensing characteristics at a relatively low operating temperature. *Appl. Surf. Sci.* **2020**, *525*, 146002. [CrossRef]

118. Li, J.; Chen, Y.; Wu, Q.; Wu, J.; Xu, Y. Synthesis of sea-urchin-like $Fe_3O_4/SnO_2$ heterostructures and its application for environmental remediation by removal of p-chlorophenol. *J. Mater. Sci.* **2018**, *54*, 1341–1350. [CrossRef]
119. Bhardwaj, N.; Satpati, B.; Mohapatra, S. Plasmon-enhanced photoluminescence from $SnO_2$ nanostructures decorated with Au nanoparticles. *Appl. Surf. Sci.* **2020**, *504*, 144381. [CrossRef]
120. Imran, M.; Riaz, S.; Sanaullah, I.; Khan, U.; Sabri, A.N.; Naseem, S. Microwave assisted synthesis and antimicrobial activity of $Fe_3O_4$-doped $ZrO_2$ nanoparticles. *Ceram. Int.* **2019**, *45*, 10106–10113. [CrossRef]
121. Khan, S.; Kim, J.; Sotto, A.; Van der Bruggen, B. Humic acid fouling in a submerged photocatalytic membrane reactor with binary $TiO_2$–$ZrO_2$ particles. *J. Ind. Eng. Chem.* **2015**, *21*, 779–786. [CrossRef]
122. Zhang, Y.; Wang, X.; Wang, C.; Zhai, H.; Liu, B.; Zhao, X.; Fang, D.; Wei, Y. Facile preparation of flexible and stable superhydrophobic non-woven fabric for efficient oily wastewater treatment. *Surf. Coat. Technol.* **2019**, *357*, 526–534. [CrossRef]
123. Elbasuney, S.; Elsayed, M.A.; Mostafa, S.F.; Khalil, W.F. $MnO_2$ nanoparticles supported on porous $Al_2O_3$ substrate for wastewater treatment: Synergy of adsorption, oxidation, and photocatalysis. *J. Inorg. Organomet. Polym. Mater.* **2019**, *29*, 827–840. [CrossRef]
124. Yaacob, N.; Sean, G.P.; Nazri, N.A.M.; Ismail, A.F.; Abidin, M.N.Z.; Subramaniam, M.N. Simultaneous oily wastewater adsorption and photodegradation by $ZrO_2$–$TiO_2$ heterojunction photocatalysts. *J. Water Process Eng.* **2021**, *39*, 101644. [CrossRef]
125. Skocaj, M.; Filipic, M.; Petkovic, J.; Novak, S. Titanium dioxide in our everyday life; is it safe? *Radiol. Oncol.* **2011**, *45*, 227–247. [CrossRef] [PubMed]
126. Sakthivel, S.; Neppolian, B.; Shankar, M.V.; Arabindoo, B.; Palanichamy, A.; Murugesan, V. Solar photocatalytic degradation of azodye: Comparison of photocatalytic efficiency of ZnO and $TiO_2$. *Sol. Energy Mater. Sol. Cells* **2003**, *77*, 65–82. [CrossRef]
127. Kobayakawa, K.; Sato, C.; Sato, Y.; Fujishima, A. Continuous-flow photoreactor packed with titanium dioxide immobilized on large silica gel beads to decompose oxalic acid in excess water. *J. Photochem. Photobiol. A Chem.* **1998**, *118*, 65–69. [CrossRef]
128. Kamat, P.V. Photochemistry on nonreactive and reactive (semiconductor) surfaces. *Chem. Rev.* **1993**, *93*, 267–300. [CrossRef]
129. Comparelli, R.; Fanizza, E.; Curri, M.L.; Cozzoli, P.D.; Mascolo, G.; Agostiano, A. UV-induced photocatalytic degradation of azo dyes by organic capped ZnO nanocrystals immobilized onto substrates. *Appl. Catal. B Environ.* **2005**, *62*, 1–11. [CrossRef]
130. Fouad, O.A.; Ismail, A.A.; Zaki, Z.I.; Mohamed, R.M. Zinc oxide thin films prepared by thermal evaporation deposition and its photocatalytic activity. *Appl. Catal. B Environ.* **2006**, *62*, 144–149. [CrossRef]
131. Yeber, M.C.; Rodriguez, J.; Freer, J.; Baeza, J.; Duran, N.H.; Mansilla, D. Advanced oxidation of a pulp mill bleaching wastewater. *Chemosphere* **1999**, *39*, 1679–1688. [CrossRef]
132. Khodja, A.A.; Sheili, T.; Pihichowski, J.F.; Boule, P. Photocatalytic degradation of 2-phenylphenol on $TiO_2$ and ZnO in aqueous suspensions. *J. Photochem. Photobiol. A* **2001**, *141*, 231–239. [CrossRef]
133. Serpone, N.; Maruthamuthu, P.; Pichat, P.; Pelizzetti, E.; Hidaka, H. Exploiting the interparticle electron transfer process in the photocatalyzed oxidation of phenol, 2-chlorophenol and pentachlorophenol chemical evidence for electron and hole transfer between coupled semiconductors. *J. Photochem. Photobiol. A Chem.* **2001**, *85*, 247–253. [CrossRef]
134. Sakthivel, S.; Shankar, M.V.; Palanichamy, M.; Arabindoo, B.; Murugesan, V. Photocatalytic decomposition of leather dye: Comparative study of $TiO_2$ supported on alumina and glass beads. *J. Photochem. Photobiol. A Chem.* **2002**, *148*, 153–159. [CrossRef]
135. Ochando-Pulido, J.M.; Pimentel-Moral, S.; Verardo, V.; Martinez-Ferez, A. A focus on advanced physico-chemical processes for olive mill wastewater treatment. *Sep. Purif. Technol.* **2017**, *179*, 161–174. [CrossRef]
136. Andreozzi, R.; Canterino, M.; Di Somma, I.; Lo Giudice, R.; Marotta, R.; Pinto, G.; Pollio, A. Effect of combined physico-chemical processes on the phytotoxicity of olive mill wastewaters. *Water Res.* **2008**, *42*, 1684–1692. [CrossRef] [PubMed]
137. Rizzo, L.; Lofrano, G.; Grassi, M.; Belgiorno, V. Pretreatment of olive mill wastewater by chitosan coagulation and advanced oxidation processes. *Sep. Purif. Technol.* **2008**, *63*, 648–653. [CrossRef]
138. Aytar, P.; Gedikli, S.; Sam, M.; Farizoglu, B.; Çabuk, A. Sequential treatment of olive oil mill wastewater with adsorption and biological and photo-Fenton oxidation. *Environ. Sci. Pollut. Res.* **2013**, *20*, 3060–3067. [CrossRef] [PubMed]
139. Hodaifa, G.; Agabo, C.; Moya, A.J.; Pacheco, R.; Mateo, S. Treatment of olive oil mill wastewater by UV-light and $UV/H_2O_2$ system. *Int. J. Green Technol.* **2015**, *1*, 46–53. [CrossRef]
140. Ruzmanova, Y.; Ustundas, M.; Stoller, M.; Chianese, A. Photocatalytic treatment of olive mill wastewater by N-doped titanium dioxide nanoparticles under visible light. *Chem. Eng. Trans.* **2013**, *32*, 2233–2238.
141. Cai, Q.; Zhu, Z.; Chen, B.; Zhang, B. Oil-in-water emulsion breaking marine bacteria for demulsifying oily wastewater. *Water Res.* **2019**, *149*, 292–301. [CrossRef] [PubMed]
142. Ahmad, T.; Guria, C.; Mandal, A. Synthesis, characterization and performance studies of mixed-matrix poly(vinyl chloride)-bentonite ultrafiltration membrane for the treatment of saline oily wastewater. *Process. Saf. Environ. Prot.* **2018**, *116*, 703–717. [CrossRef]
143. Perez-Calderon, J.; Santos, M.V.; Zaritzky, N. Optimal clarification of emulsified oily wastewater using a surfactant/chitosan biopolymer. *J. Environ. Chem. Eng.* **2018**, *6*, 3808–3818. [CrossRef]
144. Cebeci, M.S.; Gokçek, O.B. Investigation of the treatability of molasses and industrial oily wastewater mixture by an anaerobic membrane hybrid system. *J. Environ. Manag.* **2018**, *224*, 298–309. [CrossRef]
145. Elanchezhiyan, S.S.; Meenakshi, S. Encapsulation of metal ions between the biopolymeric layer beads for tunable action on oil particles adsorption from oily wastewater. *J. Mol. Liq.* **2018**, *255*, 429–438. [CrossRef]
146. Liu, B.; Chen, B.; Zhang, B. Oily wastewater treatment by nano-$TiO_2$-induced photocatalysis. *IEEE Nanotechnol.* **2017**, *11*, 2–13.

147. Yao, T.; Jia, W.; Feng, Y.; Zhang, J.; Lian, Y.; Wu, J.; Zhang, X. Preparation of reduced graphene oxide nanosheet/FexOy/nitrogen-doped carbon layer aerogel as pho-to-Fenton catalyst with enhanced degradation activity and reusability. *J. Hazard. Mater.* **2019**, *362*, 62–71. [CrossRef]
148. Rasheed, T.; Bilal, M.; Iqbal, H.M.N.; Hu, H.; Zhang, X. Reaction mechanism and degradation pathway of rhodamine 6G by photocatalytic treatment. *Water Air Soil Pollut.* **2017**, *228*, 291. [CrossRef]
149. Bilal, M.; Rasheed, T.; Iqbal, H.M.N.; Hu, H.; Wang, W.; Zhang, X. Toxicological assessment and UV/$TiO_2$-based induced degradation profile of reactive black 5 dye. *Environ. Manag.* **2018**, *61*, 171–180. [CrossRef]
150. Ma, M.; Yang, Y.; Chen, Y.; Jiang, J.; Ma, Y.; Wang, Z.; Huang, W.; Wang, S.; Liu, M.; Ma, D.; et al. Fabrication of hollow flower-like magnetic $Fe_3O_4$/C/$MnO_2$/$C_3N_4$ composite with enhanced photocatalytic activity. *Sci. Rep.* **2021**, *11*, 2597. [CrossRef] [PubMed]
151. Ma, M.; Yang, Y.; Chen, Y.; Ma, Y.; Lyu, P.; Cui, A.; Huang, W.; Zhang, Z.; Li, Y.; Si, F. Photocatalytic degradation of MB dye by the magnetically separable 3D flower- like $Fe_3O_4$/$SiO_2$/$MnO_2$/BiOBr-Bi photocatalyst. *J. Alloys Compd.* **2021**, *861*, 158256. [CrossRef]
152. Pozzo, R.L.; Baltanás, M.A.; Cassano, A.E. Supported titanium oxide as photocatalyst in water decontamination: State of the art. *Catal. Today* **1997**, *39*, 219–231. [CrossRef]
153. Zhu, Z.; Lu, Z.; Wang, D.; Tang, X.; Yan, Y.; Shi, W.; Wang, Y.; Gao, N.; Yao, X.; Dong, H. Construction of high-dispersed Ag/$Fe_3O_4$/g-$C_3N_4$ photocatalyst by selective photo-deposition and improved photocatalytic activity. *Appl. Catal. B Environ.* **2016**, *182*, 115–122. [CrossRef]
154. Zhao, X.; Wang, R.; Lu, Z.; Wang, W.; Yan, Y. Dual sensitization effect and conductive structure of $Fe_3O_4$@m$TiO_2$/C photocatalyst towards superior photodegradation activity for bisphenol A under visible light. *J. Photochem. Photobiol. A Chem.* **2019**, *382*, 111902. [CrossRef]
155. Tan, J.; Wang, X.; Hou, W.; Zhang, X.; Liu, L.; Ye, J.; Wang, D. Fabrication of $Fe_3O_4$@graphene/$TiO_2$ nanohybrid with enhanced photocatalytic activity for isopropanol degradation. *J. Alloys Compd.* **2019**, *792*, 918–927. [CrossRef]
156. Ji, H.Y.; Jing, X.C.; Xu, Y.G.; Yan, J.; Li, H.P.; Li, Y.P.; Huang, L.Y.; Zhang, Q.; Xu, H.; Li, H.M. Magnetic g-$C_3N_4$/$NiFe_2O_4$ hybrids with enhanced photocatalytic activity. *RSC Adv.* **2015**, *5*, 57960–57967. [CrossRef]
157. Shen, J.; Zhou, Y.; Huang, J.; Zhu, Y.; Zhu, J.; Yang, X.; Chen, W.; Yao, Y.; Qian, S.; Jiang, H.; et al. In-situ SERS monitoring of reaction catalyzed by multifunctional $Fe_3O_4$@$TiO_2$@Ag-Au microspheres. *Appl. Catal. B Environ.* **2017**, *205*, 11–18. [CrossRef]
158. Singh, P.; Sudhaik, A.; Raizada, P.; Shandilya, P.; Sharma, R.; Hosseini-Bandegharaei, A. Photocatalytic performance and quick recovery of BiOI/$Fe_3O_4$@graphene oxide ternary photocatalyst for photodegradation of 2,4-dintirophenol under visible light. *Mater. Today Chem.* **2019**, *12*, 85–95. [CrossRef]
159. Paul, A.; Dhar, S.S. Designing $Cu_2V_2O_7$/$CoFe_2O_4$/g-$C_3N_4$ ternary nanocomposite: A high performance magnetically recyclable photocatalyst in the reduction of 4-nitrophenol to 4-aminophenol. *J. Solid State Chem.* **2020**, *290*, 121563. [CrossRef]
160. Li, Y.; Li, L.; Hu, J.; Yan, L. A spray pyrolysis synthesis of $MnFe_2O_4$/$SnO_2$ yolk/shell composites for magnetically recyclable photocatalyst. *Mater. Lett.* **2017**, *199*, 135–138. [CrossRef]
161. Tariq, N.; Fatima, R.; Zulfiqar, S.; Rahman, A.; Warsi, M.F.; Shakir, I. Synthesis and characterization of $MoO_3$/$CoFe_2O_4$ nanocomposite for photocatalytic applications. *Ceram. Int.* **2020**, *46*, 21596–21603. [CrossRef]
162. Ma, M.; Chen, Y.; Tong, Z.; Liu, Y.; Ma, Y.; Wang, R.; Bi, Y.; Liao, Z. Research progress of magnetic bismuth-based materials in photocatalysis: A review. *J. Alloys Compd.* **2021**, *886*, 161096. [CrossRef]
163. Vaiano, V.; Sacco, O.; Sannino, D.; Stoller, M.; Ciambelli, P.; Chianese, A. Photo-catalytic Removal of Phenol by Ferromagnetic N-$TiO_2$/$SiO_2$/$Fe_3O_4$ Nanoparticles in presence of Visible Light Irradiation. *Chem. Eng. Trans.* **2016**, *47*, 235–240.
164. Hesas, R.H.; Baei, M.S.; Rostami, H.; Gardy, J.; Hassanpour, A. An investigation on the capability of magnetically separable $Fe_3O_4$/mordenite zeolite for refinery oily wastewater purification. *J. Environ. Manag.* **2019**, *241*, 525–534. [CrossRef]
165. Zhang, J.; Ma, Z. Flower-like $Ag_2MoO_4$/$Bi_2MoO_6$ heterojunctions with enhanced photocatalytic activity under visible light irradiation. *J. Taiwan Inst. Chem. Eng.* **2017**, *71*, 156–164. [CrossRef]
166. Liu, X.; Gu, S.; Zhao, Y.; Zhou, G.; Li, W. $BiVO_4$, $Bi_2WO_6$ and $Bi_2MoO_6$ photocatalysis: A brief review. *J. Mater. Sci. Technol.* **2020**, *56*, 45–68. [CrossRef]
167. Betancourt-Cantera, L.G.; Fuentes, K.M.; Bolarín-Miró, A.M.; Aldabe-Bilmes, S.; Cortés-Escobedo, C.A.; Sánchez-De Jesús, F. Enhanced photocatalytic activity of $BiFeO_3$ for water remediation via the addition of $Ni^{2+}$. *Mater. Res. Bull.* **2020**, *132*, 111012. [CrossRef]
168. Tao, R.; Li, X.; Li, X.; Liu, S.; Shao, C.; Liu, Y. Discrete heterojunction nanofibers of $BiFeO_3$/$Bi_2WO_6$: Novel architecture for effective charge separation and enhanced photocatalytic performance. *J. Colloid Interface Sci.* **2020**, *572*, 257–268. [CrossRef]
169. Cirkovic, J.; Radojkovic, A.; Golic, D.L.; Tasic, N.; Cizmic, M.; Brankovic, G.; Brankovic, Z. Visible-light photocatalytic degradation of mordant blue 9 by single-phase $BiFeO_3$ nanoparticles. *J. Environ. Chem. Eng.* **2021**, *9*, 104587. [CrossRef]
170. Haruna, A.; Abdulkadir, I.; Idris, S.O. Photocatalytic activity and doping effects of $BiFeO_3$ nanoparticles in model organic dyes. *Heliyon* **2020**, *6*, 03237. [CrossRef] [PubMed]
171. Cadenbach, T.; Benitez, M.J.; Morales, A.L.; Costa Vera, C.; Lascano, L.; Quiroz, F.; Debut, A.; Vizuete, K. Nanocasting synthesis of $BiFeO_3$ nanoparticles with enhanced visible-light photocatalytic activity. *Beilstein J. Nanotechnol.* **2020**, *11*, 1822–1833. [CrossRef] [PubMed]
172. Zhu, C.; Chen, Z.; Zhong, C.; Lu, Z. Facile synthesis of $BiFeO_3$ nanosheets with enhanced visible-light photocatalytic activity. *J. Mater. Sci. Mater. Electron.* **2017**, *29*, 4817–4829. [CrossRef]

173. Li, Y.A.; Zhang, X.; Chen, L.; Sun, H.; Zhang, H.; Si, W.; Wang, W.; Wang, L.; Li, J. Enhanced magnetic and photocatalytic properties of $BiFeO_3$ nanotubes with ultrathin wall thickness. *Vacuum* **2021**, *184*, 109867. [CrossRef]
174. Dmitriev, A.V.; Vladimirova, E.V.; Kandaurov, M.V.; Kellerman, D.G.; Kuznetsov, M.V.; Buldakova, L.U.; Samigullina, R.F. Synthesis of hollow spheres of $BiFeO_3$ from nitrate solutions with tartaric acid: Morphology and magnetic properties. *J. Alloys Compd.* **2019**, *777*, 586–592. [CrossRef]
175. Li, X.; Tang, Z.; Ma, H.; Wu, F.; Jian, R. PVP-assisted hydrothermal synthesis and photocatalytic activity of single-crystalline $BiFeO_3$ nanorods. *Appl. Phys. A* **2019**, *125*, 598. [CrossRef]
176. Bharathkumar, S.; Sakar, M.; Vinod, R.; Balakumar, S. Versatility of electrospinning in the fabrication of fibrous mat and mesh nanostructures of bismuth ferrite ($BiFeO_3$) and their magnetic and photocatalytic activities. *Phys. Chem. Chem. Phys.* **2015**, *17*, 17745–17754. [CrossRef]
177. Guo, F.; Wang, L.; Sun, H.; Li, M.; Shi, W. High-efficiency photocatalytic water splitting by a N-doped porous g-$C_3N_4$ nanosheet polymer photocatalyst derived from urea and N, N-dimethylformamide. *Inorg. Chem. Front.* **2020**, *7*, 1770–1779. [CrossRef]
178. Li, X.; Li, F.; Lu, X.; Zuo, S.; Li, Z.; Yao, C.; Ni, C. Microwave hydrothermal synthesis of $BiP_{1-x}V_xO_4$/attapulgite nanocomposite with efficient photocatalytic performance for deep desulfurization. *Powder Technol.* **2018**, *327*, 467–475. [CrossRef]
179. Shi, W.; Li, M.; Huang, X.; Ren, H.; Guo, F.; Tang, Y.; Lu, C. Construction of $CuBi_2O_4$/ $Bi_2MoO_6$ p-n heterojunction with nanosheets-on-microrods structure for improved photocatalytic activity towards broad spectrum antibiotics degradation. *Chem. Eng. J.* **2020**, *394*, 125009. [CrossRef]
180. Dieu Cam, N.T.; Pham, H.D.; Pham, T.D.; Thu Phuong, T.T.; Van Hoang, C.; Thanh Tung, M.H.; Trung, N.T.; Huong, N.T.; Thu Hien, T.T. Novel photocatalytic performance of magnetically recoverable $MnFe_2O_4/BiVO_4$ for polluted antibiotics degradation. *Ceram. Int.* **2001**, *47*, 1686–1692. [CrossRef]
181. Sakhare, P.A.; Pawar, S.S.; Bhat, T.S.; Yadav, S.D.; Patil, G.R.; Patil, P.S.; Sheikh, A.D. Magnetically recoverable $BiVO_4/NiFe_2O_4$ nanocomposite photocatalyst for efficient detoxification of polluted water under collected sunlight. *Mater. Res. Bull.* **2020**, *129*, 110908. [CrossRef]
182. Rohani Bastami, T.; Ahmadpour, A.; Ahmadi Hekmatikar, F. Synthesis of $Fe_3O_4/Bi_2WO_6$ nanohybrid for the photocatalytic degradation of pharmaceutical ibuprofen under solar light. *J. Ind. Eng. Chem.* **2017**, *51*, 244–254. [CrossRef]
183. Xiu, Z.; Cao, Y.; Xing, Z.; Zhao, T.; Li, Z.; Zhou, W. Wide spectral response photothermal catalysis-fenton coupling systems with 3D hierarchical $Fe_3O_4/Ag/Bi_2MoO_6$ ternary hetero-superstructural magnetic microspheres for efficient high-toxic organic pollutants removal. *J. Colloid Interface Sci.* **2019**, *533*, 24–33. [CrossRef] [PubMed]
184. Ren, X.; Gao, M.; Zhang, Y.; Zhang, Z.; Cao, X.; Wang, B.; Wang, X. Photocatalytic reduction of $CO_2$ on BiOX: Effect of halogen element type and surface oxygen vacancy mediated mechanism. *Appl. Catal. B Environ.* **2020**, *274*, 119063. [CrossRef]
185. Wang, Z.; Chu, Z.; Dong, C.; Wang, Z.; Yao, S.; Gao, H.; Liu, Z.; Liu, Y.; Yang, B.; Zhang, H. Ultrathin BiOX (X = Cl, Br, I) nanosheets with exposed {001} facets for photocatalysis. *ACS Appl. Nano Mater.* **2020**, *3*, 1981–1991. [CrossRef]
186. Li, Z.; Huang, G.; Liu, K.; Tang, X.; Peng, Q.; Huang, J.; Ao, M.; Zhang, G. Hierarchical BiOX (X = Cl, Br, I) microrods derived from bismuth-MOFs: In situ synthesis, photocatalytic activity and mechanism. *J. Clean. Prod.* **2020**, *272*, 122892. [CrossRef]
187. Wang, L.; Liu, G.P.; Wang, B.; Chen, X.; Wang, C.T.; Lin, Z.X.; Xia, J.X.; Li, H.M. Oxygen vacancies engineering-mediated BiOBr atomic layers for boosting visible light-driven photocatalytic $CO_2$ reduction. *Sol. RRL* **2020**, *5*, 2000480. [CrossRef]
188. Cao, L.; Ma, D.; Zhou, Z.; Xu, C.; Cao, C.; Zhao, P.; Huang, Q. Efficient photocatalytic degradation of herbicide glyphosate in water by magnetically separable and recyclable $BiOBr/Fe_3O_4$ nanocomposites under visible light irradiation. *Chem. Eng. J.* **2019**, *368*, 212–222. [CrossRef]
189. Li, X.; Wang, L.; Zhang, L.; Zhuo, S. A facile route to the synthesis of magnetically separable $BiOBr/NiFe_2O_4$ composites with enhanced photocatalytic performance. *Appl. Surf. Sci.* **2017**, *419*, 586–594. [CrossRef]
190. Li, X.; Xu, H.; Wang, L.; Zhang, L.; Cao, X.; Guo, Y. Spinel $NiFe_2O_4$ nanoparticles decorated BiOBr nanosheets for improving the photocatalytic degradation of organic dye pollutants. *J. Taiwan Inst. Chem. Eng.* **2018**, *85*, 257–264. [CrossRef]
191. Sin, J.C.; Lam, S.M.; Zeng, H.; Lin, H.; Li, H.; Tham, K.O.; Mohamed, A.R.; Lim, J.W.; Qin, Z. Magnetic $NiFe_2O_4$ nanoparticles decorated on N-doped BiOBr nanosheets for expeditious visible light photocatalytic phenol degradation and hexavalent chromium reduction via a Z-scheme heterojunction mechanism. *Appl. Surf. Sci.* **2021**, *559*, 149966. [CrossRef]
192. Ma, W.; Chen, L.; Zhu, Y.; Dai, J.; Yan, Y.; Li, C. Facile synthesis of the magnetic BiOCl/ $ZnFe_2O_4$ heterostructures with enhanced photocatalytic activity under visible- light irradiation. *Colloids Surf. A* **2016**, *508*, 135–141. [CrossRef]
193. Zhou, P.; Zhang, A.; Zhang, D.; Feng, C.; Su, S.; Zhang, X.; Xiang, J.; Chen, G.; Wang, Y. Efficient removal of Hg0 from simulated flue gas by novel magnetic $Ag_2WO_4/BiOI/CoFe_2O_4$ photocatalysts. *Chem. Eng. J.* **2019**, *373*, 780–791. [CrossRef]
194. Zhang, A.; Zhou, P.; Zhang, X.; Li, H.; Wang, Y.; Sun, Z.; Xiang, J.; Su, S.; Liu, Z. Insights into efficient removal of gaseous Hg0 using $AgIO_3$-Modified $BiOI/CoFe_2O_4$ composites through photocatalytic oxidation. *Energ. Fuel* **2019**, *33*, 12538–12548. [CrossRef]
195. Shan, B.; Zhao, Y.; Li, Y.; Wang, H.; Chen, R.; Li, M. High-quality dual-plasmonic Au@$Cu_{2-x}Se$ nanocrescents with precise $Cu_{2-x}Se$ domain size control and tunable optical properties in the second near-infrared biowindow. *Chem. Mater.* **2019**, *31*, 9875–9886. [CrossRef]
196. Shao, B.; Liu, X.; Liu, Z.; Zeng, G.; Liang, Q.; Liang, C.; Cheng, Y.; Zhang, W.; Liu, Y.; Gong, S. A novel double Z-scheme photocatalyst $Ag_3PO_4/Bi_2S_3/Bi_2O_3$ with enhanced visible-light photocatalytic performance for antibiotic degradation. *Chem. Eng. J.* **2019**, *368*, 730–745. [CrossRef]

197. Li, S.; Wang, Z.; Zhang, X.; Zhao, J.; Hu, Z.; Wang, Z.; Xie, X. Preparation of magnetic nanosphere/nanorod/nanosheet-like $Fe_3O_4/Bi_2S_3$/BiOBr with enhanced (001) and (110) facets to photodegrade diclofenac and ibuprofen under visible LED light irradiation. *Chem. Eng. J.* **2019**, *378*, 122169. [CrossRef]
198. Cong, Y.Q.; Ji, Y.; Ge, Y.H.; Jin, H.; Zhang, Y.; Wang, Q. Fabrication of 3D $Bi_2O_3$-BiOI heterojunction by a simple dipping method: Highly enhanced visible-light photoelectrocatalytic activity. *Chem. Eng. J.* **2017**, *307*, 572–582. [CrossRef]
199. Abbasi, Z.; Farrokhnia, A.; Garcia-Lopez, E.I.; Shoushtari, M.Z. Superparamagnetic recoverable flowerlike $Fe_3O_4$@$Bi_2O_3$ core-shell with g-$C_3N_4$ sheet nanocomposite: Synthesis, characterization, mechanism and kinetic study of photo-catalytic activity. *J. Mater. Sci. Mater. Electron.* **2019**, *31*, 1022–1033. [CrossRef]
200. Gao, N.; Lu, Z.; Zhao, X.; Zhu, Z.; Wang, Y.; Wang, D.; Hua, Z.; Li, C.; Huo, P.; Song, M. Enhanced photocatalytic activity of a double conductive C/$Fe_3O_4$/$Bi_2O_3$ composite photocatalyst based on biomass. *Chem. Eng. J.* **2016**, *304*, 351–361. [CrossRef]
201. Falletta, E.; Longhi, M.; Di Michele, A.; Boffito, D.C.; Bianchi, C.L. Floatable graphitic carbon nitride/alginate beads for the photodegradation of organic pollutants under solar light irradiation. *J. Clean. Prod.* **2022**, 133641. [CrossRef]
202. Wang, X.; Wang, X.; Zhao, J.; Song, J.; Zhou, L.; Wang, J.; Tong, X.; Chen, Y. An alternative to in situ photocatalytic degradation of microcystin-LR by worm-like N, P co-doped $TiO_2$/expanded graphite by carbon layer (NPT-EGC) floating composites. *Appl. Catal. B Environ.* **2017**, *206*, 479–489. [CrossRef]
203. Zhou, W.; Li, W.; Wang, J.Q.; Qu, Y.; Yang, Y.; Xie, Y.; Zhang, K.; Wang, L.; Fu, H.; Zhao, D. Ordered Mesoporous Black $TiO_2$ as Highly Efficient Hydrogen Evolution Photo-catalyst. *J. Am. Chem. Soc.* **2014**, *136*, 9280–9283. [CrossRef] [PubMed]
204. Djellabi, R.; Zhao, X.; Frias Ordonez, M.; Falletta, E.; Bianchi, C.L. Comparison of the photoactivity of several semiconductor oxides in floating aerogel and suspension systems towards the reduction of Cr(VI) under visible light. *Chemosphere* **2001**, *281*, 130839. [CrossRef] [PubMed]
205. Wang, S.; Zhang, Y.; Zhang, T.; Dong, F.; Huang, H. Readily attainable spongy foam photocatalyst for promising practical photocatalysis. *Appl. Catal. B Environ.* **2017**, *208*, 75–81. [CrossRef]
206. Chawla, H.; Saha, M.; Upadhyay, S.; Rohilla, J.; Ingole, P.P.; Sapi, A.; Szenti, I.; Yadav, M.; Lebedev, V.T.; Chandra, A.; et al. Enhanced photocatalytic activity and easy recovery of visible light active $MoSe_2$/$BiVO_4$ heterojunction immobilized on Luffa cylindrica—Experimental and DFT study. *Environ. Sci. Nano* **2021**, *8*, 3028. [CrossRef]
207. Huang, X.-H.; Hu, T.; Bu, H.; Li, W.X.; Li, Z.L.; Hu, H.J.; Chen, W.Z.; Lin, Y.; Jiang, G.B. Transparent floatable magnetic alginate sphere used as photocatalysts carrier for im-proving photocatalytic efficiency and recycling convenience. *Carbohydr. Polym.* **2021**, *254*, 117281. [CrossRef]
208. Said, K.A.M.; Ismail, A.F.; Karim, Z.A.; Abdullah, M.S.; Hafeez, A. A review of tech-nologies for the phenolic compounds recovery and phenol removal from wastewater. *Process Saf. Environ. Prot.* **2021**, *151*, 257–289. [CrossRef]
209. Rawindran, H.; Lim, J.W.; Goh, P.S.; Subramaniam, M.N.; Ismail, A.F.; Arzhandi, M.R.D. Simultaneous separation and degradation of surfactants laden in produced water using PVDF/$TiO_2$ photocatalytic membrane. *J. Clean. Prod.* **2019**, *221*, 490–501. [CrossRef]
210. Algamdi, M.S.; Alsohaimi, I.H.; Lawler, J.; Ali, H.M.; Aldawsari, A.M.; Hassan, H.M.A. Fabrication of graphene oxide incorporated polyethersulfone hybrid ultra-filtration membranes for humic acid removal. *Sep. Purif. Technol.* **2019**, *223*, 17–23. [CrossRef]
211. Zhou, K.-G.; McManus, D.; Prestat, E.; Zhong, X.; Shin, Y.; Zhang, H.-L.; Haigh, S.; Casiraghi, C. Self-catalytic membrane photo-reactor made of carbon nitride nanosheets. *J. Mater. Chem. A Mater. Energy Sustain.* **2016**, *4*, 11666–11671. [CrossRef]
212. Dzinun, H.; Othman, M.H.D.; Ismail, A.F.; Puteh, M.H.; Rahman, M.A.; Jaafar, J. Photocatalytic degradation of nonylphenol using coextruded dual-layer hollowfibre membranes incorporated with a different ratio of TiO2/PVDF. *React. Funct. Polym.* **2016**, *99*, 80–87. [CrossRef]
213. Dzinun, H.; Othman, M.H.D.; Ismail, A.F.; Puteh, M.H.; Rahman, M.A.; Jaafar, J.; Adrus, N.; Hashim, N.A. Antifouling behavior and separation performance of immobilized $TiO_2$ in dual layer hollow fiber membranes. *Polym. Eng. Sci.* **2017**, *58*, 1636–1643. [CrossRef]
214. Dzinun, H.; Othman, M.H.D.; Ismail, A.F. Photocatalytic performance of$TiO_2$/Clinoptilolite: Comparison study in suspension and hybrid photocatalytic membrane reactor. *Chemosphere* **2019**, *228*, 241–248. [CrossRef] [PubMed]
215. Sharma, N.K.; Philip, L. Combined biological and photocatalytic treatment of real coke oven wastewater. *Chem. Eng. J.* **2016**, *295*, 20–28. [CrossRef]
216. Salim, N.E.; Jaafar, J.; Ismail, A.F.; Othman, M.H.D.; Rahman, M.A.; Yusof, N.; Qtaishat, M.; Matsuura, T.; Aziz, F.; Salleh, W.N.W. Preparation and characterization of hydrophilic surface modifier macro-molecule modified poly (ether sulfone) photocatalytic membrane for phenol removal. *Chem. Eng. J.* **2018**, *335*, 236–247. [CrossRef]
217. Salim, N.E.; Nor, N.A.M.; Jaafar, J.; Ismail, A.F.; Qtaishat, M.R.; Matsuura, T.; Othman, M.H.D.; Rahman, M.A.; Aziz, F.; Yusof, N. Effects of hydrophilic surface macromolecule modifier loading on PES/O-g-$C_3N_4$ hybrid photocatalytic membrane for phenol removal. *Appl. Surf. Sci.* **2019**, *465*, 180–191. [CrossRef]

catalysts

MDPI

Article

# Synthesis, Characterization and Photoelectric Properties of $Fe_2O_3$ Incorporated $TiO_2$ Photocatalyst Nanocomposites

Atif Mossad Ali [1,2,*], Mahmoud Ahmed Sayed [1], Hamed Algarni [1], Vanga Ganesh [1], Muhd Aslam [1], Adel Ali Ismail [3] and Haitham Mohamed El-Bery [4,*]

1 Department of Physics, Faculty of Science, King Khalid University, Abha 61421, Saudi Arabia; frrag75@gmail.com (M.A.S.); halgarni@hotmail.com (H.A.); vangaganesh@gmail.com (V.G.); muhd.aslam@gmail.com (M.A.)
2 Department of Physics, Faculty of Science, Assiut University, Assiut 71516, Egypt
3 Advanced Materials Department, Central Metallurgical R&D Institute (CMRDI), P.O. Box 87, Helwan 11421, Egypt; adelali141@yahoo.com
4 Department of Chemistry, Faculty of Science, Assiut University, Assiut 71516, Egypt
* Correspondence: atifali@kku.edu.sa (A.M.A.); Haitham.El-Bery@aun.edu.eg (H.M.E.-B.)

**Abstract:** In the present work we report the sol-gel synthesis of pure $TiO_2$ and $(TiO_2)_{1-x}(Fe_2O_3)_x$ nanocomposites with different $Fe_2O_3$ contents (x = 0, 0.1, 0.5, and 1.0 for pure $TiO_2$, $Fe_2O_3$ incorporated 0.1, 0.5, and pure $Fe_2O_3$ which are denoted as PT, 0.1F, 0.5F, and PF, respectively). The structural, morphological, optical, and surface texture of the prepared nanocomposites were characterized using various techniques. The structural studies confirm the strong influence of $Fe_2O_3$ contents on the crystallite sizes and dislocation values. The size of the crystallites was increased by the increase in $Fe_2O_3$ contents. The bandgap values elucidated from DRS analysis were decreased from 3.15 eV to 1.91 eV with increasing $Fe_2O_3$ contents. The $N_2$-Physorption analysis has confirmed the mesoporous nature of the samples with a comparable specific surface area of 35 $m^2/g$. The photoelectrochemical measurements (CV, CA and EIS) were performed to assess the photoelectric properties of the prepared materials. It was found that the PT samples have the highest catalytic activity and photocurrent response compared to other composites. The reduction in current density was as follows: 2.8, 1.65, 1.5 and 0.9 $mA/cm^2$, while the photocurrent response was ca. 800, 450, 45, 35 $\mu A/cm^2$ for PT, 0.1F, 0.5F and PF samples, respectively. The EIS results showed that the $(TiO_2)_{1-x}(Fe_2O_3)_x$ nanocomposites exhibit lower charge transfer resistance than pure titania and hematite samples.

**Keywords:** $TiO_2$; $Fe_2O_3$; photoelectrochemical

**Citation:** Ali, A.M.; Sayed, M.A.; Algarni, H.; Ganesh, V.; Aslam, M.; Ismail, A.A.; El-Bery, H.M. Synthesis, Characterization and Photoelectric Properties of $Fe_2O_3$ Incorporated $TiO_2$ Photocatalyst Nanocomposites. *Catalysts* **2021**, *11*, 1062. https://doi.org/10.3390/catal11091062

Academic Editors: Ioan Balint and Monica Pavel

Received: 6 August 2021
Accepted: 27 August 2021
Published: 31 August 2021

**Publisher's Note:** MDPI stays neutral with regard to jurisdictional claims in published maps and institutional affiliations.



## 1. Introduction

Titanium dioxide ($TiO_2$) nanoparticles (NPs) exhibit interesting potential applications in various fields such as sensors, catalysts, optical filters, anti-reflection coatings, etc. [1]. These $TiO_2$ NPs are widely used in personal care products such as cosmetics, toothpaste, and sunscreen lotions [2–4]. The properties such as stability and non-toxicity make them available for a wide range of applications [5]. In addition to the properties mentioned above, $TiO_2$ is also widely studied due to its electronic, optical, and photocatalytic activity [6,7]. Incorporating with metal ions such as iron, copper, zinc, nickel, chromium has been well addressed, altering the structure, chemical composition, and optical properties of $TiO_2$ [8,9]. Amongst all, iron is the favorable ion to replace Ti in $TiO_2$ as it has the ionic radius of $Fe^{3+}$ (0.64 Å) which is comparable with $Ti^{4+}$ (0.68 Å). Thus, $Fe^{3+}$ can replace $Ti^{4+}$ comfortably in the crystal structure of $TiO_2$ with some defects.

Moreover, $Fe^{3+}$ incorporating in $TiO_2$ improves the photocatalytic activity under visible light [10] and reduces the optical bandgap [11]. Moreover, Fe-doped $TiO_2$ has applications in spintronic and magneto-optic devices [12]. Application of Fe-doped $TiO_2$ in gas sensor and photocatalytic degradation prepared from ball milling method was successfully presented by Sunil et al. [13]. Hussain et al. has studied the electrical, optical, and magnetic properties of Fe-doped $TiO_2$ nanotubes [14]. In the literature, there are a variety of approaches for synthesizing Fe-doped $TiO_2$ NPs such as the hydrothermal method [10], thermal hydrolysis [15], wet-chemical synthesis [16], and sol-gel method [17]. The influence of annealing at different temperatures was also studied by Gareso et al. [18] using the co-precipitation method. Rodriguez and co-workers [19] used the sol-gel technique for the synthesis of Fe-incorporated $TiO_2$ and observed a decrease in the optical bandgap from 3.3 eV to 2.9 eV with enhanced absorption in the visible region as the incorporating concentration of Fe increases. Fe-doped $TiO_2$ NPs prepared by the hydrothermal method were applied as an efficient electron mediator for the fabrication of hydrazine chemical sensor using glassy carbon electrode [20] which proves to be a sensitive sensor. The sol-gel routed spin coating technique was employed to synthesize Fe-doped $TiO_2$ NPs and observed that the optical transmittance is was decreased with increasing Fe dopant concentrations [21]. Here, we chose the sol-gel method to synthesize $Fe_2O_3$-doped $TiO_2$ NPs in the presence of diethanolamine. This method is easy and cost-effective and can obtain a high degree of purity, homogeneity and small particle size at low temperatures. Despite $TiO_2$ is considered one of the most extensively investigated photocatalysts due to its chemical and photostability, low cost and availability [22,23]. Still, it suffers from a high recombination rate of photogenerated charge carriers and low activity in visible light due to its wide bandgap (Eg). One of the strategies that can overcome this problem is the fabrication of heterostructure with another narrower Eg visible active semiconductor [24]. Band alignment (CB and VB) between semiconductor photocatalysts is crucial. Therefore, we chose hematite $Fe_2O_3$ as a visible active photocatalyst (Eg = ca. 2 eV) and combined it with $TiO_2$ to form the composite and investigated. Formation of internal electric field upon light irradiation and Fermi energy level alignment between $TiO_2$ and $Fe_2O_3$ will promote the separation of photogenerated electron-hole pairs. Therefore, $TiO_2$–$Fe_2O_3$ heterostructures were applied in different photocatalytic-based applications for instance as an effective photoanode for water oxidation in photoelectrochemical cells (PEC) [25], photoreduction of Cr(VI) [26], and dye degradation [27]. Some of the applications use external bias to reduce the recombination rate, as in the case of PEC applications. In view of these important and potential applications, it is worthwhile to study the different properties of Fe-doped $TiO_2$. Hence, in the present work, pure and doped $TiO_2$ with different $Fe_2O_3$ contents were prepared by the sol-gel technique. The prepared samples were denoted as PT, 0.1F, 0.5F, and PF for pure $TiO_2$ and Fe doped 0.1, 0.5, and pure $Fe_2O_3$, respectively. These materials were subjected to characterize for structural, morphological, optical, and photoelectrochemical properties.

## 2. Results and Discussion

### 2.1. Structural Analysis

X-ray diffraction (XRD) studies were carried out to determine the structural and crystallite size information of the $(TiO_2)_{1-x}(Fe_2O_3)_x$ nanoparticles. Figure 1 displays the typical XRD pattern of $Fe_2O_3$–$TiO_2$ nanocomposites, for different concentrations of $Fe_2O_3$ $(TiO_2)_{1-x}(Fe_2O_3)_x$, where x = 0, 0.1, 0.5, and 1.0 for pure $TiO_2$ and $Fe_2O_3$ contents 0.1, 0.5, and pure $Fe_2O_3$ which are denoted as PT, 0.1F, 0.5F, and PF, respectively). Pure $TiO_2$ is found in three dominant crystal structures, anatase, rutile, and brookite in nature [28]. Among them, anatase form is the most common polymorph due to its lower surface energy, especially at the nanoscale [29]. From Figure 1, it is evident that the pure $TiO_2$ is formed in the polycrystalline tetragonal anatase phase with multiple peaks marked with the asterisk (*) symbol and the planes are marked as (101), (103), (004), (112), (200), (105), (211), (204), (116), (220), (215) and (301) in Figure 1. The peaks are matching well with the ICDD pattern 21-1272. Similarly, the pure $Fe_2O_3$ is formed in the rhombohedral hematite structure matching well with the ICDD pattern# 33-0664. Figure 1 also displays the intermediate structures showing systematic changes with the $Fe_2O_3$ incorporating on $TiO_2$ lattice, which is clearly evident from the changes in the peak intensities, emergence of new peaks and peak shifts happening with different dopant levels.

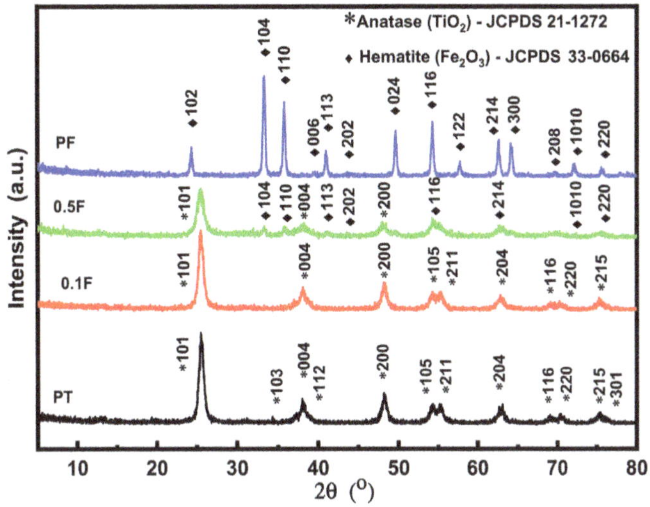


**Figure 1.** XRD analysis of $(TiO_2)_{1-x}(Fe_2O_3)_x$ nanoparticles.

Scherrer's formula was employed to calculate the mean crystal size ($D$) of the NPs from the XRD peaks using the following relation [30],

$$D = \frac{0.9\lambda}{\beta \cos(\theta)} \quad (1)$$

where $\lambda$ is the X-ray wavelength, and $\beta$ is the peak width at half maximum. The micro-strain ($\varepsilon$) and dislocation density ($\delta$) values were also calculated using the relations.

$$\varepsilon = \frac{\beta \cos(\theta)}{4} \quad (2)$$

and

$$\delta = \frac{1}{D^2} \quad (3)$$

The obtained values are tabulated in Table 1. The results clearly show that the $Fe_2O_3$ concentration strongly influences the crystallite sizes and so the dislocation values. Microstrain values are almost the same. It can be inferred from the table that the crystallite size increases in general with the $Fe_2O_3$ content. The results are matched well with the results obtained by Zhao et al. [30]. Zhao et al. [31] synthesized a set of $Fe_2O_3$, $TiO_2$ and $TiO_2/Fe_2O_3$ multilayered thin films and found that the mean crystalline size increases with the $Fe_2O_3$, which is in accord with the present results. Tang et al. [32] report that the formation of $Fe_2O_3$ and $TiO_2$ particles which are significantly affected by the concentration of hydrolysis liquid. The intensities of the characteristic peaks of both anatase and rutile phases increased with the hydrolysis liquid concentration.

**Table 1.** Crystallite size ($D$), dislocation density ($\delta$) and microstrain ($\varepsilon$).

| Sample Name | Crystallite Size ($D$) (nm) | Dislocation Density ($\delta$) | Microstrain ($\varepsilon$) |
|---|---|---|---|
| PT | 15.6 | 0.006 | 0.002 |
| 0.1F | 21.6 | 0.004 | 0.002 |
| 0.5F | 34.4 | 0.006 | 0.002 |
| PF | 33.7 | 0.001 | 0.001 |

The surface morphology studies and distribution of particles of the nanostructured material provide useful information about the utilization of the sample in various technological important applications. Figure 2 shows the scanning electron microscope (SEM) images of PT, 0.1F, 0.5F, and PF samples. It is quite evident that the Fe content has altered the surface morphology of the samples, notably. The pure $TiO_2$ sample shows an irregular distribution of spherical and rectangular-shaped particles. When Fe is added to $TiO_2$, the size of the particles reduces, and there are small particles observed throughout the surface of the sample. Apart from this, in the case of the pure Fe sample, a group of nanoparticles of larger size was observed. From morphology studies, it is clear that the decrease in grain size results in an increase in the surface area of the synthesized material and produces more active sites for adsorption of target contaminant and thus attacked by reactive oxygen species, which will be discussed later in this study. A similar type of decreasing particles size with an increasing percentage of incorporating was observed by Gareso et al. [18]. Moreover, it is also observed that the decrease in particle size with Fe dopant can make the synthesized catalyst a potential candidate for various photocatalysis and sensing applications.

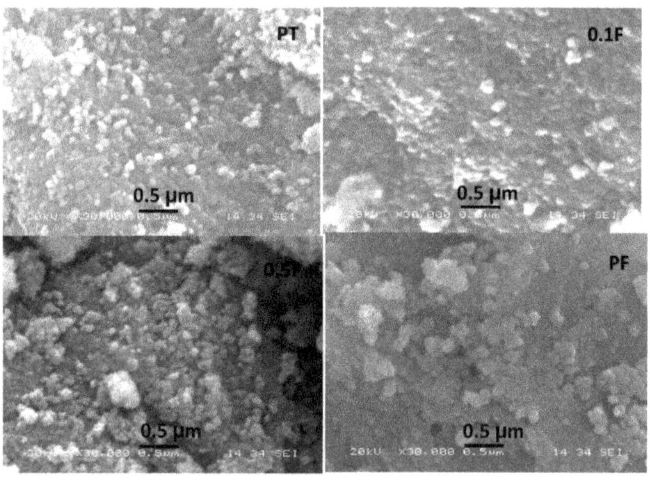


**Figure 2.** SEM images of $(TiO_2)_{1-x}(Fe_2O_3)_x$ nanoparticles.

Figure 3 shows the Raman spectra of PT, 0.1F, 0.5F, and PF samples. In Raman spectra, four main peaks were observed at 145 $cm^{-1}$, 396 $cm^{-1}$, 514 $cm^{-1}$, and 637 $cm^{-1}$, which belong to Eg, B1g, A1g + B1g, Eg mode, respectively. The main peaks observed in Raman spectra were well correlated to earlier reports confirms the phase of the prepared samples [20,33]. The position of the main peak at 145 $cm^{-1}$ is slightly shifted towards a higher wavenumber when Fe was doped in $TiO_2$. The shift in Raman bands is attributed to the incorporation of Fe that causes changes in defect structure and particle size. Moreover, a continuous decrease in the intensity of the Raman band (145 $cm^{-1}$) was also observed at higher concentrations, suggesting a decrease in the particle size of synthesized photocatalysts with $Fe_2O_3$ incorporating. The vibrational properties of materials are significantly affected when the grain size decreases to the nanometer scale. Due to the size-induced radial strain, a volume contraction occurs primarily within the nanoparticles, which leads to an increase in force constants due to the decrease in interactor pressure. In the case of pure Fe material, the new peak at 1310 $cm^{-1}$ reveals the hematite group of $\alpha$-$Fe_2O_3$.

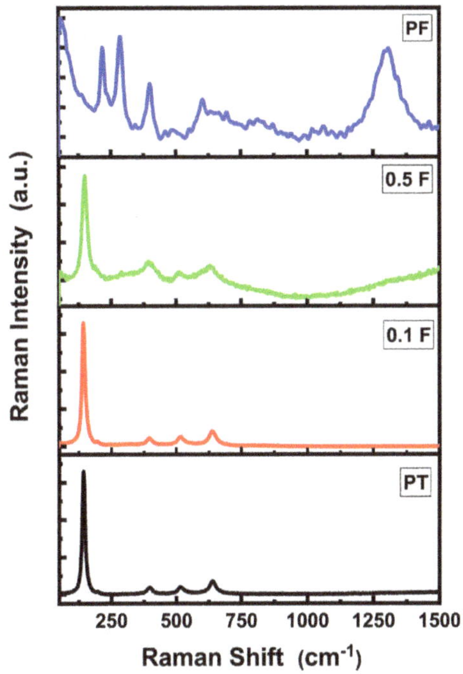


**Figure 3.** Raman spectra for $(TiO_2)_{1-x}(Fe_2O_3)_x$ nanoparticles.

Figure 4 shows the Fourier-transform infrared spectroscopy (*FT-IR*) spectra of PT, 0.1F, 0.5F, and PF samples. From Figure 4, the band observed at around 3412 $cm^{-1}$ was attributed to the presence of the stretching vibrations of the O-H groups of $H_2O$ molecules physically adsorbed on the surface of $TiO_2$. This band is gradually shifted to a lower wavelength due to incorporating concentration, suggesting the crystal structure of $Fe_2O_3$ was distorted [34]. It is also observed that at higher incorporating concentration and in pure Fe material, the intensity of band is increased. Some weak bands are also observed at around 2919 $cm^{-1}$ attributed to different vibrational modes of $TiO_2$. Zhang et al. [34] observed peaks at 2928 $cm^{-1}$, 2845 $cm^{-1}$, 1502 $cm^{-1}$, 1421 $cm^{-1}$ and 1364 $cm^{-1}$. These peaks are assigned to $sp^3$ and $sp^2$ C-H, C=O, unsaturated C-H and C-OH bonds, indicating the existence of carbon quantum dots (CQDs) in the composites [34]. In pure Fe, this band disappears, clearly suggesting the other vibration modes of $TiO_2$. The stretching vibrations of the O-H groups were also observed at around 1645 $cm^{-1}$. Wu et al. also observed the same peaks close to 1630 $cm^{-1}$ [35]. The first one is attributed to the stretching vibration of the corresponding –OH derived from the hydroxyl radical or the adsorbed water on the $TiO_2$ surface. The second peak close to 1630 $cm^{-1}$ corresponds to the bending vibration of the H–O–H bond of the adsorbed water on the $TiO_2$ surface. These results are a common feature of semiconductor oxides and a basic condition for photocatalysis. There are some other bands such as 1033, 1439, 557, 469 $cm^{-1}$ also observed in the case of pure Fe and incorporating are attributed to vibration modes of $Fe_2O_3$ and can indicate the iron oxide formation at the structure of $Fe_2O_3$–$TiO_2$ nanocomposites [36].

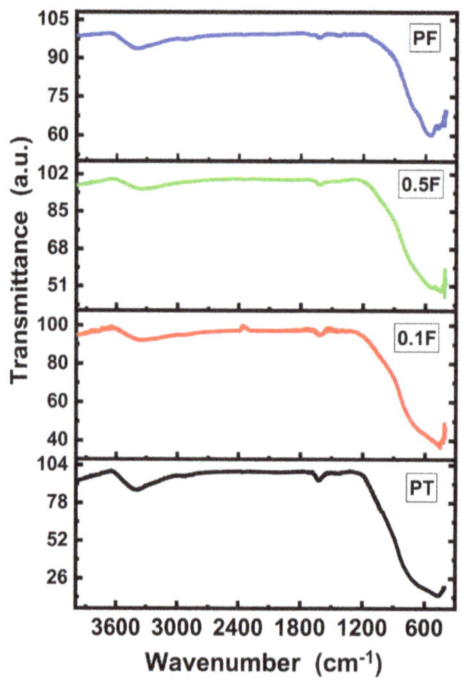


**Figure 4.** IR transmittance spectra for $(TiO_2)_{1-x}(Fe_2O_3)_x$ nanoparticles.

## 2.2. Optical Properties

Diffuse reflectance spectroscopy (DRS) is a unique technique to study the electronic structure of nanostructured materials. The non-destructive method of this technique allows us to measure exact values of the bandgap of powdered materials by a mirror-like reflection from the loaded samples by diffuse illumination. In literature, this method is well studied by the incident light is partially absorbed and scattered [37]. In the present case, pure $TiO_2$, pure $Fe_2O_3$, and different concentrations of $Fe_2O_3$ doped $TiO_2$ were subjected to DRS analysis and the corresponding spectra are shown in Figure 5. The optical bandgap of the present nanostructured material was determined from the following Kubelka–Munk model equations [38]:

$$F(R) = \frac{\left(1 - R^2\right)}{2R} \tag{4}$$

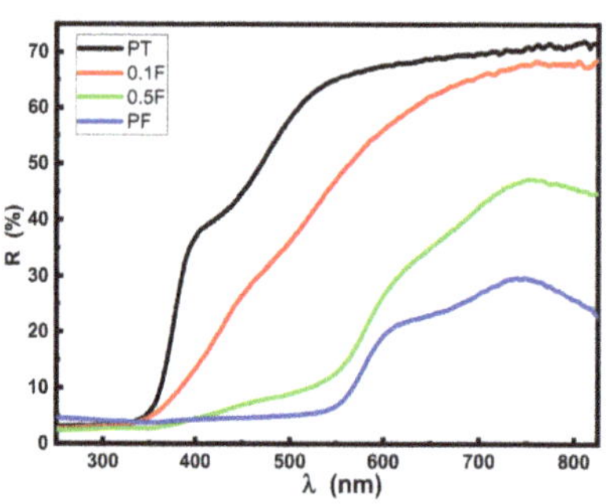


**Figure 5.** Diffuse reflectance UV–visible spectra of $(TiO_2)_{1-x}(Fe_2O_3)_x$ nanoparticles.

F(R) is the Kubelka–Munk function, and R is the absolute reflectance. For calculating the ($\alpha$), the Equation (4) is modified in terms of F(R) as [39,40]:

$$\alpha = \frac{\text{absorbance}}{t} = \frac{F(R)}{t} \quad (5)$$

where t is the height of the sample holder, which is equal to 2 mm, and the optical bandgap is calculated from Equation (6)

$$\alpha h\upsilon = \left(\frac{(F(R)h\upsilon)}{t}\right)^n = A\left(h\upsilon - E_g\right)^n \quad (6)$$

where $\alpha$ absorption coefficient, $E_g$ is bandgap, $h\nu$ is the absorbed energy, A is the parameter that is related to the effective mass associated with the valence and conduction bands, and n ($n = \frac{1}{2}$ for direct bandgap) is an optical transition. From Figure 6, with increasing the $Fe_2O_3$ incorporating concentration, the bandgap values are decreasing from 3.15 eV to 1.91 eV (see Table 2) as a result of incorporating $Fe_2O_3$ into the $TiO_2$ lattice. The decreasing band gap with increasing incorporating concentration indicates that the present samples found huge applications in optoelectronic devices.

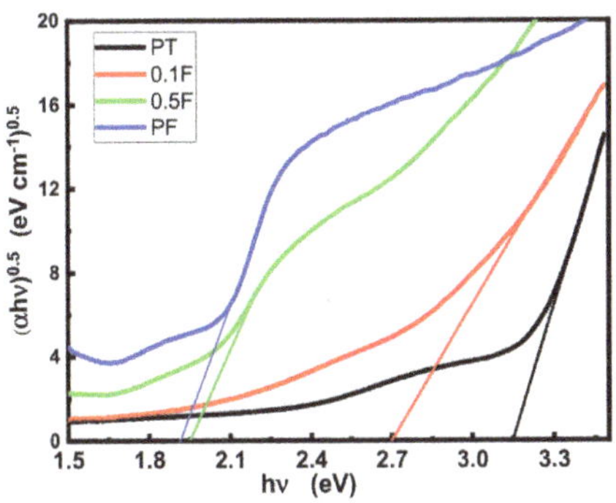


**Figure 6.** A plot of transferred Kubelka–Munk versus the energy of the light absorbed of $(TiO_2)_{1-x}(Fe_2O_3)_x$ nanoparticles.

**Table 2.** The bandgap values of $(TiO_2)_{1-x}(Fe_2O_3)_x$ nanoparticles.

| Sample Name | The Bandgap (eV) |
|---|---|
| PT | 3.15 |
| 0.1F | 2.71 |
| 0.5F | 1.95 |
| PF | 1.91 |

### 2.3. $N_2$-Physorption Analysis

Specific surface area, surface texture, and pore size distribution were acquired via the $N_2$-physisorption technique. $N_2$ adsorption–desorption isotherms, Barrett, Joyner, and Halenda method (BJH) for calculating pore size distribution, and t-plot of the prepared composites are shown in Figure 7. As it can be seen, all the obtained isotherms are of Type III. The reversible Type III isotherm is represented as a convex to the x-axis over its entire range. It also reveals the multilayer formation process. It forms because of strong lateral interactions between adsorbed molecules in comparison to interactions between adsorbate and the adsorbent surface.

The BJH pore size distribution indicates that all samples are mesoporous in nature with an average pore radius of ca. 5 nm as shown in Table 3. Moreover, the t-plot method is a widely used procedure to estimate the total surface area and the external surface area of materials. For mesoporous materials, the adsorbed volume (V) varies linearly with the film thickness (t) and passes through the origin. The total surface area ($S_{tot}$) is directly given by the slope of the first linear fit at low pressure. Once all pores are filled, adsorption occurs only on the external surface of the material and leads to a second linear fit at high pressure, the slope of which allows the determination of the external surface ($S_{ext}$). The external surface area obtained for our samples was similar to the value of the BET surface area because of the mesoporous nature of the prepared composites. The values of BET-specific surface area, total pore volume and average pore radius are given in Table 3.

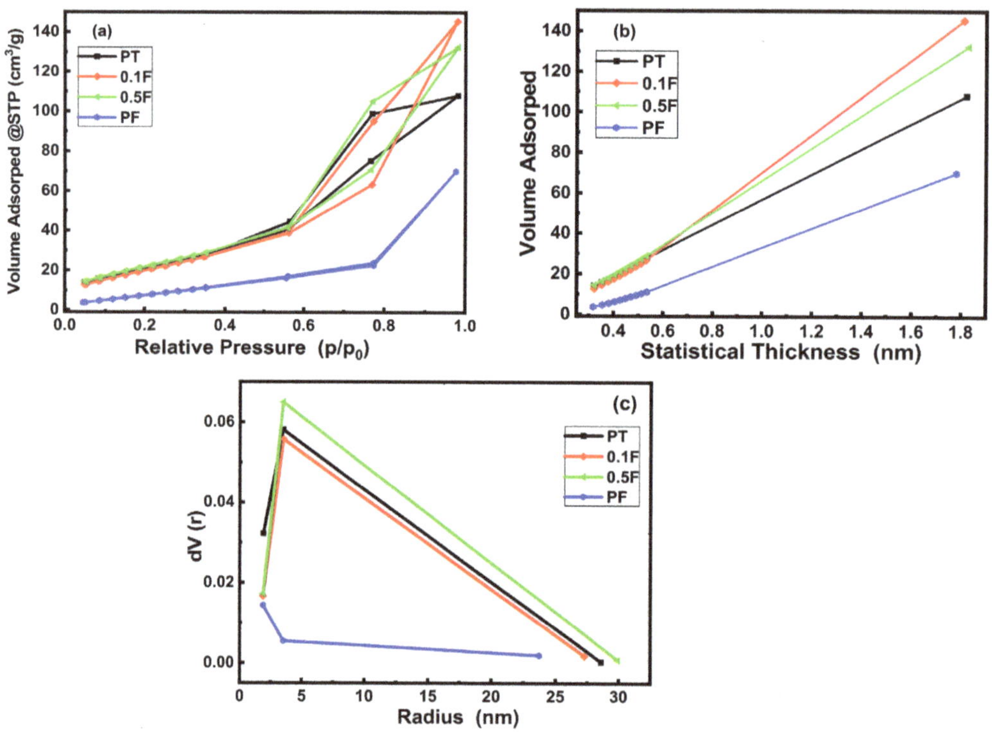


**Figure 7.** (**a**) $N_2$ adsorption–desorption isotherm, (**b**) t-plot of the as-prepared samples, and (**c**) BJH pore size.

**Table 3.** $N_2$-Physorption analysis results.

| Sample/Parameter | $S_{BET}$ $m^2/g$ | Total Pore Volume $cm^3/g$ | Average Pore Radius nm |
|---|---|---|---|
| PT | 82 | 0.1675 | 4.1 |
| 0.1 F | 80.3 | 0.2258 | 5.6 |
| 0.5 F | 86 | 0.2058 | 4.8 |
| PF | 35.6 | 0.1088 | 6.1 |

## 2.4. Photoelectrochemical Properties

In order to evaluate the photocatalytic performance of the pure titania (PT), pure hematite (PF) and the prepared hematite incorporated titania nanocomposites, we conducted photoelectrochemical measurements as indicated in Figure 8. Cyclic voltammetry, chronoamperometry and EIS were performed in a three-electrode cell configuration where dissolved air free 0.1 M $Na_2SO_4$ electrolyte was employed. The working electrode was prepared from the powder materials via a drop-casting technique using isopropanol solvent and fluorine-doped tin oxide (FTO) substrate without any post-treatment of the electrodes. The oxidation and reduction current values displayed in the cyclic voltammetry measurement can be used to judge the catalytic activity of the prepared materials. All samples exhibited only reduction current, except the PT sample exhibited both oxidation and reduction current. The reduction current density of the electrodes was 2.8, 1.65, 1.5 and 0.9 mA for PT, 0.1F, 0.5F and PF samples, respectively, while the oxidation current of PT samples was 1.5 mA. Therefore, the pure titania sample has the highest catalytic activity, while incorporating it with hematite its catalytic activity decreases. Moreover, to investigate the effective separation of the photogenerated charge carriers, we examined the photocurrent response under fixed potential (0.6 V versus Ag/AgCl) using the chronoamperometry technique. The photoresponse under light illumination demonstrates the rate of electrons transport as the majority charge carriers of n-type semiconductor from the sample to FTO as the collecting electrode. It was found that the PT sample shows the highest photocurrent magnitude reached ca. 800 μA/cm$^2$ while 0.1F sample reached almost half this value and, 0.5F and PF samples give negligible photocurrent response. This could be explained based on the low bandgap value of the hematite, which increases the recombination rate of the photogenerated electron/hole pair and the existence of a high density of surface electron traps. Furthermore, an EIS test was performed to assess the charge transfer resistance under dark conditions of the prepared samples. Since the semicircle radius shown in the Nyquist plots of the EIS data indicates the conductivity at the interface between the electrode/electrolyte and electrode/FTO substrate. Thus, the smaller semicircle radius suggests improvement of the charge transfer of the prepared composite. Accordingly, the conductivity of the prepared samples was found to be in the following order: 0.1F > 0.5F > PT > PF.

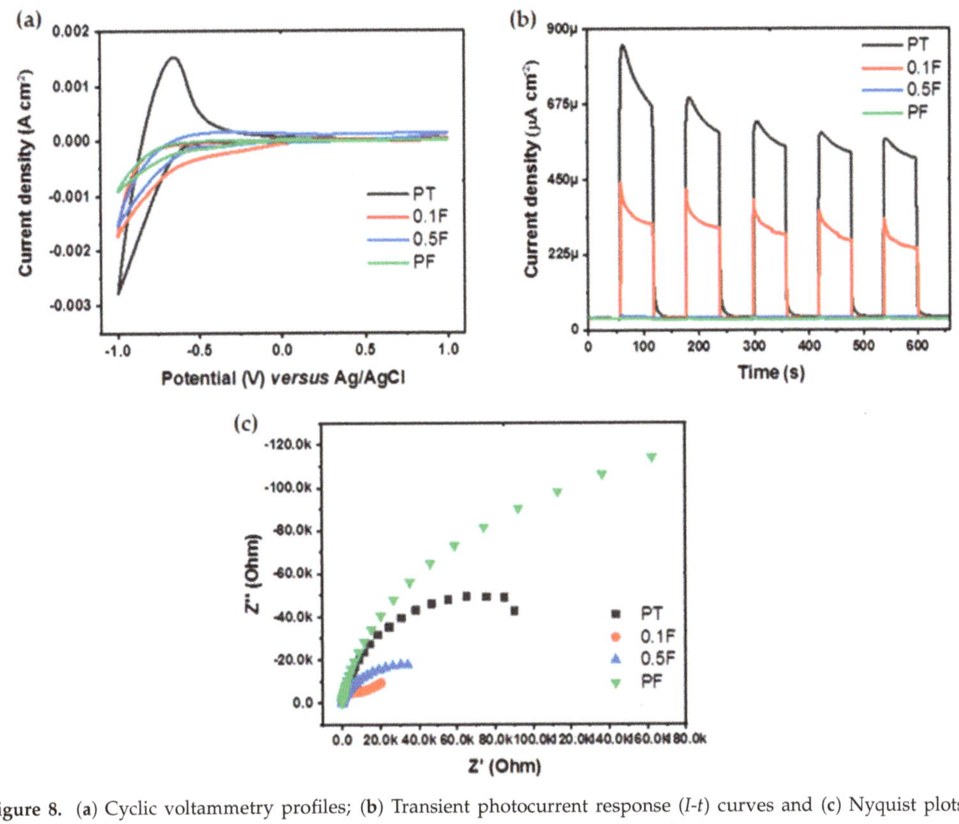


**Figure 8.** (**a**) Cyclic voltammetry profiles; (**b**) Transient photocurrent response (*I-t*) curves and (**c**) Nyquist plots of Electrochemical Impedance Spectroscopy (EIS) data for PT, 0.1F, 0.5F, and PF samples.

### 2.5. Photocatalytic Degradation of Rose Bengal Dye

The photocatalytic degradation of RB dye at different time intervals for 60 min is measured in the presence of $Fe_2O_3$ incorporated $TiO_2$ photocatalysts nanocomposites, $(TiO_2)_{1-x}(Fe_2O_3)_x$, where x = 0, 0.1, 0.5, and 1.0) as shown in Figure 9. In this work, 100 ppm aqueous solution of RB dye is taken without pH adjustment, and 150 mg of catalyst was added in 150 mL of dye solution to test photocatalytic activity. The as-prepared nanocomposites were suspended in dye solution (this refers to all the photocatalysts set that were prepared) via ultrasonication for 5 min, then was stirred in the dark for 1 h to reach adsorption–desorption equilibrium. Then the solution was irradiated with UV light (25 W, UV-LED, 365 nm) to perform the experiments. The concentration of RB dye solution was continuously monitored via UV-Vis spectrophotometer at stipulated time intervals. The absorbance of RB dye decreased with time during the photocatalytic degradation, clearly observable at 546 nm.

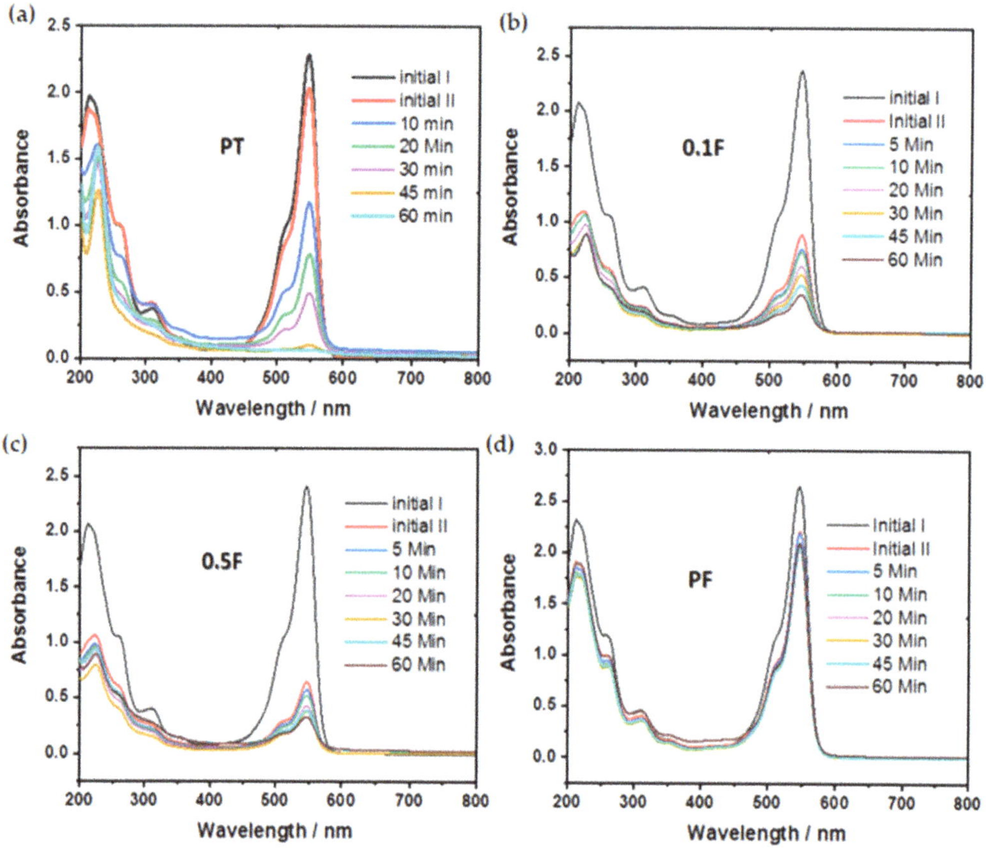


**Figure 9.** UV-Vis absorbance of the photocatalytic degradation of 100 ppm of Rose Bengal dye (60 min) over 1 g $L^{-1}$ of (**a**) PT, (**b**) 0.1F, (**c**) 0.5F and (**d**) PF samples.

Based on previous reports, the conduction band position of $TiO_2$ and $Fe_2O_3$ was estimated to be −0.22 and +0.30 V (versus NHE at pH = 0) [25]. While the valence band position could be calculated based on the bandgap values estimated from the DRS measurements (3.15 eV for $TiO_2$ and 1.91 eV for $Fe_2O_3$). Formation of the internal electric field upon light irradiation and Fermi energy level alignment between $TiO_2$ and $Fe_2O_3$ has promoted the separation of photogenerated electron-hole pairs. Therefore, as the UV light falls on the suspended photocatalyst on the dye solution, photogenerated electron/hole pairs are formed on the conduction and valence bands. The photogenerated electrons are known for their ability to react with the dissolved oxygen forming $O_2^{\bullet-}$ radicals while the holes react with $OH^-$ groups forming $OH^\bullet$ radicals. Those oxidation species react efficiently with dye resulting in the degradation of the dye molecules. The following equations could explain the possible degradation mechanism using our photocatalyst composite:

$$((TiO_2)\text{-}(Fe_2O_3)\text{ photocatalyst} + \text{UV irritation} \rightarrow (TiO_2)\text{-}(Fe_2O_3) \times (h_{VB}{}^+ + e_{CB}{}^-)$$

$$OH^- + h_{VB}{}^+ \rightarrow OH^\bullet$$

$$O_2 + e_{CB}{}^- \rightarrow O_2{}^{\bullet-}$$

$$O_2{}^{\bullet-} + H^+ \rightarrow HO_2{}^\bullet$$

$$HO_2{}^\bullet + H^+ + e_{CB}{}^- \rightarrow H_2O_2$$

$$H_2O_2 + e_{CB}{}^- \rightarrow OH^- + OH^\bullet$$

$$H_2O_2 \rightarrow H^+ + HO_2{}^\bullet$$

$$H_2O_2 + \text{UV irritation} \rightarrow 2[OH^\bullet]$$

$$OH^\bullet \text{ and } h_{VB}{}^+ + \text{RB dye} \rightarrow CO_2 + H_2O$$

The maximum photocatalytic degradation of the RB dye reaches 97% in the PT sample, which is pure $TiO_2$ NPs. Pure $TiO_2$ is widely known for its high oxidation potential compared to other photocatalysts under UV irradiation. In contrast, the photocatalytic activity of the other samples decreased with increasing the iron oxide content. Interestingly, the RB dye removal via adsorption increased remarkably with increasing the iron oxide content to almost 80% with the 0.5F catalyst. However, both adsorption and photocatalytic removal activity decreased to the minimum in the case of the pure iron oxide sample PF as shown in Table 4 and Figure 9. The enhanced adsorption capacity of the samples can be explained based on the surface texture analysis (surface area and pore size distribution). The terms related to degradation were calculated using the following formulae and briefed in Table 4:

$$\%\text{ Removal efficiency} = [C_o - C] \times 100/C \quad (7)$$

$$-\ln (C/C_o) = kt \quad (8)$$

where $C_o$ and C are the initial and final concentration of RB dye, respectively, k is the 1st order rate constant, and t is the irradiation time. All the RB dye degradation parameters are briefed in Table 4.

**Table 4.** Photocatalytic degradation parameters.

| Parameter/Sample | * Removal Efficiency % (Dark Adsorption) | * Removal Efficiency % (Photocatalysis) | Rate Constant K /$min^{-1}$ | ** $R^2$ |
|---|---|---|---|---|
| PT | 11.2 | 97 | 0.058 | 0.97 |
| 0.1F | 62.5 | 61 | 0.016 | 0.98 |
| 0.5F | 79 | 53 | 0.014 | 0.94 |
| PF | 16.3 | 7.3 | 0.0014 | 0.98 |

* Both dark adsorption and photocatalytic reaction were conducted for 60 min where the Photocatalytic degradation (removal efficiency) was calculated based on the initial concentration of the Rose Bengal dye after the adsorption step (Initial II). ** $R^2$ is a correlation coefficient.

## 3. Experimental Details

### 3.1. Synthesis of $(TiO_2)_{1-x}$ $(Fe_2O_3)_x$ Nanoparticles (NPs)

$(TiO_2)_{1-x}$ $(Fe_2O_3)_x$ NPs were prepared by the sol-gel process. In brief, 150 mL of ethanol, 3 mL of acetic acid, 5.26 mL of diethanolamine, 10.9 mL of titanium tetrachloride were mixed with vigorously stirring for 15 min, followed by adding 1.8 mL of deionized water. The desired amount of iron (III) acetate hydroxide was gradually added at X = 0.0, 0.1,

0.5, and 1.0 to obtain $(TiO_2)_{1-x}(Fe_2O_3)_x$ nanocomposites. The solution was magnetically stirred for 4 h, and the pH value was adjusted at~8. The obtained wet powder was washed three times by water and ethanol, respectively. The wet $(TiO_2)_{1-x}(Fe_2O_3)_x$ NPs were dried for 24 h at 100 °C, and then it was annealed for 3 h at 450 °C. All chemicals applied were purchased from Sigma-Aldrich, Burlington, MA, USA and used without further purification.

### 3.2. Characterization of $(TiO_2)_{1-x}(Fe_2O_3)_x$ Nanocomposites

The crystal structures and phases of the samples were recorded on Bruker D8 advance equipped with Cu Kα radiation (40 kV,40 mA) at a wavelength of 1.5406 Å. JEOL 6360 LA Japan SEM was employed to analyze the surface topography. DXR FT-Raman system from Fisher-Scientific was operated for vibrational study under 532 nm excitation unit at 5 mW power. The UV-Visible diffuse reflectance spectroscopy (DRS) analysis of powder samples was acquired from 200–800 nm using Evolution 220 spectrophotometer coupled with ISA-220 integrating sphere (Thermo Scientific, Waltham, MA, USA). The specific surface areas were determined by $N_2$ adsorption–desorption isotherm (Quantachrom Instrument Corporation, NOVATOUCH LX4, Boynton Beach, FL, USA) using Brunauer–Emmett-Teller (BET) calculations, whereas the samples were pretreated at 150 °C for 2 h, and the pore size distributions were recorded by desorption branch using BJH method. FTIR spectra of the samples were acquired from NICOLET IS 10 (Thermo Scientific, USA).

### 3.3. Photoelectrochemical Measurements

Various electrochemical techniques such as cyclic voltammetry (CV), chronoamperometry (CAM) and electrochemical impedance spectroscopy (EIS) were performed to evaluate the photoelectric properties of the prepared composites. Electrodes of the prepared powder materials were first prepared over fluorine-doped tin oxide (FTO) substrate using the drop-casting technique. A slurry of the powder materials and isopropanol solvent was prepared via ultrasonication. 50 µL was withdrawn and uniformly deposited over FTO substrate, and this step is repeated three times while the amount of the powder was controlled to be 4 mg over an area of ca. 1 cm$^2$. Then the electrodes were dried naturally at room temperature without any posttreatment. A rectangular quartz cell (50 mL capacity) with a three-electrode cell configuration was employed in the measurements, where a platinum wire was used as the counter electrode, sat. Ag/AgCl electrode was used as reference electrode and FTO act as the working electrode. All three electrodes were immersed in a dissolved air-free electrolyte of 0.1M Na2SO4 (purge with Ar gas) then connected to the electrochemical workstation (CS350, CorrTest Instruments, Wuhan, China). UV-LED (25 W, 365 nm, NVMUR020A, NICHIA, Japan) was used as a light source in the chronoamperometry test. Multiple cyclic voltammetry was measured with a scan rate of 50 mV/s and potential window from −1.0 to 1.0 V. Chronoamperometry was performed for 660 s at a fixed potential of 0.6 V. Electrochemical impedance spectroscopy (EIS) measurement was conducted at 20 mV amplitude with a frequency range from $10^5$ to 0.01 Hz under dark conditions.

### 3.4. Photocatalytic Dye Degradation

The photocatalytic activity of the as-prepared samples was evaluated toward RB degradation. 100 ppm of RB dye (4,5,6,7-tetrachloro-2′,4′,5′,7′-tetraiodofluorescein, 973.67 g/mol) was employed, while the reaction conditions were 1 g/L, 20 °C, 6.5 for catalyst concentration, reaction temperature, and pH (without adjustment), respectively. UV-LED (25 W, 365 nm, NVMUR020A, NICHIA, Tokushima, Japan) was used as the light source. First, the catalyst was suspended in 150 mL of dye aqueous solution inside irradiation made of Pyrex glass via ultrasonication and stirred in the dark for 1h to establish the adsorption–desorption equilibrium. Then the sample was irradiated with UV-LED for another 1 h while the concentration of the dye was monitored via UV-Vis. spectrophotometer (Perkin

Elmer, Lamda-40, Waltham, MA, USA). Dye samples were withdrawn, and catalysts were separated via centrifugation prior to absorption measurement.

## 4. Conclusions

Highly oriented nanostructures of pure and $TiO_2$ incorporated with different $Fe_2O_3$ contents $(TiO_2)_{1-x}(Fe_2O_3)_x$, where x = 0, 0.1, 0.5, and 1.0 for pure $TiO_2$ and $Fe_2O_3$ incorporated 0.1, 0.5, and pure $Fe_2O_3$ which are denoted as PT, 0.1F, 0.5F, and PF, respectively) were prepared by a sol-gel approach. X-ray diffraction studies revealed the polycrystalline tetragonal anatase phase, and the pure $Fe_2O_3$ is formed in the rhombohedral hematite structure. From SEM images, all the individual particles are uniformly distributed throughout the surface of the sample. FT-IR studies revealed that at higher incorporating concentration and in pure Fe incorporating, the intensity of band is increased. The decreasing bandgap with increasing incorporating concentration indicates that the present samples have huge applications in optoelectronic devices. The BET surface areas are high because of the mesoporous nature of the prepared composites. The synthesized $TiO_2$ NPs in the present work are very active under UV irradiation with high photonic efficiency. We continue our efforts to examine the synthesized photocatalysts under visible light irradiation employing different pollutants. All the promising results of the prepared samples clearly indicate their huge applications in the field of wastewater treatment.

**Author Contributions:** Methodology, M.A.S. and A.M.A.; investigation, A.M.A., M.A.S. and H.M.E.-B.; writing—original draft preparation, V.G., M.A. and A.M.A.; writing—review and editing, A.M.A., H.A., H.M.E.-B. and A.A.I.; supervision, A.M.A. and H.A. All authors have read and agreed to the published version of the manuscript.

**Funding:** This research was funded by the Deputyship for Research and Innovation, Ministry of Education, Saudi Arabia (Research Capability-Project number 412).

**Data Availability Statement:** Data are contained within the article.

**Acknowledgments:** The authors extend their appreciation to the Deputyship for Research and Innovation, Ministry of Education in Saudi Arabia, for funding this research work through project number 412.

**Conflicts of Interest:** The authors declare no conflict of interest.

## References

1. Reddy, B.M.; Ganesh, I.; Khan, A. Stabilization of nanosized titania-anatase for high temperature catalytic applications. *J. Mol. Catal. A Chem.* **2004**, *223*, 295–304. [CrossRef]
2. Heringa, M.B.; Geraets, L.; van Eijkeren, J.C.; Vandebriel, R.J.; de Jong, W.H.; Oomen, A.G. Risk assessment of titanium dioxide nanoparticles via oral exposure, including toxicokinetic considerations. *Nanotoxicology* **2016**, *10*, 1515–1525. [CrossRef]
3. Chang, X.; Zhang, Y.; Tang, M.; Wang, B. Health effects of exposure to nano-$TiO_2$: A meta-analysis of experimental studies. *Nanoscale Res. Lett.* **2013**, *8*, 1–10. [CrossRef]
4. Buzea, C.; Pacheco, I.I.; Robbie, K. Nanomaterials and nanoparticles: Sources and toxicity. *Biointerphases* **2007**, *2*, MR17–MR71. [CrossRef] [PubMed]
5. Ismail, A.A.; Bahnemann, D.W. Mesoporous titania photocatalysts: Preparation, characterization and reaction mechanisms. *J. Mater. Chem.* **2011**, *21*, 11686–11707. [CrossRef]
6. Ismail, A.A.; Ibrahim, I.A. Impact of supercritical drying and heat treatment on physical properties of titania/silica aerogel monolithic and its applications. *Appl. Catal. A Gen.* **2008**, *346*, 200–205. [CrossRef]
7. Fujishima, A.; Rao, T.N.; Tryk, D.A. Titanium dioxide photocatalysis. *J. Photochem. Photobiol. C Photochem. Rev.* **2000**, *1*, 1–21. [CrossRef]
8. Nasralla, N.; Yeganeh, M.; Astuti, Y.; Piticharoenphun, S.; Shahtahmasebi, N.; Kompany, A.; Karimipour, M.; Mendis, B.G.; Poolton, N.R.; Šiller, L. Structural and spectroscopic study of Fe-doped $TiO_2$ nanoparticles prepared by sol-gel method. *Sci. Iran.* **2013**, *20*, 1018–1022.
9. Assadi, M.H.N.; Hanaor, D.A.H. The effects of copper doping on the photocatalytic activity at (101) planes of anatase $TiO_2$: A theoretical study. *Appl. Surf. Sci.* **2016**, *387*, 682–689. [CrossRef]
10. Zhu, J.; Zheng, W.; He, B.; Zhang, J.; Anpo, M. Characterization of Fe-$TiO_2$ photocatalysts synthesized by hydrothermal method and their photocatalytic reactivity for photodegradation of XRG dye diluted in water. *J. Mol. Catal. A Chem.* **2004**, *216*, 35–43. [CrossRef]

11. Zhou, M.; Yu, J.; Cheng, B. Effects of Fe-doping on the photocatalytic activity of mesoporous $TiO_2$ powders prepared by an ultrasonic method. *J. Hazard. Mater.* **2006**, *137*, 1838–1847. [CrossRef] [PubMed]
12. Calle, A.M.; Sanchez, L.C.; Arboleda, J.D.; Beltran, J.J.; Barrero, C.A.; Osorio, J.; Nomura, K. Mixtures of iron and anatase $TiO_2$ by mechanical alloying. *Microelectron. J.* **2008**, *39*, 1322–1323. [CrossRef]
13. Eadi, S.B.; Kim, S.; Jeong, S.W.; Jeon, H.W. Novel preparation of Fe doped $TiO_2$ nanoparticles and their application for gas sensor and photocatalytic degradation. *Adv. Mater. Sci. Eng.* **2017**, *2017*, 2191659. [CrossRef]
14. Hussain, S.T.; Siddiqa, A.; Siddiq, M.; Ali, S. Iron-doped titanium dioxide nanotubes: A study of electrical, optical, and magnetic properties. *J. Nanopart. Res.* **2011**, *13*, 6517–6525. [CrossRef]
15. Hirano, M.; Joji, T.; Inagaki, M.; Iwata, H. Direct Formation of Iron (III)-Doped Titanium Oxide (Anatase) by Thermal Hydrolysis and Its Structural Property. *J. Am. Ceram. Soc.* **2004**, *87*, 35–41. [CrossRef]
16. Wang, C.Y.; Böttcher, C.; Bahnemann, D.W.; Dohrmann, J.K. A comparative study of nanometer sized Fe (III)-doped $TiO_2$ photocatalysts: Synthesis, characterization and activity. *J. Mater. Chem.* **2003**, *13*, 2322–2329. [CrossRef]
17. Mohamed, R.M.; Kadi, M.W.; Ismail, A.A. A Facile synthesis of mesoporous $\alpha$-$Fe_2O_3/TiO_2$ nanocomposites for hydrogen evolution under visible light. *Ceram. Int.* **2020**, *46*, 15604–15612. [CrossRef]
18. Gareso, P.L.; Sampe, N.; Palentek, V.; Taba, P. Influence of annealing on Fe-doped $TiO_2$ powders using co-precipitation technique. *AIP Conf. Proc.* **2017**, *1801*, 020002. [CrossRef]
19. Rodríguez, P.A.; Pecchi, G.A.; Casuscelli, S.G.; Elías, V.R.; Eimer, G.A. A simple synthesis way to obtain iron-doped $TiO_2$ nanoparticles as photocatalytic surfaces. *Chem. Phys. Lett.* **2019**, *732*, 136643. [CrossRef]
20. Umar, A.; Harraz, F.A.; Ibrahim, A.A.; Almas, T.; Kumar, R.; Al-Assiri, M.S.; Baskoutas, S. Iron-doped titanium dioxide nanoparticles as potential scaffold for hydrazine chemical sensor applications. *Coatings* **2020**, *10*, 182. [CrossRef]
21. Lourduraj, S.; Williams, R.V. Effect of iron doping on structural and optical properties of $TiO_2$ thin film by sol-gel routed spin coating technique. *J. Adv. Dielectr.* **2017**, *7*, 1750024. [CrossRef]
22. El-Bery, H.M.; Salah, M.R.; Ahmed, S.M.; Soliman, S.A. Efficient non-metal based conducting polymers for photocatalytic hydrogen production: Comparative study between polyaniline, polypyrrole and PEDOT. *RSC Adv.* **2021**, *11*, 13229–13244. [CrossRef]
23. Nabil, S.; Hammad, A.S.; El-Bery, H.M.; Shalaby, E.A.; El-Shazly, A.H. The $CO_2$ photoconversion over reduced graphene oxide based on $Ag/TiO_2$ photocatalyst in an advanced meso-scale continuous-flow photochemical reactor. *Environ. Sci. Pollut. Res.* **2021**, *28*, 36157–36173. [CrossRef]
24. El-Bery, H.M.; Abdelhamid, H.N. Photocatalytic hydrogen generation via water splitting using ZIF-67 derived $Co_3O_4@C/TiO_2$. *J. Environ. Chem. Eng.* **2021**, *9*, 105702. [CrossRef]
25. Singh, A.P.; Wang, R.B.; Tossi, C.; Tittonen, I.; Wickman, B.; Hellman, A. Hydrogen induced interface engineering in $Fe_2O_3$–$TiO_2$ heterostructures for efficient charge separation for solar-driven water oxidation in photoelectrochemical cells. *RSC Adv.* **2021**, *11*, 4297–4307. [CrossRef]
26. Mei, Q.; Zhang, F.; Wang, N.; Yang, Y.; Wu, R.; Wang, W. $TiO_2/Fe_2O_3$ heterostructures with enhanced photocatalytic reduction of Cr(VI) under visible light irradiation. *RSC Adv.* **2019**, *9*, 22764–22771. [CrossRef]
27. Cao, Y.Q.; Zi, T.Q.; Zhao, X.R.; Liu, C.; Ren, Q.; Fang, J.P.; Li, W.M.; Li, A.D. Enhanced visible light photocatalytic activity of $Fe_2O_3$ modified $TiO_2$ prepared by atomic layer deposition. *Sci. Rep.* **2020**, *10*, 13437. [CrossRef]
28. Hanaor, D.A.; Sorrell, C.C. Review of the anatase to rutile phase transformation. *J. Mater. Sci.* **2011**, *46*, 855–874. [CrossRef]
29. Da Silva, A.L.; Hotza, D.; Castro, R.H. Surface energy effects on the stability of anatase and rutile nanocrystals: A predictive diagram for $Nb_2O_5$-doped-$TiO_2$. *Appl. Surf. Sci.* **2017**, *393*, 103–109. [CrossRef]
30. Aslam, M.M.; Ali, S.M.; Fatehmulla, A.; Farooq, W.A.; Atif, M.; Al-Dhafiri, A.M.; Shar, M.A. Growth and characterization of layer by layer CdS–ZnS QDs on dandelion like $TiO_2$ microspheres for QDSSC application. *Mater. Sci. Semicond. Process.* **2015**, *36*, 57–64. [CrossRef]
31. Mahadik, M.A.; Shinde, S.S.; Mohite, V.S.; Kumbhar, S.S.; Moholkar, A.V.; Rajpure, K.Y.; Ganesan, V.; Nayak, J.; Barman, S.R.; Bhosale, S.H. Visible light catalysis of rhodamine B using nanostructured $Fe_2O_3$, $TiO_2$ and $TiO_2/Fe_2O_3$ thin films. *J. Photochem. Photobiol. B Biol.* **2014**, *133*, 90–98. [CrossRef] [PubMed]
32. Tang, W.; Zhang, Y.; Chen, X.; Zeng, X. $Fe_2O_3/TiO_2$ film electrodes prepared by the forced hydrolysis method and their photoelectrocatalytic performance. *Mater. Lett.* **2018**, *217*, 109–112. [CrossRef]
33. Barkhade, T.; Mishra, S.; Chander, H.; Mahapatra, S.K.; Banerjee, I. Effect of $TiO_2$ and Fe doped $TiO_2$ nanoparticles on mitochondrial membrane potential in HBL-100 cells. *Biointerphases* **2019**, *14*, 041003. [CrossRef] [PubMed]
34. Zhang, J.; Kuang, M.; Wang, J.; Liu, R.; Xie, S.; Ji, Z. Fabrication of carbon quantum dots/$TiO_2/Fe_2O_3$ composites and enhancement of photocatalytic activity under visible light. *Chem. Phys. Lett.* **2019**, *730*, 391–398. [CrossRef]
35. Wu, L.; Yan, H.; Xiaoc, J.; Li, X.; Wang, X.; Zhao, T. Characterization and photocatalytic properties of nano-$Fe_2O_3$–$TiO_2$ composites prepared through the gaseous detonation method. *Ceram. Int.* **2017**, *43*, 14334–14339. [CrossRef]
36. Abbas, N.; Shao, G.N.; Haider, M.S.; Imran, S.M.; Park, S.S.; Kim, H.T. Sol-gel synthesis of $TiO_2$-$Fe_2O_3$ systems: Effects of $Fe_2O_3$ content and their photocatalytic properties. *J. Ind. Eng. Chem.* **2016**, *39*, 112–120. [CrossRef]
37. Morales, A.E.; Mora, E.S.; Pal, U. Use of diffuse reflectance spectroscopy for optical characterization of un-supported nanostructures. *Rev. Mex. Fís.* **2007**, *53*, 18–22.

38. Barton, D.G.; Shtein, M.; Wilson, R.D.; Soled, S.L.; Iglesia, E. Structure and electronic properties of solid acids based on tungsten oxide nanostructures. *J. Phys. Chem. B* **1999**, *103*, 630–640. [CrossRef]
39. Shkir, M.; AlFaify, S.; Yahia, I.S.; Ganesh, V.; Shoukry, H. Microwave-assisted synthesis of $Gd^{3+}$ doped $PbI_2$ hierarchical nanostructures for optoelectronic and radiation detection applications. *Phys. B Condens. Matter* **2017**, *508*, 41–46. [CrossRef]
40. Shkir, M.; Yahia, I.S.; Ganesh, V.; Algarni, H.; AlFaify, S. Facile hydrothermal-assisted synthesis of $Gd^{3+}$ doped $PbI_2$ nanostructures and their characterization. *Mater. Lett.* **2016**, *176*, 135–138. [CrossRef]

catalysts

Article

# One-Pot Synthesis of $TiO_2$/Hectorite Composite and Its Photocatalytic Degradation of Methylene Blue

Dingqing Yang *, Jinyang Chen *, Xiaomin Hong, Jingying Cui and Lingzhen Li

School of Environmental and Chemical Engineering, Shanghai University, Shanghai 200444, China; hxm1102@shu.edu.cn (X.H.); qtaacui@shu.edu.cn (J.C.); 123abcabc@shu.edu.cn (L.L.)

* Correspondence: ydq@shu.edu.cn (D.Y.); chenjy@shu.edu.cn (J.C.); Tel.: +86-21-66137729 (J.C.); Fax: +86-21-66137725 (J.C.)

**Abstract:** $TiO_2$/hectorite composite photocatalysts with different molar ratios of lithium, magnesium, and silicon were synthesized by a one-pot hydrothermal method. The samples were characterized by X-ray diffraction (XRD), Fourier transform infrared (FTIR) spectroscopy, scanning electron microscopy (SEM), transmission electron microscopy (TEM), X-ray photoelectron spectroscopy (XPS), $N_2$ adsorption-desorption isotherms, and ultraviolet-visible diffuse reflectance spectra (UV-Vis DRS). When the molar ratio of lithium, magnesium, and silicon was 1.32:5.34:8 (TH-2), the composite showed the highest UV photocatalytic degradation of methylene blue (MB). The apparent rate constant of TH-2 was 0.04361 $min^{-1}$, which was about 3.12 times that of EVONIK Degussa commercial $TiO_2$ of AEROXIDE P25. The improvement of photocatalytic efficiency of the composite was mainly due to its high specific surface area, light trapping ability, and effective separation of electrons ($e^-$) and holes ($h^+$). At the same time, the F element of hectorite is beneficial to the formation of $Ti^{3+}$ in $TiO_2$, thus enhancing the photocatalytic activity. After five cycles, the removal rate of MB with TH-2 still reached 87.9%, indicating its excellent reusability.

**Keywords:** photocatalysis; $TiO_2$; hectorite; one-pot synthesis; degradation of organic dyes

**Citation:** Yang, D.; Chen, J.; Hong, X.; Cui, J.; Li, L. One-Pot Synthesis of $TiO_2$/Hectorite Composite and Its Photocatalytic Degradation of Methylene Blue. *Catalysts* **2022**, *12*, 297. https://doi.org/10.3390/catal12030297

Academic Editor: Ewa Kowalska

Received: 25 January 2022
Accepted: 3 March 2022
Published: 6 March 2022



## 1. Introduction

The development of the chemical industry has resulted in large-scale pollutant emission, which has brought about a series of environmental problems [1]. Organic dyes, in particular, are highly toxic and chemically stable, potentially teratogenic, and carcinogenic to humans [2]. As a "green" technology, photocatalysis has attracted widespread attention because it can efficiently degrade dyes to avoid its pollution to the environment [3,4].

Titanium dioxide ($TiO_2$) is considered one of the most promising photocatalysts because of its advantages of good chemical stability, nontoxicity, and low cost to degrade organic pollutants in the field of printing and dyeing [5,6]. However, $TiO_2$ has a wide band gap (3.0–3.2 eV) and excitation light is limited to ultraviolet light (4%), which greatly decreases its utilization efficiency of solar energy [7]. In addition, $TiO_2$ nanoparticles exhibit a low specific surface area, easy aggregation, and poor recycling, which limit its application range [8]. Therefore, improving the adsorption capacity and photocatalytic performance of $TiO_2$ is important.

Smectite clay is a layered silicate mineral with a high adsorption capacity and specific surface area, and some composite clay materials can change the phase of semiconductors or improve the separation of $e^-$ and $h^+$, which can be used to support $TiO_2$ [9]. The interlayer cations in the composite clays tend to capture electrons and oxidize the holes, thus reducing the charge recombination rate and improving the photocatalytic performance. Hectorite belongs to 2:1-type layered clay composed of a Si–O tetrahedron and a Mg–O octahedron arranged in 2:1 order in the vertical direction, and its molecular structure is $M_x[Li_xMg_{6-x}Si_8O_{20}(OH)_yF_z]$ (M = Na, Li) [10,11]. Hectorite can facilitate the separation of composites and improve the recycling of catalysts [12]. Therefore, $TiO_2$/hectorite composite material with high photocatalytic performance can be prepared by utilizing the adsorption and ion exchange properties of hectorite to enhance the degradation rate of dye wastewater.

The composite of clay and $TiO_2$ has the advantages of small titanium crystal size, large porosity, and large specific surface area [13]. However, it is usually difficult for $TiO_2$ to enter the interlayer. Moreover, due to the wide band gap of $TiO_2$, $TiO_2$/clay composites have photocatalytic activity only under ultraviolet light [14]. Up until now, $TiO_2$/hectorite has been synthesized mainly by the hydrothermal method, sol-gel method, and hydrolysis method [15–17]. In previous studies, most of them reacted by mixing $TiO_2$ precursor solution with the suspension of hectorite directly. The process is complicated and takes a long time. Furthermore, it is difficult for $TiO_2$ to enter the interlayer of hectorite, which limits the photocatalytic activity [18]. Ma et al. synthesized $TiO_2$/hectorite and found the influence of titanium content on photocatalytic performance [19]. However, $TiO_2$ was usually only supported on the mineral surface, not in the interlayer space of hectorite. It is worth studying $TiO_2$/hectorite composite materials by utilizing the exchange ability of the interlayer cation to introduce titanium ions into the interlayer during the synthesis of $TiO_2$/hectorite. Therefore, we studied the synthesis of $TiO_2$/hectorite nanocomposites as photocatalysts at different molar ratios of lithium, magnesium, and silicon prepared by a simple one-pot hydrothermal method and utilization for the photodegradation of MB under UV light irradiation.

## 2. Results and Discussion

### 2.1. Characterization

#### 2.1.1. X-ray Diffraction (XRD)

Figure 1 shows the XRD patterns of hectorite and $TiO_2$/hectorite (TH-1, TH-2, TH-3, TH-4, and TH-5). As for the hectorite, characteristic diffraction peaks of 2θ approximately appear at 19.6, 28.0, 35.1, 53.3, 61.0, and 72.3° [20,21]. It can be observed that hectorite has low crystallinity and small particle size from the widened peaks. Some weak peaks of 2θ appear at 29.0, 32.0, 38.7, 48.9, and 54.7°, which may be attributed to the residual starting materials and some intermediate products, such as lithium fluoride, lithium, and silicate [10].

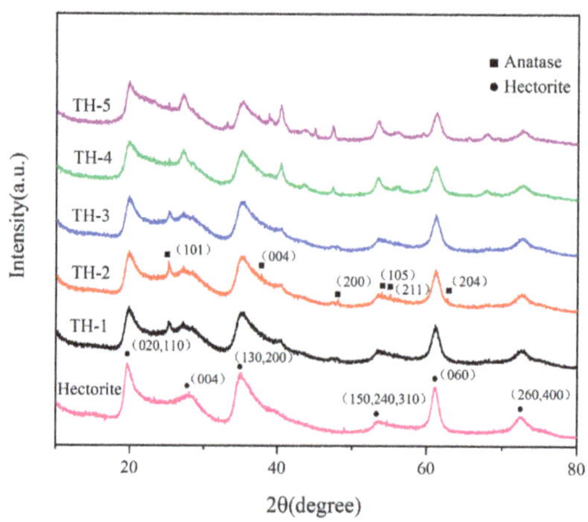


**Figure 1.** XRD patterns of hectorite: TH-1, TH-2, TH-3, TH-4, and TH-5.

$TiO_2$ shows the characteristic diffraction peaks of the anatase phase (Joint Committee on Powder Diffraction Standards Card NO. 21-1272) at 25.3, 37.8, 48.06, 54.0, 55.07, and 62.8°, which correspond to the diffraction planes of (101), (004), (200), (105), (211), and (204) [10]. The same diffraction peaks are found for the $TiO_2$/hectorite photocatalysts with different molar ratios of Li, Mg, and Si. However, the characteristic peaks of hectorite become slightly weaker in comparison to the synthesized $TiO_2$/hectorite, indicating that the layered structure of hectorite is partially destroyed, but its skeleton structure is still maintained [22]. The crystallite size of $TiO_2$ can be calculated according to the Debye–Scherrer equation (Equation (1)) [23]:

$$D = K\lambda/\beta cos\theta, \quad (1)$$

where the $K$ constant is the shape factor (value as 0.89), $\lambda$ is the wavelength, and $\beta$ is the half-peak width of the signal. The crystallite sizes of $TiO_2$ in TH-1, TH-2, TH-3, TH-4, and TH-5 are 11.3, 10.8, 11.2, 11.4, and 11.9 nm, respectively, according to the (101) peak by Equation (1), which are all smaller than that of P25 (21 nm) (Table 1). The appropriate particle size of $TiO_2$ is beneficial to increase the surface area, providing more photocatalytic active sites to improve the photocatalytic efficiency.

**Table 1.** Specific surface area ($S_{BET}$), pore size, pore volume, and crystal size of samples.

| Sample | $S_{BET}$ ($m^2/g$) | Pore Size (nm) | Pore Volume ($cm^3/g$) | Crystal Size of $TiO_2$ (nm) |
|---|---|---|---|---|
| TH-5 | 457.83 | 3.21 | 0.32 | 11.9 |
| TH-4 | 412.34 | 3.48 | 0.29 | 11.4 |
| TH-3 | 353.69 | 3.52 | 0.20 | 11.2 |
| TH-2 | 491.97 | 4.65 | 0.39 | 10.8 |
| TH-1 | 429.55 | 3.69 | 0.19 | 11.3 |
| Hectorite | 260.27 | 3.16 | 0.13 | - |
| P25 | 50 | - | - | 21 |

It has been found that for hectorite, magnesium from the magnesium oxide octahedron is more easily replaced by lithium ions, while silicon from silica the tetrahedron is difficult [20]. The number of lithium ions and magnesium ions have some effect on $TiO_2$/hectorite in the hydrothermal synthesis. With the increase in $Li^+$, the crystal size of $TiO_2$ decreases first and then increases gradually. $Li^+$ can accelerate the conversion of lithium silicate into hectorite, as well as increase the number of $Li^+$ replacing $Mg^{2+}$ on the lamella, enhancing the layer charge, so that more titanium ions have more chances to intercalate the interlayer to balance the negative charge. As for the samples, $TiO_2$ in TH-2 shows the best crystallinity. With the increase in the amount of lithium, diffraction peaks of TH-4 and TH-5 appear at 38.70, 44.99, and 65.48°, corresponding to the diffraction of (111), (200), and (220) crystal planes of the cubic LiF crystal, suggesting the increase in residual LiF. In addition, the peaks at 48.9 and 54.7° are caused by the intermediate product lithium silicate.

### 2.1.2. Fourier Transform Infrared (FTIR) Spectroscopy

Figure 2 shows the FTIR spectra of the samples in the range of 4000 to 400 $cm^{-1}$. It can be obviously observed that there are similar peaks between hectorite and $TiO_2$/hectorite, indicating that the basic skeleton structure of hectorite is not damaged during the introduction of $TiO_2$. The absorption bands such as -OH stretching vibrations of crystalline water at 3620 $cm^{-1}$, Si–O stretching at around 1030 $cm^{-1}$, Mg–O at 650–670 $cm^{-1}$, and the vibration peak of Si–O–Si bonds at 470 $cm^{-1}$ can be observed, which are characteristics of hectorite [24,25]. The peaks at 1640 and 3460 $cm^{-1}$ are attributed to the bound water and the vibration of structural hydroxyl groups in the samples [26,27]. The bands of all the composite photocatalysts centered on 640 and 605 $cm^{-1}$ are typical for $TiO_2$ (anatase phase), corresponding to the stretching modes of Ti–OH and Ti–O bonds, respectively [28]. The bands between 700 and 500 $cm^{-1}$ are usually assigned to the stretching vibration modes characteristic of Ti–OH and Ti–O–Ti bonds. All samples have these peaks and thus all have $TiO_2$.

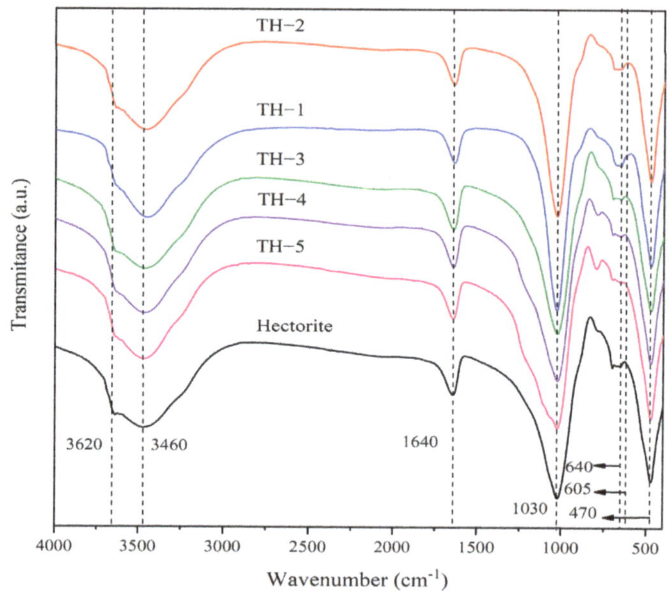


**Figure 2.** FTIR spectra for hectorite, TH-1, TH-2, TH-3, TH-4, and TH-5.

#### 2.1.3. Scanning Electron Microscopy (SEM)

The SEM images of hectorite, TH-1, TH-2, TH-3, TH-4, and TH-5 are shown in Figure 3. Figure 3a shows the typical structure of hectorite, which is a nanoscale layered structure and forms blocky particles with various sizes, rough surfaces, and different thicknesses [15]. SEM images of the five $TiO_2$/hectorite with different initial material ratios are given in Figure 3b–f. After the introduction of $TiO_2$, the structure of $TiO_2$/hectorite is different from that of hectorite. The particle size of hectorite pillared by $TiO_2$ is significantly reduced, showing a looser structure and disordered accumulation of flake structure, which can increase the specific surface area, enhance the adsorption ability, and improve the photocatalytic effect [10].

The lamellar structure of $TiO_2$/hectorite also shows that the ion leaching degree in the structure of hectorite is small and the skeleton is damaged to some degree, which is consistent with the XRD analysis results. The SEM images of TH-4 and TH-5 show a higher degree of agglomeration and lower nucleation of agglomerates in comparison with the TH-2 sample. In particular, the pattern of TH-5 gives an increase in irregular shape compared with other samples, which may be due to the excessive content of lithium ions inhibiting the entry of titanium ions into the layers of hectorite and agglomerating on the surface.

#### 2.1.4. Transmission Electron Microscopy (TEM)

To further understand the structure of the composites, TEM was applied to characterize the microstructure. The TEM images of TH-1, TH-2, TH-3, TH-4, and TH-5 are shown in Figure 4. The black areas show that $TiO_2$ is embedded in the interlayer of hectorite or supported on the surface, which is consistent with the XRD analysis results, indicating that some $TiO_2$ enter the interlayer of hectorite. It can be seen from these images that $TiO_2$ nanoparticles in TH-2 are well distributed in comparison with other samples, which is conducive to photocatalysis. Figure 4c shows that the distance (d) between two crystal faces of $TiO_2$ is 0.353 nm, indicating the existence of an anatase phase structure of $TiO_2$, corresponding to the (101) crystal plane of anatase [29]. The results suggest that the basic structure of $TiO_2$ cannot be influenced by the hectorite, and the lattice plane of $TiO_2$ in contact with the hectorite surface is the (101) lattice plane.

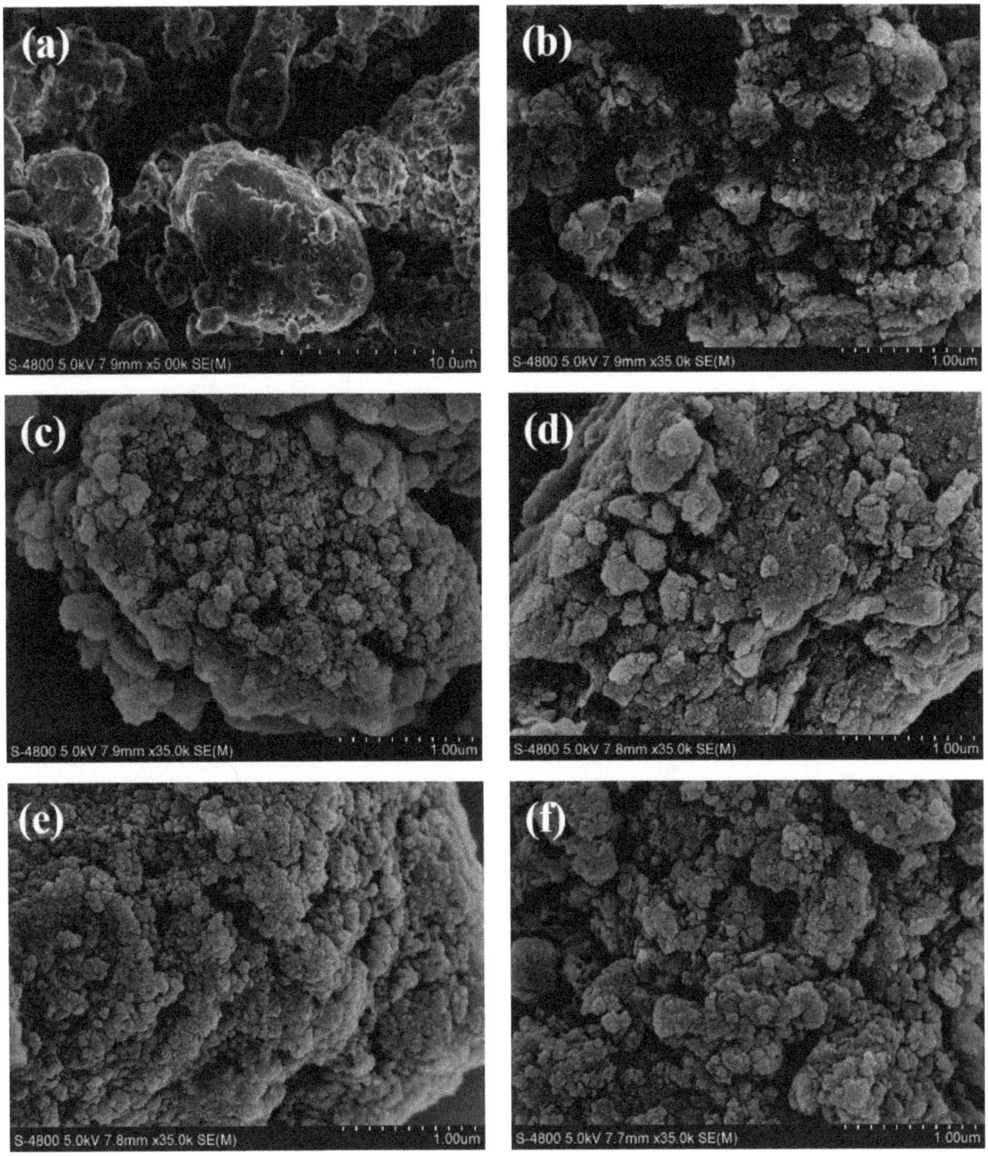


**Figure 3.** Scanning electron microscopy (SEM) images: (**a**) hectorite, (**b**) TH-1, (**c**) TH-2, (**d**) TH-3, (**e**) TH-4, and (**f**) TH-5.

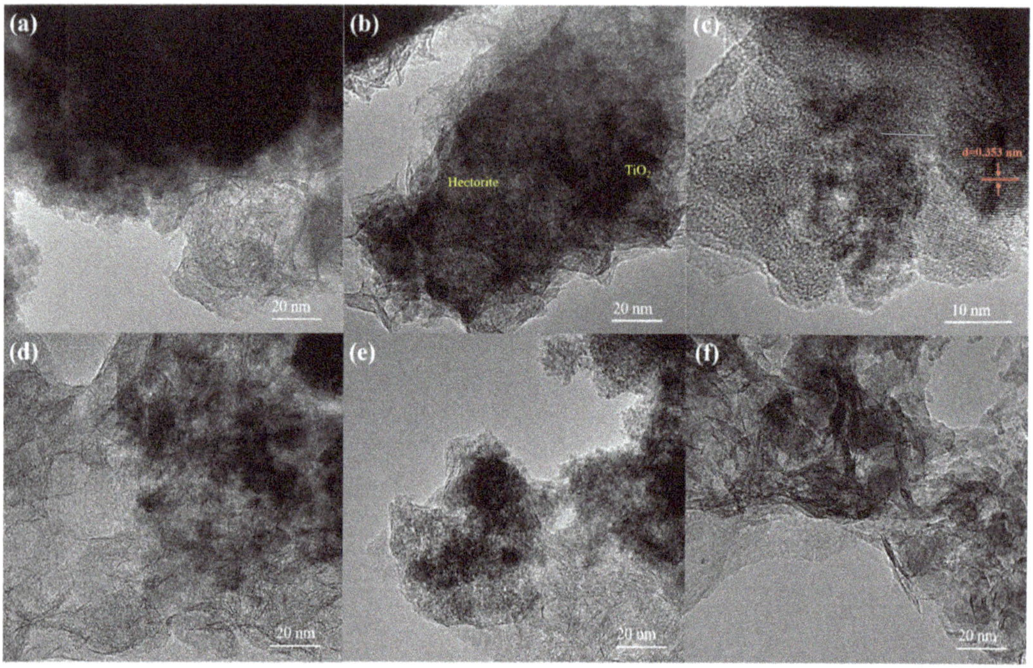


**Figure 4.** Transmission electron microscopy (TEM) images: (**a**) TH-1, (**b**,**c**) TH-2, (**d**) TH-3, (**e**) TH-4, and (**f**) TH-5.2.1.5. $N_2$ adsorption-desorption isotherms.

### 2.1.5. $N_2$ Adsorption-Desorption Isotherms

Figure 5 shows the $N_2$ adsorption-desorption isotherms and pore size distribution. According to the classification of IUPAC, the isotherms of all materials are regarded as type IV, corresponding to the mesoporous structure [30]. At the initial stage, $N_2$ adsorption capacities are greater than 0, indicating abundant micropores in the samples. Furthermore, the hysteresis loops of these samples belong to type H3, which represent that the samples are lamellar particle materials with fissure structures [31].

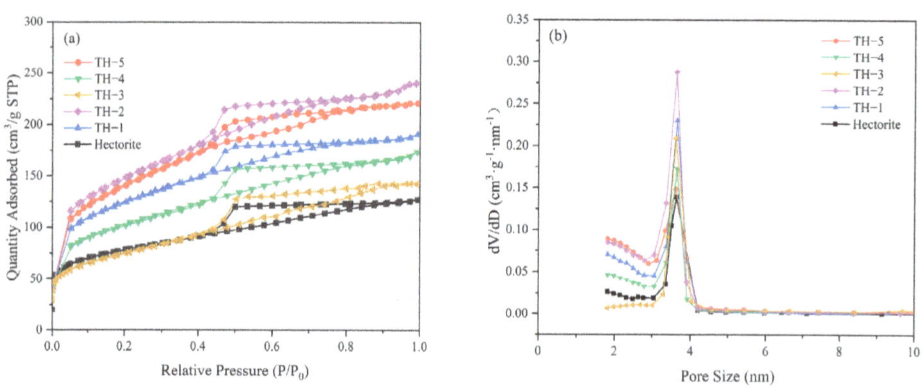


**Figure 5.** $N_2$ adsorption-desorption isotherms and pore size distribution: (**a**) isotherms and (**b**) pore size.

The pore structure parameters including specific surface area ($S_{BET}$), pore volume, pore size, and crystal size are summarized in Table 1. Compared with hectorite, the surface area, pore volume, and diameter of all composite samples increase, which would be beneficial to the adsorption and degradation of organic dyes. Among them, the TH-2 sample has the largest specific surface area of 491.97 $m^2/g$, which is about twice that of hectorite and ten times that of commercial P25. The average pore size of TH-1~5 is larger than that of hectorite (3.16 nm), which proves that $TiO_2$ is successfully introduced into the interlayer of hectorite. At the same time, when the molar ratio of $Li^+$ to $Mg^{2+}$ is 1.32:5.34, most ions exchange between titanium ions and interlayer water molecules, $Li^+$ is observed, and the interlayer spacing is largest. A higher specific surface area is beneficial to adsorption performance because it can provide more active sites, which can increase the contact area between organic pollutants and catalysts, thus increasing the photodegradation rate and improving the pollutant removal rate effectively ultimately.

Figure 5b is the pore size distribution curves calculated from Figure 5a. It can be observed that the pore size distribution curves of these samples are similar and the pore size is concentrated between 2 and 5 nm, which indicates again that hectorite and $TiO_2$/hectorite belong to the mesoporous structure. TH-2 displays a larger pore size distribution and the largest pore size, indicating that titanium ions have more chances to intercalate in the interlayer of hectorite under the condition. The pore size of TH-5 is larger than that of TH-4, which could be because of the increase in $TiO_2$ particles and their inhomogeneity [15].

#### 2.1.6. Ultraviolet-Visible Diffuse Reflectance Spectra (UV-Vis DRS)

Figure 6 gives the UV-Vis DRS, and the optical absorption capacity of hectorite is weak in the wavelength range of 200–800 nm, which is probably attributed to its composition and particle size [10]. All the composite photocatalysts present an intense optical absorption under 425 nm compared with the single hectorite, indicating the electron Ti–O transformation of $TiO_2$ after the intercalation of titanium cations [32]. The insertion of $TiO_2$ inside the layers of hectorite may modify the electronic structure and thus the band gap energy, promoting a shift in the absorption spectra [15,32]. The band gap values were obtained with Equation (2):

$$(\alpha h\nu)^{1/2} = A(h\nu - E_g), \tag{2}$$

where $\alpha$, $h\nu$, $A$, and $E_g$ are the optical absorption coefficient, photon energy, A constant, and band gap, respectively [33].

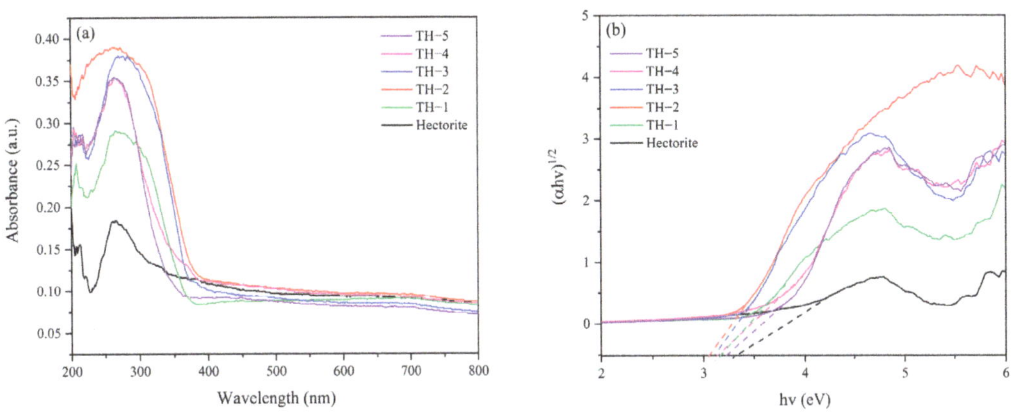


**Figure 6.** UV-Vis diffuse reflectance spectra and band gaps: (**a**) reflectance spectra and (**b**) band gaps.

The values of the band gap energies calculated from Equation (2) (Figure 6b) of hectorite, TH-1, TH-2, TH-3, TH-4, and TH-5, are 3.33, 3.17, 3.06, 3.11, 3.14 and 3.22 eV, respectively. Except for TH-5, the band gaps of other composite photocatalysts are smaller than that of pure $TiO_2$ (3.2 eV). The TH-2 sample exhibits the strongest UV absorption capacity, thus providing a good opportunity to broaden the absorption band and improve the photocatalytic performance.

2.1.7. X-ray Photoelectron Spectroscopy (XPS)

The elemental states and surface components of the TH-2 sample were analyzed by XPS and are given in Figure 7. It mainly contains Ti, Si, Mg, F, C, and O, while the Li element in hectorite is not observed in TH-2 (Figure 7a). The phenomenon indicates that the ion exchange reaction has occurred in which Li ions are likely to be replaced by hydrogen ions or hydrate titanium ions during the reaction [10]. In addition, the leaching of Mg from the magnesia octahedron has a great influence on the skeleton structure.

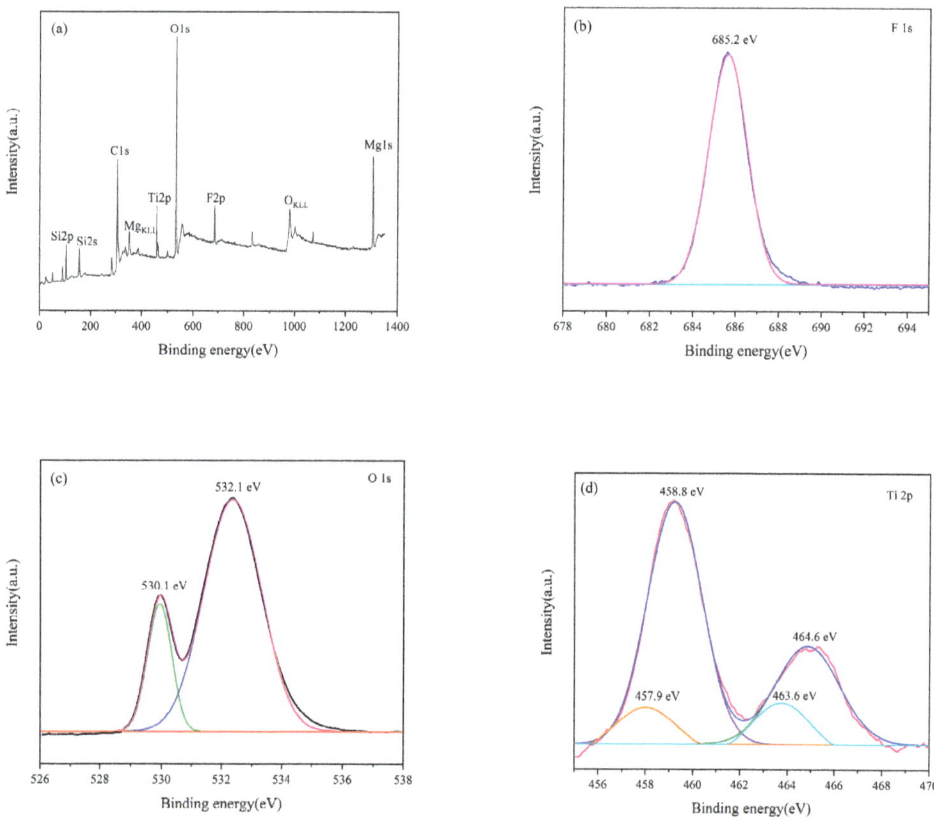


**Figure 7.** XPS spectra of TH-2 composite sample: (**a**) survey spectrum, (**b**) F 1s, (**c**) O 1s, and (**d**) Ti 2p.

The presence of a peak of F 1s for TH-2 with a binding energy at 685.2 eV is shown in Figure 7b, which is associated with the physical surface adsorption of F [34]. The UV-Vis DRS of TH-2 exhibits intense absorption in the UV region without an obvious redshift, which is consistent with the previous reports of F-doped $TiO_2$ [35,36]. In addition, the doping of the F element may contribute to the formation of oxygen holes.

According to the spectrum of O 1s for TH-2 (Figure 7c), the peak at 530.1 eV demonstrates the existence of crystal lattice oxygen $O_2^-$ and the peak at 532.1 eV is mainly related to the Si–O bond from silicon lattice [32,37]. The surface adsorption of F (≡Ti–F) and surface –OH can proceed as follows (Equation (3)) [34], while it is difficult to see the peak of –OH as F occupies more space on the surface of the $TiO_2$/hectorite.

$$\equiv \mathrm{Ti}-\mathrm{OH}+\mathrm{F}^- \leftrightarrow \equiv \mathrm{Ti}-\mathrm{F}+\mathrm{OH}^-\mathrm{pK_F}=6.2 \quad (3)$$

Two strong peaks of Ti 2p at 458.8 eV and 464.6 eV are attributed to the Ti $2p_{3/2}$ and Ti $2p_{1/2}$ (Ti–O bond), respectively, indicating the existence of the $Ti^{4+}$ chemical state in the TH-2 photocatalyst [32]. Two peaks at 457.9 eV and 463.6 eV correspond to the $Ti^{3+}$ $2p_{3/2}$ and $Ti^{3+}$ $2p_{1/2}$, respectively, which confirms the formation of $Ti^{3+}$ [38,39]. As is well understood, $Ti^{3+}$ is generally considered to be beneficial for improving photocatalytic activity.

The electron cloud density decreases with the binding energy [40,41]. Compared with pure $TiO_2$ (458.4 and 464.1 eV), the binding energy of Ti $2p_{3/2}$ (458.8 eV) and Ti $2p_{1/2}$ (464.6 eV) of TH-2 is much higher, which should be the formation of the Ti–O–Si bond between $TiO_2$ and hectorite [35,41,42]. The result indicates that the hectorite helps to promote the separation of $e^-$-$h^+$ pairs.

## 2.2. Photocatalytic Study

The photocatalytic performance of composite samples was studied in the degradation of MB with a 125 W high-pressure mercury lamp. The blank control group and P25 were used as comparisons. At low concentration, the photocatalytic degradation process conforms to the first-order kinetic equation, which can be fitted by the Langmuir-Hinshelword (L-H) model, as shown by Equation (4) [43,44].

$$ln\left(\frac{C_0}{C}\right) = kt, \quad (4)$$

where $C_0$ is the adsorption and desorption equilibrium concentration of MB, $C$ is the concentration of MB at time $t$, and $k$ is the photocatalytic kinetic constant. $k$ can be used to evaluate the photocatalytic performance, and the higher the value, the higher the catalytic efficiency.

The photocatalytic performance of TH-1, TH-2, TH-3, TH-4, TH-5, and P25 was evaluated for MB degradation under UV light irradiation in Figure 8. On the one hand, Figure 8a shows the degradation curves at 10 ppm of MB for the samples TH-1, TH-2, TH-3, TH-4, TH-5, and P25, where their corresponding removal rates are 57.5%, 97.8%, 94.5%, 80.1%, 78%, and 60.7%, respectively. Without catalysis, the removal rate of MB is only 8.2%. The result of the photolysis does not show a significant reduction in MB, indicating that the irradiation with UV light by itself is not capable of degrading the dye. The composition of $TiO_2$ and hectorite clearly enhances the photocatalytic activity. Specifically, TH-2 and TH-3 show excellent photocatalytic activity, and photocatalysts are able to degrade 97.8% and 94.5% of MB within 60 min, while P25 and TH-1 only degrade MB by 60.7% and 57.5%, respectively. Although TH-1 has a higher surface area value than P25, its degradation is lower than that achieved by P25. The lower value from TH-1 can be attributed to the excess of defects present in the TH-1 sample, and P25 probably has a higher recombination of $e^-/h^+$ pairs by affecting the photodegradation. Therefore, its degradation is lower than those of TH-2, TH-3, TH-4, and TH-5. In particular, P25 (composed of ~30% rutile and 70% anatase) presents a better performance than TH-1, attributed to the transfer of electrons from Cb of anatase to those of rutile $TiO_2$ [32].

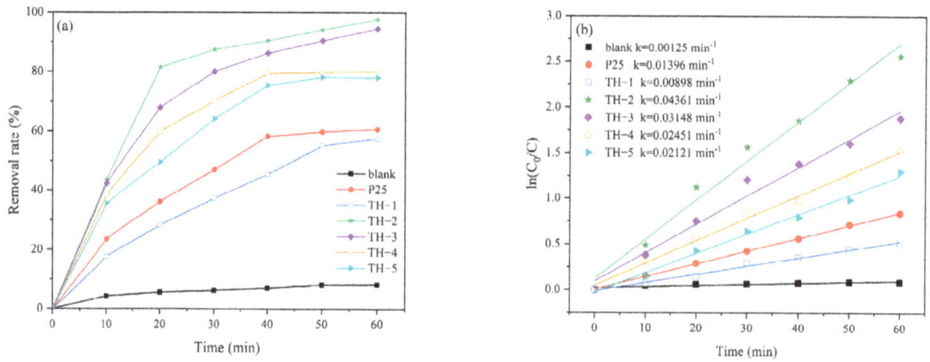


**Figure 8.** (**a**) Photocatalytic degradation of MB under UV light; (**b**) the first-order kinetic fitting curve of the photocatalytic MB degradation.

On the other hand, the kinetic constants (*k*), derived from the first-order kinetic fitting curve (Figure 8b), for the photodegradation of MB from highest to lowest, are shown in the order of TH-2 (0.04361 $min^{-1}$), TH-3 (0.03148 $min^{-1}$), TH-4 (0.0245 $min^{-1}$), TH-5 (0.02121 $min^{-1}$), P25 (0.01396 $min^{-1}$), and TH-1 (0.00898 $min^{-1}$) catalysts. The apparent rate constant of TH-2 is about 3.12 times that of P25. Based on these results, the removal rates of MB are increased by adding a quantity of $TiO_2$ in hectorite, and the highest degradation rate is achieved when the molar ratio of lithium, magnesium, and silicon is 1.32:5.34:8 (TH-2). At the same time, the combination of $TiO_2$ and hectorite is beneficial to the generation of oxygen vacancies. These oxygen vacancies can trap the photogenerated species and prolong their life, which improves the photoactivity. However, TH-1 could produce a defect excess and the formation of polycrystals by increasing the recombination of $e^-/h^+$ pairs, affecting the degradation of MB.

Based on the above result, the photocatalytic mechanism of TH-2 is given in Figure 9. First, $TiO_2$ in the interlayer of hectorite can increase the specific surface area to increase the adsorption capacity and contact of MB and the photocatalyst. Secondly, $TiO_2$ is uniformly dispersed on the hectorite to provide more active sites for the reaction. Thirdly, it is the most important that during the photocatalytic process, the electrons are stimulated from the valence band to the oxygen vacancy and $Ti^{3+}$ in $TiO_2$ under UV light irradiation, and the oxygen vacancy defect is beneficial to the adsorption of $O_2$, while the positive charge holes are left in the conduction band. At the same time, the negatively charged interlayer surface of hectorite is conducive to improving the photocatalytic activity, and the photogenerated holes can migrate to the surface quickly under the electronic attraction on the negatively charged interlayer surface.

In addition, photocatalytic activity is closely related to the behavior of photocarriers [45,46]. Subsequently, electrons ($e^-$) gathered on the surface of $TiO_2$ react with dissolved oxygen molecules in water to generate the superoxide radical anion $\cdot O_2^-$ and other high oxidation groups. As the conduction band has a more positive potential than that of $\cdot OH/H_2O$, the holes in the valence band react with water molecules (or surface hydroxyl groups) adsorbed on the surface of $TiO_2$ to form $\cdot OH$ [34]. Furthermore, due to the adsorption of F on the surface of $TiO_2$, $\cdot OH_{free}$ is generated in the bulk solution. Finally, MB is oxidized into decomposed products by these active substances $\cdot O_2^-$ and $\cdot OH$.

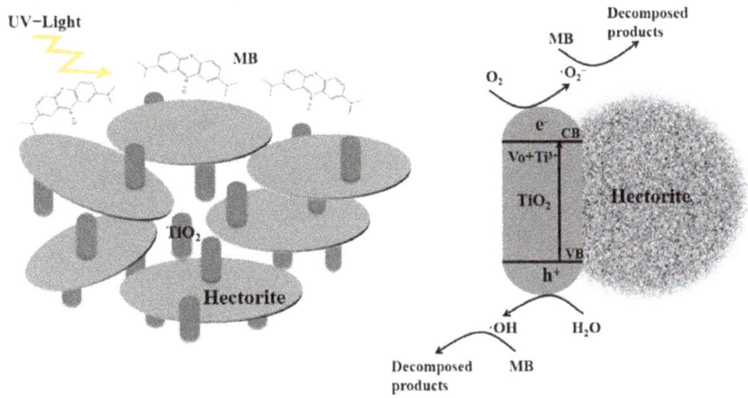


**Figure 9.** Schematic illustration of a photocatalytic mechanism for the degradation process of MB under UV light toward TH-2 composite.

## 2.3. *Reusability*

Stability and reutilization are important to industrial application [47,48]. The reusability of the TH-2 composite is exhibited in Figure 10. After 5 cycles, the results show that the photoactivity can still achieve 87.9%, indicating the outstanding stability of TH-2. The decrease in catalytic activity may be mainly caused by the loss of surface area and the deposition of byproducts at the surface of the photocatalyst. It can result in the loss of active sites, agglomeration of nanoparticles, and reduction in crystallinity, which has been suggested by previous studies [49,50].

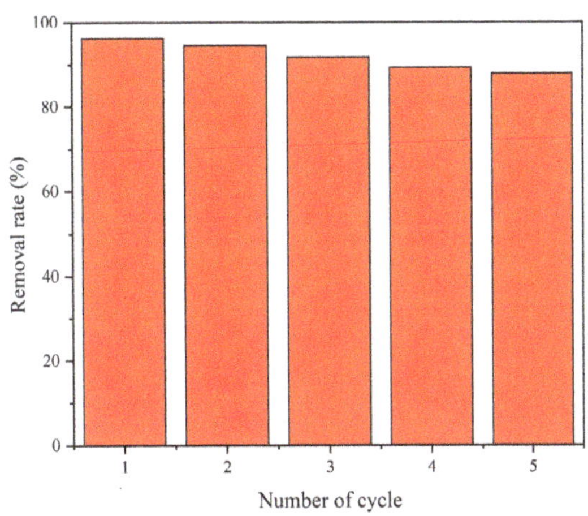


**Figure 10.** Reusability of TH-2 composite to the degradation of MB.

In addition, $TiO_2$/hectorite can be recovered from the mixture only by sedimentation, while P25 is difficult to precipitate due to its strong suspension force, indicating that the composite has good recyclability. It can be concluded that $TiO_2$/hectorite is an excellent catalyst material with stable activity and recoverability.

## 3. Materials and Methods

### 3.1. Materials

Tetrabutyl titanate ($C_{16}H_{36}O_4Ti$, TBOT, ≥99%) and anhydrous ethanol ($C_2H_6O$, ≥99.8%) were purchased from Aladdin Biochemical Technology Co., Ltd., Shanghai, China. Lithium fluoride (LiF), magnesium sulfate ($MgSO_4$), hydrochloric acid (HCl), sodium silicate ($Na_2O{\cdot}nSiO_2$), sodium hydroxide (NaOH), and methylene blue ($C_{16}H_{18}N_3ClS$, MB) were all obtained from Sinopharm Chemical Reagent Co., Ltd., Shanghai, China. The Evonik Degussa P25 was provided by Evonik Industries AG, Frankfurt, Germany. Deionized water was used.

### 3.2. Synthesis of Photocatalysts

The $TiO_2$/hectorite nanocomposites were prepared through the one-pot hydrothermal method. First, a certain amount of LiF was added into 200 mL of distilled water with stirring to obtain an LiF dispersion. Secondly, a certain amount of $MgSO_4$ was dissolved in 100 mL of distilled water, adding sodium hydroxide solution to obtain $Mg(OH)_2$ precipitation. Then, the $Mg(OH)_2$ was transferred to the LiF dispersion and stirred to form a uniform mixed slurry after being filtered and washed with deionized water. Subsequently, 18.46 g of sodium silicate ($SiO_2$ 26% wt, $Na_2O$ 8.2% wt) was added into 100 mL of distilled water with the dripping of hydrochloric acid to obtain $SiO_2$. Then, the $SiO_2$ was transferred to the $Mg(OH)_2$ and LiF mixed slurry and stirred to form a uniform mixed slurry (A) after being filtered and washed with deionized water.

Meanwhile, 1.1295 g of TBOT was dissolved in 10 mL of anhydrous ethanol with stirring for about 30 min to obtain a light yellow solution (B). Then, the dispersion mix A and solution B were put into the hydrothermal reaction kettle and reacted at 180 °C for 12 h. After the hydrothermal reaction, the solution was cooled to room temperature, centrifuged, washed, dried, ground, and finally sifted through 400 mesh to obtain a $TiO_2$/hectorite composite photocatalyst.

A starting mixture with different molar ratios of Li:Mg:Si (Si is 8) was prepared, as shown in Table 2. The five synthesized groups of catalysts were named TH-1, TH-2, TH-3, TH-4, and TH-5.

**Table 2.** The molar ratio and mass of the samples.

| Catalysts | Li:Mg:Si Molar Ratio of the Starting Mixture | Mass (g) | | |
|---|---|---|---|---|
| | | LiF | $MgSO_4$ | Sodium Silicate |
| TH-1 | 0.72:5.64:8 | 0.1872 | 6.768 | 18.46 |
| TH-2 | 1.32:5.34:8 | 0.3432 | 6.408 | 18.46 |
| TH-3 | 1.92:5.04:8 | 0.4992 | 6.048 | 18.46 |
| TH-4 | 2.52:4.74:8 | 0.6552 | 5.688 | 18.46 |
| TH-5 | 3.12:4.44:8 | 0.8112 | 5.328 | 18.46 |

### 3.3. Characterizations

The crystallinity and structure of $TiO_2$/hectorite were obtained by the X'pert PRO Empyrean X-ray diffractometer (PANalytical, Almelo, The Netherlands) equipped with Cu-K$\alpha$ radiation ($\lambda$ = 0.15418 nm) at 45 kV and 40 mA. The scanning speed was 5°/min and the scanning range of 2$\theta$ was 10–80°. The scanning electron microscopy (SEM) images were captured by a S-4800 scanning electron microscope (Hitachi, Tokyo, Japan) with an accelerated working voltage of 5 kV. TEM images were observed using a Tecnai F20 instrument (FEI Corp, Waltham, MA, USA) at 200 kV. The FTIR spectrum of these samples was recorded with a Nicolet iS5 FTIR spectrometer (Thermo Scientific, Madison, WI, USA) from 4000 to 400 $cm^{-1}$. The $N_2$ adsorption-desorption isotherms were captured on a Micromeritics, ASAP 2460 nitrogen volumetric adsorption facility (Norcross, GA, USA) at liquid nitrogen temperature (77 K). The specific surface area and pore size were calculated by the Brunauer-Emmett-Teller (BET) and Barret-Joyner-Halenda methods, respectively.

Ultraviolet-visible diffuse reflectance spectra (UV-Vis DRS) were recorded by using a UV-3600 spectrophotometer (Shimadzu, Tokyo, Japan) equipped with an integrating sphere, and $BaSO_4$ was used as the reference standard. X-ray photoelectron spectroscopy (XPS) was measured on a Thermo Scientific K-Alpha electron spectrometer (Thermo Fisher Scientific, Hillsboro, OR, USA) by using 12 kV Al-K$\alpha$ X-ray radiation.

*3.4. Photocatalytic Degradation*

The photocatalysis of $TiO_2$/hectorite was studied by the degradation of MB in a photocatalytic reactor [10]. The photocatalytic degradation was performed under a 125 W high-pressure mercury lamp. The Mercury lamp was preheated for 30 min before reaction to ensure stable luminescence. The lamp was kept approximately 30 cm away from the solution.

Firstly, 3 mg of the $TiO_2$/hectorite catalyst was dispersed in 100 mL of 10 ppm MB solution. Then, the mixture was stirred magnetically for 30 min in the dark to achieve adsorption-desorption equilibrium. After that, with underexposure of UV light, 3 mL of suspension was taken every 10 min and filtered by a 0.45 μm membrane to remove photocatalyst. The absorbance of the filtrate was determined by a WF Z UV-2800H UV-vis spectrophotometer (Unico, Suite E Dayton, NJ, USA) at 664 nm to evaluate the photocatalytic efficiency. The removal rate of MB was calculated by Equation (5):

$$R = \frac{C_0 - C_t}{C_0} \times 100, \quad (5)$$

where $C_0$ (mg/L) is the initial concentration of MB and $C_t$ (mg/L) corresponds to its concentration at time $t$.

After the photocatalytic degradation, the MB adsorbed on the material was completely degraded by irradiation under UV light. Then, the MB solution and photocatalytic materials were separated by centrifuge and dried at 120 °C for 2 h for recycling.

## 4. Conclusions

$TiO_2$ was introduced into the interlayer of hectorite by the one-pot hydrothermal method and the synthesized $TiO_2$/hectorite composites exhibited a higher UV photocatalytic activity than commercial P25. The number of titanium ions entering into the hectorite layer was changed by adjusting the molar ratio of lithium and magnesium in the raw material. The materials synthesized in this study showed anatase phase $TiO_2$, and the appropriate amount of lithium ion was beneficial to improve the crystallinity of the products according to XRD and FTIR analysis. SME and TEM studies indicated that $TiO_2$/hectorite showed a looser structure after being pillared by $TiO_2$ in comparison with hectorite, and the layer spacing of $TiO_2$ was 0.353 nm. In the $N_2$ adsorption-desorption analysis, TH-2 presented the largest specific surface area, the strongest adsorption capacity, and the best photocatalytic effect. UV-Vis DRS studies showed that the absorption band edge of $TiO_2$/hectorite was redshifted and its UV absorption capacity became stronger after $TiO_2$ incorporation, indicating the electron Ti–O transformation of $TiO_2$. XPS analysis indicated that Li or Na ions of hectorite were easily replaced by titanium ions or hydrogen ions during the preparation process, which can promote the separation of $e^-$-$h^+$ pairs.

The results showed that a suitable ratio of lithium to magnesium is beneficial to the improvement of the photocatalytic effect. When the molar ratio of lithium, magnesium, and silicon was TH-2 of 1.32:5.34:8, the $TiO_2$/hectorite photocatalyst had the highest removal rate of MB dye (97.8%). In addition, the TH-2 sample could be easily recycled and the removal rate of the MB still achieved 87.9% after five cycles, indicating good reusability. High specific surface area, strong light capture ability, and great $e^-$-$h^+$ separation efficiency are favorable for the promotion of photoactivity. The formation of oxygen vacancies and $Ti^{3+}$ can also promote the separation of photogenic $e^-$ and $h^+$. Therefore, $TiO_2$/hectorite has good photoactivity and reusability as a UV photocatalyst to be used in the field of photodegradation of organic pollutants.

**Author Contributions:** Literature search and study design, D.Y. and J.C. (Jinyang Chen); investigation and data collection, D.Y., X.H. and J.C. (Jingying Cui); artwork and figures, D.Y.; writing—original draft preparation, D.Y.; writing—review and editing, J.C. (Jinyang Chen); data curation, D.Y. and L.L. All authors have read and agreed to the published version of the manuscript.

**Funding:** The research was supported by the Program for Innovative Research Team in University (No. IRT13078).

**Data Availability Statement:** The data presented in this study are available on request from the corresponding author.

**Acknowledgments:** The authors would like to thank Shiyanjia Lab (www.shiyanjia.com) for the SEM and TEM tests.

**Conflicts of Interest:** The authors declare no conflict of interest.

## References

1. Wang, Q.H.; Lai, Z.Y.; Luo, C.M.; Zhang, J.; Cao, X.D.; Liu, J.A.; Mu, J. Honeycomb-like activated carbon with microporous nanosheets structure prepared from waste biomass cork for highly efficient dye wastewater treatment. *J. Hazard. Mater.* **2021**, *416*, 125896. [CrossRef] [PubMed]
2. Yu, Y.; Qiao, N.; Wang, D.J.; Zhu, Q.Z.; Fu, F.; Cao, R.Q.; Wang, R.; Liu, W.; Xu, B. Fluffy honeycomb-like activated carbon from popcorn with high surface area and well-developed porosity for ultra-high efficiency adsorption of organic dyes. *Bioresour. Technol.* **2019**, *285*, 121340. [CrossRef] [PubMed]
3. Xu, J.; Wang, W.Z.; Shang, M.; Gao, E.P.; Zhang, Z.J.; Ren, J. Electrospun nanofibers of Bi-doped $TiO_2$ with high photocatalytic activity under visible light irradiation. *J. Hazard. Mater.* **2011**, *196*, 426–430. [CrossRef] [PubMed]
4. Chen, D.J.; Cheng, Y.L.; Zhou, N.; Chen, P.; Wang, Y.P.; Li, K.; Huo, S.H.; Cheng, P.F.; Peng, P.; Zhang, R.C.; et al. Photocatalytic degradation of organic pollutants using $TiO_2$-based photocatalysts: A review. *J. Clean. Prod.* **2020**, *268*, 121725. [CrossRef]
5. Ge, Y.H.; Luo, H.; Huang, J.R.; Zhang, Z.M. Visible-light-active $TiO_2$ photocatalyst for efficient photodegradation of organic dyes. *Opt. Mater.* **2021**, *115*, 111058. [CrossRef]
6. Ijaz, M.; Zafar, M. Titanium dioxide nanostructures as efficient photocatalyst: Progress, challenges and perspective. *Int. J. Energ. Res.* **2020**, *45*, 3569–3589. [CrossRef]
7. Zhang, X.K.; Li, L.; Zeng, Y.Q.; Liu, F.; Yuan, J.J.; Yu, Y.; Zhu, X.R.; Xiong, Z.Z.; Yu, H.J.; Xie, Y.M. $TiO_2$/graphitic carbon nitride nanosheets for the photocatalytic degradation of Rhodamine B under simulated sunlight. *ACS Appl. Nano. Mater.* **2019**, *2*, 7255–7265. [CrossRef]
8. Yuan, L.L.; Huang, D.D.; Guo, W.N.; Yang, Q.X.; Yu, J. $TiO_2$/montmorillonite nanocomposite for removal of organic pollutant. *Appl. Clay Sci.* **2011**, *53*, 272–278. [CrossRef]
9. Auta, M.; Hameed, B.H. Acid modified local clay beads as effective low-cost adsorbent for dynamic adsorption of methylene blue. *J. Ind. Eng. Chem.* **2013**, *19*, 1153–1161. [CrossRef]
10. You, R.; Chen, J.Y.; Hong, M.H.; Li, J.R.; Hong, X.M. Facile Synthesis of g-$C_3N_4$/$TiO_2$/Hectorite Z-Scheme Composite and Its Visible Photocatalytic Degradation of Rhodamine B. *Materials* **2020**, *13*, 5304. [CrossRef]
11. Vinci, D.; Dazas, B.; Ferrage, E.; Lanson, M.; Magnin, V.; Findling, N.; Lanson, B. Influence of layer charge on hydration properties of synthetic octahedrally-charged Na-saturated trioctahedral swelling phyllosilicates. *Appl. Clay Sci.* **2020**, *184*, 105404. [CrossRef]
12. Mishra, A.; Mehta, A.; Basu, S. Clay supported $TiO_2$ nanoparticles for photocatalytic degradation of environmental pollutants: A review. *J. Environ. Chem. Eng.* **2018**, *6*, 6088–6107. [CrossRef]
13. Belessi, V.; Lambropoulou, D.; Konstantinou, I.; Katsoulidis, A.; Pomonis, P.; Petridis, D.; Albanis, T. Structure and photocatalytic performance of $TiO_2$/clay nanocomposites for the degradation of dimethachlor. *Appl. Catal. B Environ.* **2007**, *73*, 292–299. [CrossRef]
14. Daniel, L.M.; Frost, R.L.; Zhu, H.Y. Synthesis and characterisation of clay-supported titania photocatalysts. *J. Colloid Interface Sci.* **2007**, *316*, 72–79. [CrossRef]
15. Huo, M.Y.; Guo, H.G.; Jiang, Y.S.; Ju, H.; Xue, B.; Li, F.F. A Facile Method of Preparing Sandwich layered $TiO_2$ in between Montmorillonite Sheets and Its Enhanced UV-light Photocatalytic Activity. *J. Photochem. Photobiol. A Chem.* **2018**, *358*, 121–129. [CrossRef]
16. Jin, J.; Chen, B.Z.; Liu, L.; Liu, R.H.; Qian, G.P.; Wei, H.; Zheng, J.L. A study on modified bitumen with metal doped nano-$TiO_2$ pillared montmorillonite. *Materials* **2019**, *12*, 1910. [CrossRef]
17. Djellabi, R.; Ghorab, M.F.; Cerrato, G.; Morandi, S.; Gatto, S.; Oldani, V.; Di Michele, A.; Bianchi, C.L. Photoactive $TiO_2$-montmorillonite composite for degradation of organic dyes in water. *J. Photochem. Photobiol. A Chem.* **2015**, *295*, 57–63. [CrossRef]
18. Chen, D.M.; Zhu, H.L.; Wang, X. A facile method to synthesize the photocatalytic $TiO_2$/montmorillonite nanocomposites with enhanced photoactivity. *Appl. Surf. Sci.* **2014**, *391*, 158–166. [CrossRef]
19. Ma, J.; Jia, Y.Z.; Jing, Y.; Sun, J.H.; Yao, Y. Synthesis and photocatalytic activity of $TiO_2$-hectorite composites. *Appl. Clay Sci.* **2009**, *46*, 114–116. [CrossRef]

20. Qi, Y.; AlMukhtar, M.; Alcover, J.F.; Bergaya, F. Coupling analysis of macroscopic and microscopic behaviour in highly consolidated Na-laponite clays. *Appl. Clay Sci.* **1996**, *11*, 185–197. [CrossRef]
21. Gombos, E.D.; Krakko, D.; Zaray, G.; Illes, A.; Dobe, S.; Szegedi, A. Laponite immobilized $TiO_2$ catalysts for photocatalytic degradation of phenols. *J. Photochem. Photobiol. A Chem.* **2020**, *387*, 112045. [CrossRef]
22. An, T.C.; Chen, J.X.; Li, G.Y.; Ding, X.J.; Sheng, G.Y.; Fu, J.M.; Mai, B.X.; O'Shea, K.E. Characterization and the photocatalytic activity of $TiO_2$ immobilized hydrophobic montmorillonite photocatalysts. *Catal. Today.* **2008**, *139*, 69–76. [CrossRef]
23. Zhu, H.Y.; Li, J.Y.; Zhao, J.C.; Churchman, G.J. Photocatalysts prepared from layered clays and titanium hydrate for degradation of organic pollutants in water. *Appl. Clay Sci.* **2005**, *28*, 79–88. [CrossRef]
24. Ogawa, M.; Matsutomo, T.; Okada, T. Preparation of hectorite-like swelling silicate with controlled layer charge density. *J. Ceram. Soc. Jpn.* **2008**, *116*, 1309–1313. [CrossRef]
25. Yuan, P.; Annabi-Bergaya, F.; Tao, Q.; Fan, M.D.; Liu, Z.W.; Zhu, J.X.; He, H.P.; Chen, T.H. A combined study by XRD, FTIR, TG and HRTEM on the structure of delaminated Fe-intercalated/pillared clay. *J. Colloid Interface Sci.* **2008**, *324*, 142–149. [CrossRef]
26. Li, S.Z.; Wu, P.X.; Li, H.L.; Zhu, N.W.; Li, P.; Wu, J.H.; Wang, X.D.; Dang, Z. Synthesis and characterization of organo-montmorillonite supported iron nanoparticles. *Appl. Clay Sci.* **2010**, *50*, 330–336. [CrossRef]
27. Manjanna, J. Preparation of Fe(II)-montmorillonite by reduction of Fe(III)-montmorillonite with ascorbic acid. *Appl. Clay Sci.* **2008**, *42*, 32–38. [CrossRef]
28. López, R.; Gómez, R.; Oros-Ruiz, S. Photophysical and photocatalytic properties of $TiO_2$-Cr sol-gel prepared semiconductors. *Catal. Today.* **2011**, *166*, 159–165. [CrossRef]
29. Dai, S.X.; Wu, Y.Q.; Sakai, T.; Du, Z.L.; Sakai, H.; Abe, M. Preparation of Highly Crystalline $TiO_2$ Nanostructures by Acid-assisted Hydrothermal Treatment of Hexagonal-structured Nanocrystalline Titania/Cetyltrimethyammonium Bromide Nanoskeleton. *Nanoscale Res. Lett.* **2010**, *5*, 1829–1835. [CrossRef]
30. Pérez-Verdejo, A.; Sampieri, A.; Pfeiffer, H.; Ruiz-Reyes, M.; Santamaría, D.J.; Fetter, G. Nanoporous composites prepared by acombination of SBA-15 with Mg-Al mixed oxides. Water vapor sorption properties. *Beilstein J. Nanotechnol.* **2014**, *5*, 1226–1234. [CrossRef]
31. Chen, D.M.; Du, G.X.; Zhu, Q.; Zhou, F.S. Synthesis and characterization of $TiO_2$ pillared montmorillonites: Application for methylene blue degradation. *J. Colloid Interface Sci.* **2013**, *409*, 151–157. [CrossRef] [PubMed]
32. Ruiz-Santoyo, V.; Maranon-Ruiz, V.F.; Romero-Toledo, R.; Vargas, O.A.G.; Perez-Larios, A. Photocatalytic Degradation of Rhodamine B and Methylene Orange Using $TiO_2$-$ZrO_2$ as Nanocomposite. *Catalysts* **2021**, *11*, 1035. [CrossRef]
33. Mishra, A.; Mehta, A.; Kainth, S.; Basu, S. Effect of g-$C_3N_4$ loading on $TiO_2$/Bentonite nanocomposites for efficient heterogeneous photocatalytic degradation of industrial dye under visible light. *J. Alloys Compd.* **2018**, *764*, 406–415. [CrossRef]
34. Park, H.; Choi, W. Effects of $TiO_2$ surface fluorination on photocatalytic reactions and photoelectrochemical behaviors. *J. Phys. Chem. B* **2004**, *108*, 4086–4093. [CrossRef]
35. Wang, Q.; Rhimi, B.; Wang, H.; Wang, C. Efficient photocatalytic degradation of gaseous toluene over F-doped $TiO_2$/exfoliated bentonite. *Appl. Surf. Sci.* **2020**, *530*, 147286. [CrossRef]
36. Li, D.; Haneda, H.; Labhsetwar, N.K.; Hishita, S.; Ohashi, N. Visible-light-driven photocatalysis on fluorine-doped $TiO_2$ powders by the creation of surface oxygen vacancies. *Chem. Phys. Lett.* **2005**, *401*, 579–584. [CrossRef]
37. Kuźniarska-Biernacka, I.; Silva, A.R.; Carvalho, A.P.; Pires, J.; Freire, C. Organo-Laponites as novel mesoporous supports for manganese (III) salen catalysts. *Langmuir* **2005**, *21*, 10825–10834. [CrossRef]
38. Zhang, Y.; Xing, Z.P.; Liu, X.F.; Li, Z.Z.; Wu, X.Y.; Jiang, J.J.; Li, M.; Zhu, Q.; Zhou, W. $Ti^{3+}$ Self-Doped Blue $TiO_2$(B) Single-Crystalline Nanorods for Efficient Solar-Driven Photocatalytic Performance. *Appl. Mater. Inter.* **2016**, *8*, 26851–26859. [CrossRef]
39. Ren, R.; Wen, Z.H.; Cui, S.M.; Hou, Y.; Guo, X.R.; Chen, J.H. Controllable Synthesis and Tunable Photocatalytic Properties of $Ti^{3+}$-doped $TiO_2$. *Sci. Rep.* **2015**, *5*, 10714. [CrossRef]
40. Fajrina, N.; Tahir, M. 2D-montmorillonite-dispersed g-$C_3N_4$/$TiO_2$ 2D/0D nanocomposite for enhanced photo-induced $H_2$ evolution from glycerol-water mixture. *Appl. Surf. Sci.* **2019**, *471*, 1053–1064. [CrossRef]
41. Tang, Q.; Meng, X.F.; Wang, Z.F.; Zhou, J.W.; Tang, H. One-step electrospinning synthesis of $TiO_2$/g-$C_3N_4$ nanofibers with enhanced photocatalytic properties. *Appl. Surf. Sci.* **2018**, *430*, 253–262. [CrossRef]
42. Li, M.; Li, Y.J.; Chen, F.T.; Lin, X.; Feng, Q.J. Electrically enhanced photocatalysis for gas-phase benzaldehyde degradation by ordered mesoporous titania/conductive carbon felts. *Electrochim. Acta.* **2016**, *216*, 517–527. [CrossRef]
43. Li, C.; Sun, Z.; Zhang, W.; Yu, C.; Zheng, S. Highly efficient g-$C_3N_4$/$TiO_2$/kaolinite composite with novel three-dimensional structure and enhanced visible light responding ability towards ciprofloxacin and *S. aureus*. *Appl. Catal. B Environ.* **2018**, *220*, 272–282. [CrossRef]
44. Tang, S.P.; Fu, Z.H.; Li, Y.; Li, Y.J. Study on boron and fluorine-doped $C_3N_4$ as a solid activator for cyclohexane oxidation with $H_2O_2$ catalyzed by 8-quinolinolato iron(III) complexes under visible light irradiation. *Appl. Catal. A Gen.* **2020**, *590*, 117342. [CrossRef]
45. Xiang, H.C.; Tuo, B.Y.; Tian, J.W.; Hu, K.M.; Wang, J.L.; Cheng, J.G.; Tang, Y. Preparation and photocatalytic properties of Bi-doped $TiO_2$/montmorillonite composite. *Opt. Mater.* **2021**, *117*, 111137. [CrossRef]
46. Yuan, L.; Jiang, S.M.; Li, Z.Z.; Zhu, Y.; Yu, J.; Li, L.; Li, M.Z.; Tang, S.; Sheng, R.R. Photocatalyzed cascade Meerwein addition/cyclization of N-benzylacrylamides toward azaspirocycles. *Org. Biomol. Chem.* **2018**, *16*, 2406–2410. [CrossRef] [PubMed]

47. Cao, W.X.; Lin, L.; Qi, H.F.; He, Q.; Wu, Z.J.; Wang, A.Q.; Luo, W.H.; Zhang, T. In-situ synthesis of single-atom Ir by utilizing metal-organic frameworks: An acid-resistant catalyst for hydrogenation of levulinic acid to gamma-valerolactone. *J. Catal.* **2019**, *373*, 161–172. [CrossRef]
48. Xiong, S.F.; Yin, Z.L.; Zhou, Y.J.; Peng, X.Z.; Yan, W.B.; Liu, Z.X.; Zhang, X.Y. The Dual-frequency (20/40 kHz) Ultrasound Assisted Photocatalysis with the Active Carbon Fiber-loaded $Fe^{3+}$-$TiO_2$ as Photocatalyst for Degradation of Organic Dye. *Bull. Korean Chem. Soc.* **2013**, *34*, 3039–3045. [CrossRef]
49. Yaacob, N.; Pei, G.; Aina, N.; Nazri, M.; Fauzi, A.; Nidzhom, M.; Abidin, Z.; Naidu, M. Simultaneous oily wastewater adsorption and photodegradation by $ZrO_2$-$TiO_2$ heterojunction photocatalysts. *J. Water Process Eng.* **2021**, *39*, 101644. [CrossRef]
50. Cheng, T.T.; Gao, H.J.; Liu, G.R.; Pu, Z.S.; Wang, S.F.; Yi, Z.; Wu, X.W.; Yang, H. Preparation of core-shell heterojunction photocatalysts by coating CdS nanoparticles onto $Bi_4Ti_3O_{12}$ hierarchical microspheres and their photocatalytic removal of organic pollutants and Cr(VI) ions. *Colloids Surf. A Physicochem. Eng. Asp.* **2022**, *233*, 127918. [CrossRef]

*catalysts*

MDPI

*Article*

# Solochrome Dark Blue Azo Dye Removal by Sonophotocatalysis Using $Mn^{2+}$ Doped ZnS Quantum Dots

**Jyoti Patel [1], Ajaya K. Singh [1,*], Bhawana Jain [1], Sushma Yadav [1], Sónia A. C. Carabineiro [2] and Md. Abu Bin Hasan Susan [3]**

1 Department of Chemistry, Govt. V. Y. T. PG Autonomous College, Durg 491001, Chhattisgarh, India; jyotibhilai17@gmail.com (J.P.); bhawanajain123@gmail.com (B.J.); sushmabhilai80@gmail.com (S.Y.)
2 LAQV-REQUIMTE, Department of Chemistry, NOVA School of Science and Technology, Universidade NOVA de Lisboa, 2829-516 Caparica, Portugal; sonia.carabineiro@fct.unl.pt
3 Department of Chemistry, University of Dhaka, Dhaka 1000, Bangladesh; susan@du.ac.bd
* Correspondence: ajayaksingh_au@yahoo.co.in; Tel.: +91-940-620-7572

**Abstract:** This work investigates the degradation of the azo dye solochrome dark blue (SDB) by measurement of the photocatalytic, sonocatalytic and sonophotocatalytic activities, under low ultrasonic frequency (40 kHz) and UV-C (254 nm) light, using Mn-doped ZnS semiconductor quantum dots ($Mn^{2+}$:ZnS Qds) as catalysts, prepared by a simple chemical precipitation procedure. In order to study the different morphological and optical crystal properties, various characterization techniques were used, such as high resolution transmission electron microscopy, scanning electron microscopy, energy dispersive X-ray analysis, X-ray diffraction, $N_2$ adsorption-desorption at −196 °C and ultraviolet-visible spectroscopy. The average particle size of the semiconductor Qds was in the range of 3–4 nm. The optimal parameters affecting dye degradation, such as the catalyst loading, solution pH, time of irradiation, initial concentration of dye, dopant concentration, ultrasonic power and frequency effect were evaluated. The synthesized catalytic material exhibited a high activity for sonophotocatalytic degradation of SDB (89%), larger than that observed for sonocatalysis (69.7%) or photocatalysis (55.2%) alone, which was due to the improved electron-holes separation, formation of more reactive radicals and enhancement of the active surface area. Qds showed good stability and reusability after five repeated cycles. Finally, the degradation products were identified by liquid chromatography-mass spectrometry (LC-MS).

**Keywords:** semiconductors; quantum dots; doping; optical properties; sonocatalysis; photocatalysis

**Citation:** Patel, J.; Singh, A.K.; Jain, B.; Yadav, S.; Carabineiro, S.A.C.; Susan, M.A.B.H. Solochrome Dark Blue Azo Dye Removal by Sonophotocatalysis Using $Mn^{2+}$ Doped ZnS Quantum Dots. *Catalysts* **2021**, *11*, 1025. https://doi.org/10.3390/catal11091025

Academic Editors: Ioan Balint and Monica Pavel

Received: 27 July 2021
Accepted: 20 August 2021
Published: 24 August 2021



## 1. Introduction

From the viewpoint of environmental sustainability, textile processing and production release a huge quantity of polluted wastewaters, which enter the environment with unknown effects on health. Those wastewaters contain various synthetic dyes used in coloring of textile fabrics [1]. These textile dyes are a matter of concern, due to their high stability and low degradation capacity in wastewater treatment plants. A variety of techniques can be used for the degradation of such dyes [2].

Solochrome dark blue (SDB) is an essential azo dye used for dyeing nylon, wool, silk and other fibers. Solochrome dark blue (SDB) is mainly used as an indicator in complexometric titrations for the determination of total hardness of water due to elements such as calcium, zinc, magnesium, and to a lesser extent for other metal ions, including manganese. These metal ions readily undergo oxidation in alkaline media to form products of uncertain stoichiometry. It is a hazardous dye, and its degradation intermediates may be carcinogenic. Therefore, it is highly desirable to develop an effective method for removal of such dye pollutants from wastewater effluents, even at trace levels.

In recent years, quantum dots (Qds) gained substantial interest due to their exceptional properties [3]. Their dimension is comparable to the excitonic Bohr radius and they are well-known for their tunable size, broad absorption and narrow emission spectra, high photo-stability and strong signal intensity [4]. In recent years, much interest was shown towards semiconductor-based photocatalytic degradation of hazardous organic pollutants [5,6]. To enlarge the light absorption of these nanomaterials, modifications were made using semiconductor coupling, dye sensitization, impurity doping, and metal deposition, using coordination metal complexes, etc. [7,8]. Efforts to improve the efficiency of photocatalysis by combination with other methods were also made [9]. Among these, sonochemical oxidation was successfully applied for the degradation of several recalcitrant pollutants in water treatment methodologies [10,11]. By introducing ultrasonic waves of defined frequencies (20–1000 kHz), bubbles can be generated, followed by subsequent collapse action [12], generating very high temperatures up to ~4700 °C inside the core of bubbles [13,14], producing free radicals, such as $H^•$ and $HO^•$, due to homolytic cleavage of $H_2O$. The generated free radicals and the pyrolysis in cavitation bubbles lead to the degradation of organic compounds into a range of short-chain by-products.

In sonophotocatalysis, a photocatalyst is used, together with ultrasonic waves and UV irradiation. It is considered a useful method for enhancing the efficiency of photocatalysis, as hydrophilic components are degraded by photocatalysis and hydrophobic components by sonolysis [15]. It is also considered as a more effective procedure than photocatalysis or sonolysis alone [16–20]. Due to the extended processes taking place during the adsorption of pollutants on the catalyst surface, the photocatalytic efficiency is reduced, thus blocking the active sites. The use of ultrasound avoids the build-up of contaminants and intermediates formed during degradation, by cleaning the catalyst surface and generating effective radicals for degradation. Moreover, there is an improvement in the mass transportation of contaminants over the surface of the catalyst [21] and the $H_2O_2$ sonochemically produced is more stable, at low pollutant levels [22], being cleaved to $HO^•$ radicals during photolysis.

A wide variety of semiconductor materials can act as photocatalysts [23–25]. Among these, zinc sulfide (ZnS) Qds, having a low dimension, i.e., 1–10 nm, are well known wide band gap II-VI semiconductor materials, extensively used in photocatalytic decomposition of organic dyes and water purification, due to the high photocatalytic activity, high photochemical stability, non-toxicity and low cost [26]. Moreover, zinc based Qds have no toxic elements, show higher surface area than their bulk counterparts, have wider gap energy, and are excellent hosts for the doping of a huge variety of metals [27].

Doping or using intentional impurity atoms in Qds gives rise to further discrete energy levels in the intrinsic dots controlling the behavior of materials and enhancing the energy dynamics of excitons. It is an efficient technique to tune the energy levels and surface states, in addition to tuning optical, structural, electrical, and the magnetic behavior of the semiconducting nanocrystals [28–30]. This leads to extensive applications of Qds to light emitting devices [31–33], spintronics [34,35], solar cells [36,37], bioimaging [38–40] and sensing [41,42]. Doping allows an efficient transfer of energy from absorbed photons to the impurity, rapidly confining the excitation by restraining unwanted reactions at the crystal surface [43]. ZnS incorporated with transition elements such as chromium, manganese, iron, nickel and copper show a positive impact on their structural, magnetic and optical property [44]. ZnS nanoparticles doped with nickel, manganese, cobalt [45–47], copper [45] and iron [47–49], can be successfully prepared by simple and efficient methods, even at room temperature.

For Qds to become useful for clinical purposes, it is essential to obtain them without toxic elements. Thus, it is necessary to reduce the cytotoxicity of Qds [50,51]. Pyridine is a very good capping agent with significant photoluminescence properties. In the present work, we used nicotinic acid, which belongs to the group of the pyridine carboxylic acids, as a capping agent. It is an organic compound and a form of vitamin $B_3$. These Qds offer a good candidate system for evading toxic elements in traditional ones. Thus, good quality and noxious-element free aqueous Qds will generate materials able to be used for in vivo bio-applications.

The aim of this study was to prepare $Mn^{2+}$:ZnS Qds and to explore their sonophotocatalytic activity in SDB degradation. The goal was to improve the efficiency of the ultrasonic-based process for low-cost degradation of organic pollutants. This combination of ultrasound with a $Mn^{2+}$:ZnS Qds photocatalyst was rarely investigated. In fact, this work is the first report on sonophotocatalysis of $Mn^{2+}$:ZnS Qds for SDB dye removal from wastewater. The results show that the photocatalytic activity of Qds remarkably improved with the use of ultrasound, compared to another study dealing with sonophotocatalysis, as well as the other conventional methods for the removal of SDB molecules from an aqueous solution [52–57].

## 2. Results and Discussion

### *2.1. Catalyst Characterization*

The pure and doped Qds were characterized by different techniques. To study the sample morphology, detailed imaging investigations were carried out. Figure 1a–c show TEM images of $Mn^{2+}$:ZnS Qds. Figure 1c shows some particle sizes of the Qds sample. The micrographs reveal monodispersion of particles, with an average size of 3–4 nm (Figure 1d). Larger aggregates of small particles could also be identified.

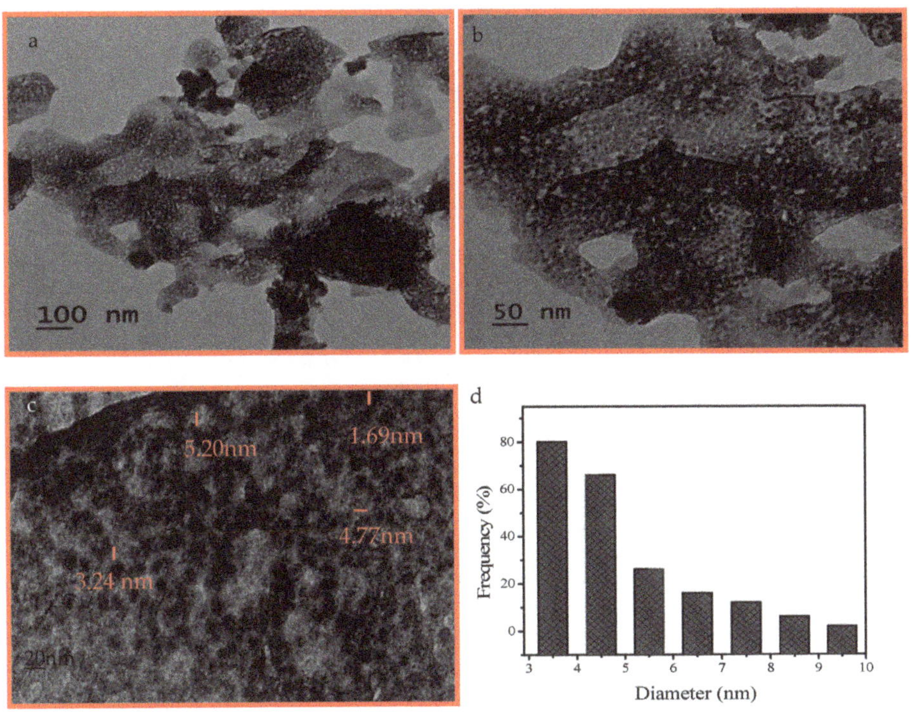


**Figure 1.** (**a–c**) TEM images, (**d**) distribution of particle sizes $Mn^{2+}$:ZnS Qds.

The morphology of the Qds was also studied by SEM, as shown in Figure 2. The images illustrate the formation of agglomerated nanoparticles with a smooth surface. The morphology shows that aggregates of Qds are formed by primary units of varied orientations. Arbitrary aggregations among the small particles occur, directing the development of disordered crystallites [58]. The micrographs (Figure 2a) at lower magnification illustrate inhomogeneities about the size distribution of the crystallites. However, an image at higher magnification (Figure 2b) shows that the smaller sized particles agglomerate, originating larger ones.

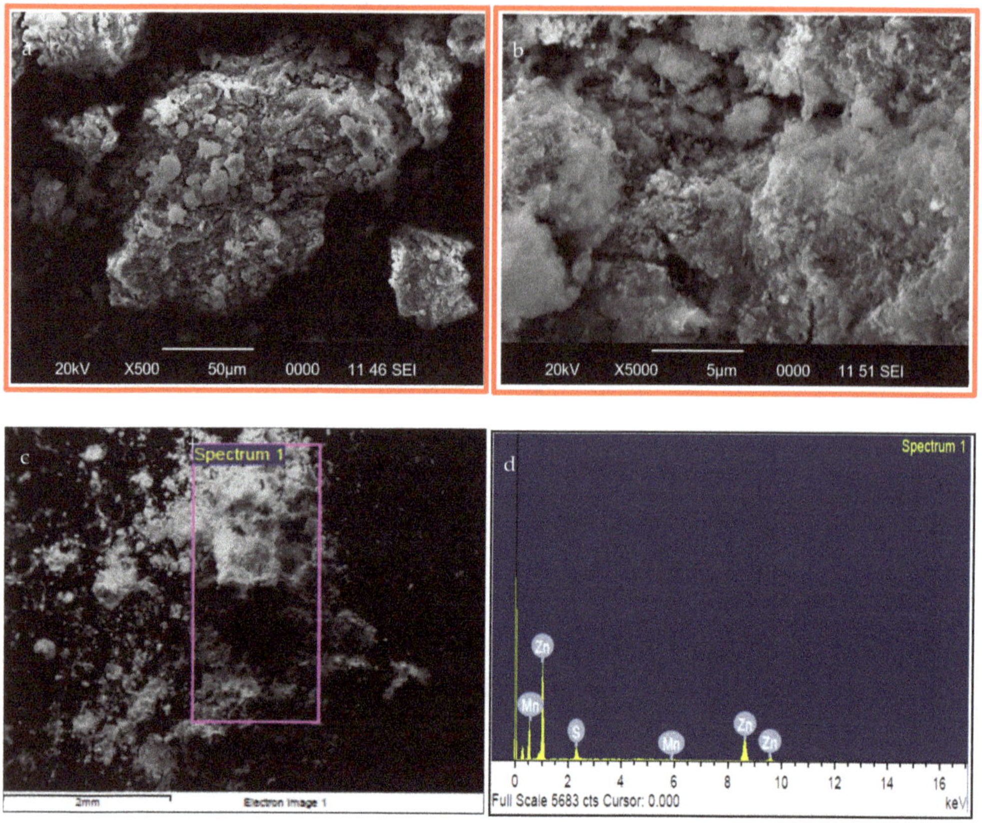


**Figure 2.** (**a**–**c**) SEM images of $Mn^{2+}$:ZnS Qds; (**d**) EDX spectra of the marked area (spectrum 1).

The elemental composition was studied by energy dispersive X-ray analysis (EDX). Figure 2c shows the synthesized Qds and Figure 2d displays the respective EDX image of the marked area (spectrum 1). The percentages of various elements present in the material are 91.19 wt.% Zn, 8.05 wt.% S and 0.77 wt.% Mn; molar: 70.78% Zn, 28.37% S and 0.85% Mn.

A small amount of S was used for the synthesis. It is known that the presence of large amounts of S during synthesis causes accumulation of S atoms at the Qds surface. These S atoms are easily oxidized and may cause quenching of luminescence intensity [29,59]. Nevertheless, the expected molar ratio Zn:S in the catalyst should be 1:1, but the amount of Zn found is larger. It is possible that, along with ZnS, $Zn(OH)_2$ is also formed, in alkaline conditions, as also shown by other authors [60]. We also used a small amount of $Mn^{2+}$, as we found that for higher concentrations a reduction in the luminescence intensity was observed. This phenomenon is ascribed to the fact that a larger $Mn^{2+}$ concentration causes a larger number of $Mn^{2+}$ emission centers per nanocrystal, and the interaction between the $Mn^{2+}$-$Mn^{2+}$ pairs intensifies the non-radiative decay of the $Mn^{2+}$ excited state, thus causing a decrease in $Mn^{2+}$ emission intensity [59]. The EDX image (Figure 2d) shows the $Mn^{2+}$ doped ZnS crystals. A theoretical value of 1% Mn was expected, but a smaller amount was found by EDX. It is possible that the prepared sample was not totally homogeneous and the small part analyzed contained a little less Mn than the bulk. Additionally, the prepared sample was washed with distilled water, which could also lead to some loss of Mn ions.

Ultraviolet-visible (UV-Vis) spectroscopy was used to evaluate the usefulness of the capping agent in stabilizing the growth of the Qds in aqueous solution. The UV-Vis spectra and curve of the band gap of ZnS and $Mn^{2+}$:ZnS Qds are shown in Figure 3a,b.

As found in Figure 3a, the absorption shoulders of the Qds are placed between 290 to 310 nm, while for bulk ZnS, a band is noted near 350 nm, having a band gap energy ($E_g$) value of 3.6 eV [61–63]. The blue shift in the absorption edge with doping is due to the decrease in the particle size. The absorption spectra are steadily blue shifted with increasing Mn content; this shift in the absorption edge can be attributed to the reduction of particle size through the doping process, which is due to the quantum confinement of the excitons, ensuing a more discrete energy spectrum of the discrete nanoparticles. Figure 3b shows the band gap energy curve of pure ZnS and $Mn^{2+}$:ZnS QDs calculated from the Tauc's relation [63], as: $\alpha h\upsilon = \alpha_o (h\upsilon - E_g)^{1/2}$; where $h\upsilon$ is photon energy, $E_g$ is optical bandgap of the nanoparticles, and $\alpha_o$, a constant. The $E_g$ values for pure and $Mn^{2+}$:ZnS Qds samples were found between 3.87 and 3.96 eV. The increment in the $E_g$ value along with the shifting of the curve may be due to the decrease in the nanoparticle size, attributed to the quantum size confinement effect [61,64]. Thus, transformations in the optical and electronic properties of nanoparticles occur when their dimensions are reduced below threshold values.

The shift in the band gap can be explained by the effective mass approximation model with the particle in a box approach. Based on the first order approximation of Brus equation, the relationship between the particle radius ($r$) and band gap ($E_g$) in ZnS nanocrystal [59,65–67] is given by:

$$r(E_g) = \frac{0.32 - 2.9\sqrt{E_g - 3.49}}{2(3.50 - E_g)} \tag{1}$$

Using Equation (1), the size of the pure and doped ZnS nanocrystal (considering $E_g$ = 3.87–3.96 eV) was calculated to be 3.62–3.96 nm, which is in agreement with the size obtained from TEM measurements.

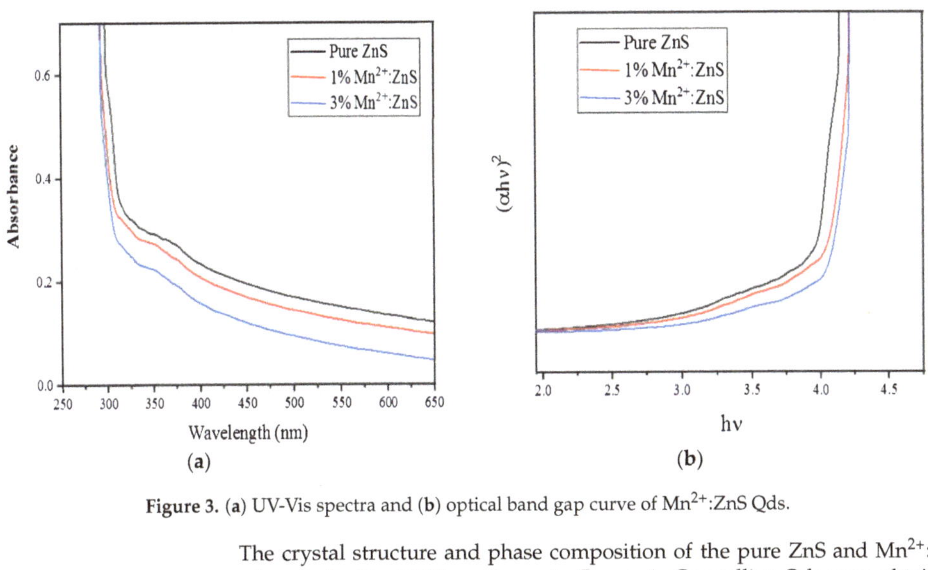


**Figure 3.** (**a**) UV-Vis spectra and (**b**) optical band gap curve of $Mn^{2+}$:ZnS Qds.

The crystal structure and phase composition of the pure ZnS and $Mn^{2+}$:ZnS Qds were also studied by XRD, as shown in Figure 4. Crystalline Qds were obtained. The peaks at 29.7°, 48.4° and 56.16° for $Mn^{2+}$:ZnS Qds correspond to the (111), (220), and (311) crystallographic planes of cubic crystalline ZnS, respectively [27,68–70]. The dopants are well inserted in the ZnS structure and $Mn^{2+}$ ions did not alter the phase [27,70,71]. A slight shift for the peak corresponding to the 220 plane is observed, which indicates the inclusion of Mn in the crystal lattice. The average size was calculated by the Debye Scherrer equation using the full width at half maximum of the XRD peaks. The average crystallite size for pure ZnS and $Mn^{2+}$:ZnS Qds were found to be 1.32 and 1.13 nm, respectively.

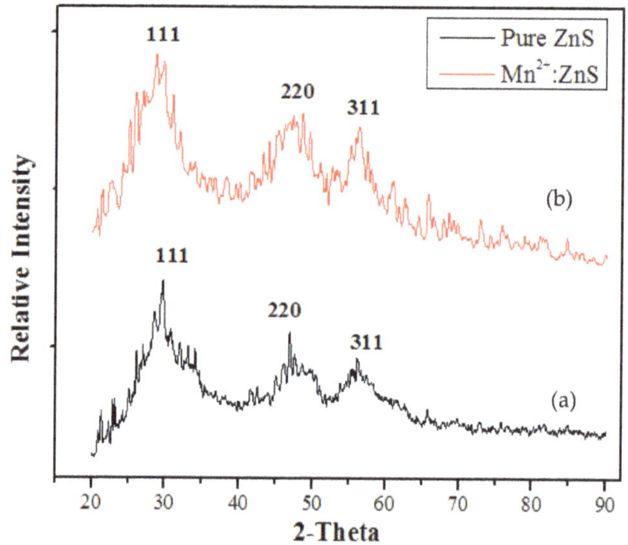


**Figure 4.** XRD patterns of (**a**) pure ZnS Qds and (**b**) $Mn^{2+}$:ZnS Qds.

Figure 5 shows the FTIR scan of $Mn^{2+}$:ZnS Qds. Several features can be found in the range of 500–4000 $cm^{-1}$, characteristic of various functional groups [5,71]. The absorption peaks above 3000 $cm^{-1}$ are due to the Ar-H or =C-H stretching. The bands between 1600–1200 $cm^{-1}$ are ascribed to the ring stretching vibrations [5,72]. The broad band at 3335.4 $cm^{-1}$ is attributed to the O-H stretching, whereas the band at 930.4 $cm^{-1}$ is due to the O-H out of plane bending. The peaks between 1630–1540 $cm^{-1}$ and around 1410 $cm^{-1}$ are due to the C=C stretching, while the peaks between 1330–1240 $cm^{-1}$ are ascribed to C=N stretching. This may be attributed to the coordinate bond formed between $Zn^{2+}$ ions and the N atoms of the pyridine moiety in nicotinic acid [73,74]. Thus, the FTIR study strongly supports the formation of nicotinic acid capped $Mn^{2+}$:ZnS Qds. The band at 1036 $cm^{-1}$ is attributed to the Zn-OH vibrations [60]. As explained above, the formation of $Zn(OH)_2$ is possible under alkaline conditions, as reported by other authors [60].

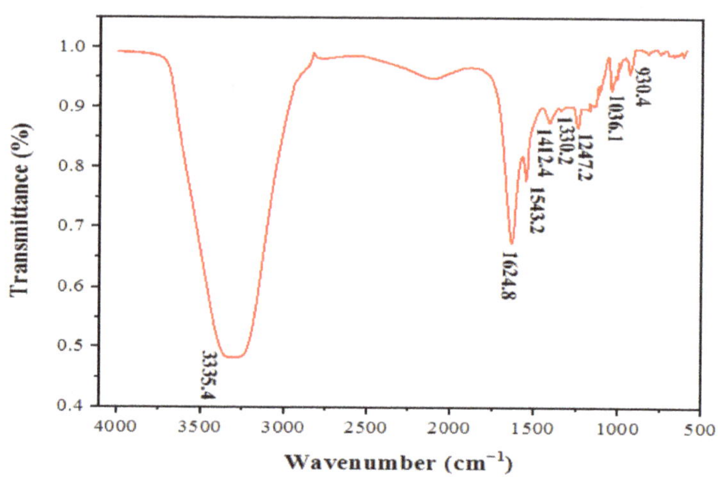


**Figure 5.** FTIR of $Mn^{2+}$:ZnS Qds.

For determining the specific surface area and pore size distribution of solid and porous materials, gas adsorption is a prevailing analysis technique. BJH calculation is a pore size distribution determination method, typically applied to $N_2$ adsorption data. Evaluation of the adsorption and desorption isotherm branches reveals information about the pore volume and pores size distribution. BET is mainly used for the surface area analysis of the prepared nanomaterials. The surface area can be calculated from the quantity of gas required to form a monolayer. In order to confirm the porous structure of Qds, $N_2$ adsorption–desorption experiments were performed at −196 °C. Figure S1a shows a type IV adsorption–desorption isotherm with a hysteresis loop, while Figure S1b displays distribution of the pore size. The pores are likely to show a very narrow slit or bottle shaped configuration or a distribution of randomly-shaped micro and mesopores. The Qds surface area value was 10 $m^2/g$, determined using the BET equation. The pores have an average diameter of 1.5 nm (determined by the BJH adsorption procedure), whereas the total pore volume was 0.01 $cm^3/g$.

Thermal analysis was carried out for studying the disintegration, strength and temperatures of phase development of the nanoparticles. To determine the thermal activity of the prepared materials, differential thermal analysis (DTA) and thermogravimetric analysis (TGA) were performed under $N_2$ atmosphere, as shown in Figure 6a,b. The data were recorded with a heating rate of 10 °C/min, up to 650 °C. Owing to the loss of absorbed water, the endothermic peak occurred near 185 °C. Different steps of weight loss are noted in the TGA plot. The initial weight loss shown by the sample up to 230 °C (12.5%) is connected to the water molecules physically adsorbed at the surface. The next peak, located at 240–370 °C (10.35%), is correlated to the disintegration of organic moieties from the precursors. The next loss of weight (11.43%) is due to the discharge of Mn ions from the sample. Further decrease (5.72%) is attributed to the loss of the S ions, as also reported by other authors [75–77]. Figure 6a indicates that the thermal study started at about 2.8 mg sample weight, which after heating to 600 °C remained at a weight of 1.96 mg. Thus, the sample loses 40% of its weight up to 700 °C. Figure 6b shows an exothermic progression up to about 460 °C, which may possibly be attributed to a shift in phase or crystallinity of the sample.

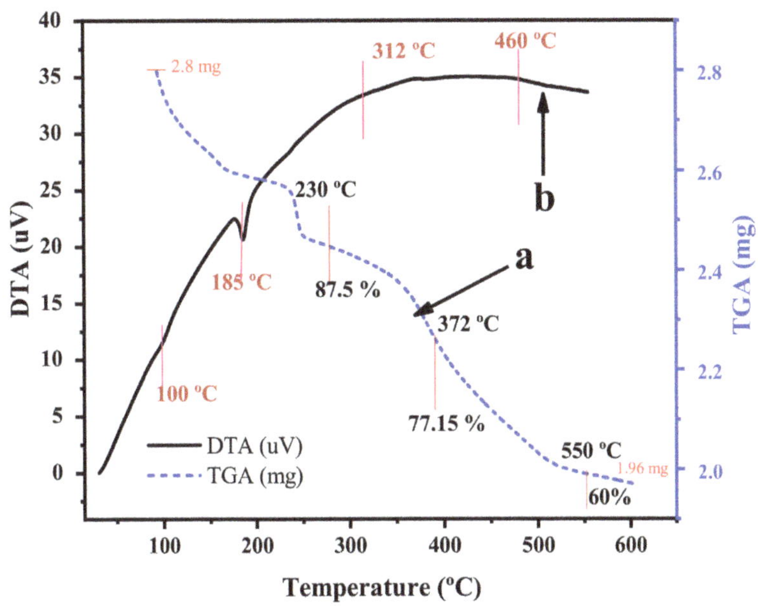


**Figure 6.** (**a**) Thermogravimetric analysis (TGA) and (**b**) differential thermal analysis (DTA) of $Mn^{2+}$:ZnS Qds.

The fluorescence spectra of pure ZnS QDs and Mn:ZnS QDs (doped with various concentrations) were measured, using 320 nm as the excitation wavelength optimal for the ZnS QDs [59]. The integrated fluorescence intensities of the emission peak versus absorbances are plotted in Figure 7. With the increase in $Mn^{2+}$ concentration from 0 to 3%, the florescence intensity of $Mn^{2+}$ ($^4T_1$-$^6A_1$) is found to increase steadily.

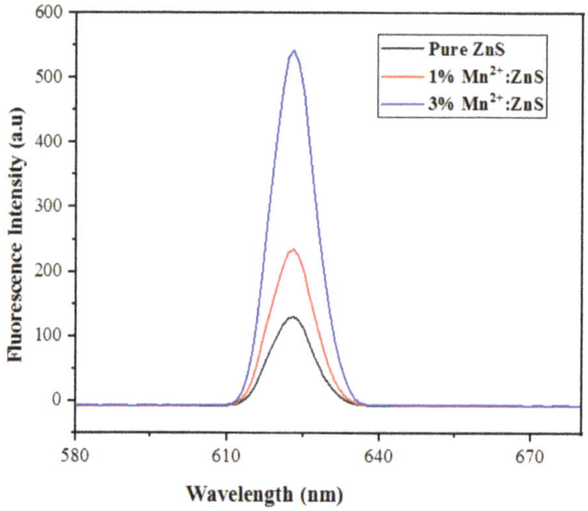


**Figure 7.** Fluorescence spectra of $Mn^{2+}$:ZnS QDs with various percentages of $Mn^{2+}$ ions ($\lambda_{ex}$ = 320 nm).

### *2.2. Efficiencies of Various Methods for SDB Degradation*

Various types of nanostructures [46,47,78–85], carbon nanotubes [54], nanocubes [55], semiconductors [56], microorganisms [86,87], electric discharge methods [88], as well as conventional sorbents [57,89–91] have been utilized for the removal of SDB dye molecules from aqueous solutions. These findings are summarized in Table 1, together with our data. When comparing the contact time for degradation, and the initial concentration of dye and dosage of catalyst, the results indicate that the results obtained with $Mn^{2+}$:ZnS Qds and sonophotocatalysis are better than those reported in other works for different nanostructures and adsorbents, as shown in Table 1.

**Table 1.** Comparison of the results obtained in the present work with literature studies of other researchers for the removal of Solochrome dark blue dye.

| Catalyst | % Degradation | Dye Conc., Catalyst Dose for Volume of Dye Solution | pH | Time | Ref. |
|---|---|---|---|---|---|
| Ti Grafted SBA-15 | 88.7% | Conc: 50 mg/L<br>Catalyst: 300 mg in 300 mL | - | 70 min | [52] |
| $TiO_2$ Nanoparticles | Anatase 65%<br>Rutile 55.65% | Conc: 100 mg/L<br>Catalyst: 1000 mg | - | 150 min | [53] |
| Iron Oxide Nanoparticles | 67% | Conc: 200 mg/L<br>Catalyst: 2000 mg/L | 5.0 | 20 min | [78] |
| Anatase $TiO_2$ Nanoparticles | 82% | Conc: 25 mg/L<br>Catalyst: 25 mg in 100 mL | - | 90 min | [79] |
| Undoped and Zn Doped Nano $TiO_2$ Nanoparticles | Undoped 88.73%<br>Doped 99.4% | Catalyst: 100 mg in 100 mL | - | 6 h | [80] |
| Nanostructured $Ho_2O_3$ | 80% | Conc: 2 mg<br>Catalyst: 40 mg | - | 100 min | [81] |

**Table 1.** *Cont.*

| Catalyst | % Degradation | Dye Conc., Catalyst Dose for Volume of Dye Solution | pH | Time | Ref. |
|---|---|---|---|---|---|
| Nanostructured $Nd_2O_3$ | 79% | Conc: 2 mg<br>Catalyst: 80 mg | - | 100 min | [82] |
| ZnO Nanoparticles | 90% | Conc: 0.08 mM<br>Catalyst: 1000 mg/L in 100 mL | 3 | 90 min | [83] |
| $TiO_2$ Nanoparticles | 92% | Conc: 50 mg/L<br>Catalyst: 250 mg/L | 5 | - | [84] |
| $TiO_2$ Nanoparticles | 100% | Conc: $10^{-4}$ M<br>Catalyst: 300 mg in 100 mL | 7 | 240 min | [85] |
| **MWCNT/Nd, N, Se$TiO_2$,<br>SWCNT/ Nd, N, Se$TiO_2$** | 49.0%,<br>54.1% | Conc: 20 ppm<br>Catalyst:100 mg in 100 mL | - | 4 h | [54] |
| Potassium Zinc Hexacyano-ferratenanocubes | 76.13% | Conc: 5 mg/L,<br>Catalyst: 15 mg in 10 mL | 7 | 120 min | [55] |
| Ammonium Phosphomolybdate Semiconductor | 93.9% | Conc: $10^{-3}$ M<br>Catalyst: 300 mg in 50 mL | 11 | 240 min | [56] |
| Bacillus Lentus Bacteria SG-7 Strain | 98% | Conc: 100 mg/L | 7 | 30 h | [86] |
| Microalga Coellastrella | 85% | Conc: 10 mg/L<br>Catalyst: 10% inoculum | 8 | 20 days | [87] |
| Non-Thermal Plasma Generated by Electric Discharges | 88.4% | - | - | 60 min | [88] |
| ZnO Nanoparticles | 75, 83, and 88% at pH 4, 8, 11 | Conc: 20 mg/L<br>Catalyst: 50 mg in 100 mL | 4, 8, 11 | 180 min | [57] |
| Adsorption on Eucalyptus Bark | 77.33% | Conc: 250 mg/L<br>Catalyst: 125 mg in 100 mL | 8 | 120 min | [89] |
| Sludge-Derived Activated Carbon | 90.66% | Conc: 500 mg/L,<br>Catalyst: 50 mL in 3000 mg/L | 6 | 720 min | [90] |
| Perlite Modified with Orthophenanthroline | 63.74% | Conc: $2 \times 10^{-4}$ M<br>Catalyst: 50 mg in 20 mL | 5 | 20 min | [91] |
| $Mn^{2+}$:ZnS Qds | 88% | Conc: 70 mg/L<br>Catalyst: 40 mg in 15 mL | 6 | 75 min | This Work |

The efficiency of the Qds catalyst, UV irradiation and ultrasonication in the SDB degradation process was studied by performing some preliminary tests. A very small dye removal rate was observed with photolysis (4%), sonolysis (5%) and ultrasound with UV (7%) without any catalysts after 75 min as shown in Figure S2. For the sonophotocatalytic experiments, the Qds solution was stirred in the dark for 20 min, to reach the adsorption–desorption equilibrium in the presence of catalysts with different Mn doping concentrations.

With UV light and Qds catalyst, the degradation was higher (47%) compared to the sole photolysis. An improvement in the photodegradation rate was found up to 3% doping with Mn and a maximum degradation of 55.2% was accomplished within 75 min of irradiation. As shown in Figure 8, the photocatalytic activity of $Mn^{2+}$:ZnS Qds was larger than that of pure ZnS. The photocatalytic absorption under UV light increased from 0 to 3% doping as the energy-gap of Qds increased due to a decrease of the particle sizes (from 1.32 to 1.13 nm, for pure ZnS and $Mn^{2+}$:ZnS Qds, respectively, as confirmed by XRD).

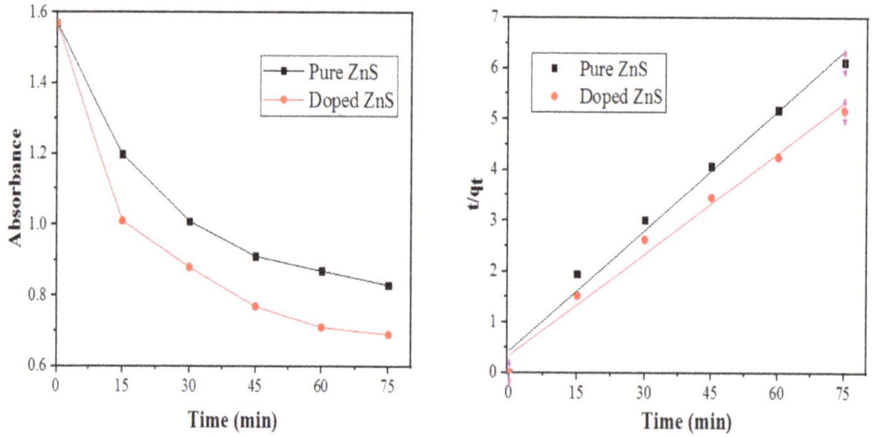


(a) (b)

**Figure 8.** (**a**) SDB absorbance changes at the maximum wavelength ($\lambda_{max}$ = 560 nm) and (**b**) the kinetic model for the photodegradation in the presence of pure and $Mn^{2+}$:ZnS Qds under optimum conditions (15 mL of 70 ppm SDB, pH 6, 75 min irradiation, 40 mg Qds).

During sonocatalytic degradation of pollutants, the ultrasonic frequency is a significant factor which greatly influences size and duration of cavitation bubbles, affecting the production of $HO^{\bullet}$ [92]. Sonication produces a kind of turbulence causing a mass transfer from the solution to the surface of the catalyst [93].

The SDB degradation in aqueous solutions was investigated at two different frequencies (40 and 80 kHz). Figure 9 shows the decline in SDB concentration with irradiation time. The degradation efficiency is directly related to the $H_2O_2$ generation, depending on the applied ultrasound frequencies [94]. From Figure 9, it was found that the differences in the degradation rates at two different frequencies were minor; still, the decrease in SDB concentration was larger at 40 kHz (62.9%) compared to 80 kHz (53%) due to the higher production of $H_2O_2$, which caused more $HO^{\bullet}$ generation, leading to more degradation at this frequency [95]. Thus, 40 kHz was chosen for further study.

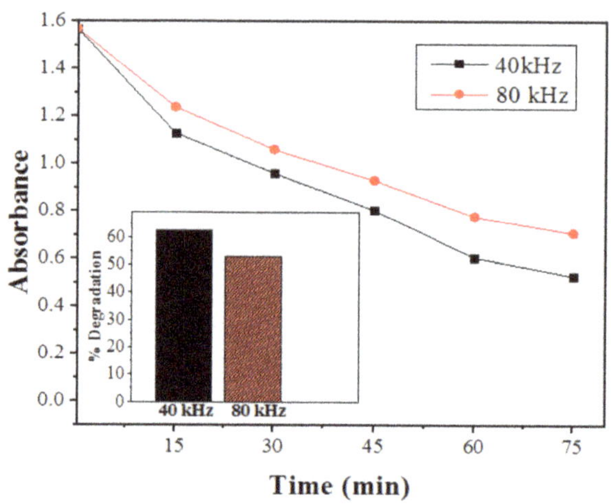


**Figure 9.** Effect of ultrasound frequency on degradation of SDB at optimum conditions (15 mL of 70 ppm SDB, pH 6, 75 min irradiation, 40 mg Qds).

With ultrasound and catalyst, the degradation obtained was 62.3%, which was higher, compared to sonolysis. The SDB removal rate increased with Mn concentration, accomplishing maximum degradation of 69.7%, after 75 min of ultrasonication. Figure 10a,b show the higher sonocatalytic removal rate of $Mn^{2+}$:ZnS Qds compared to that of pure ZnS.

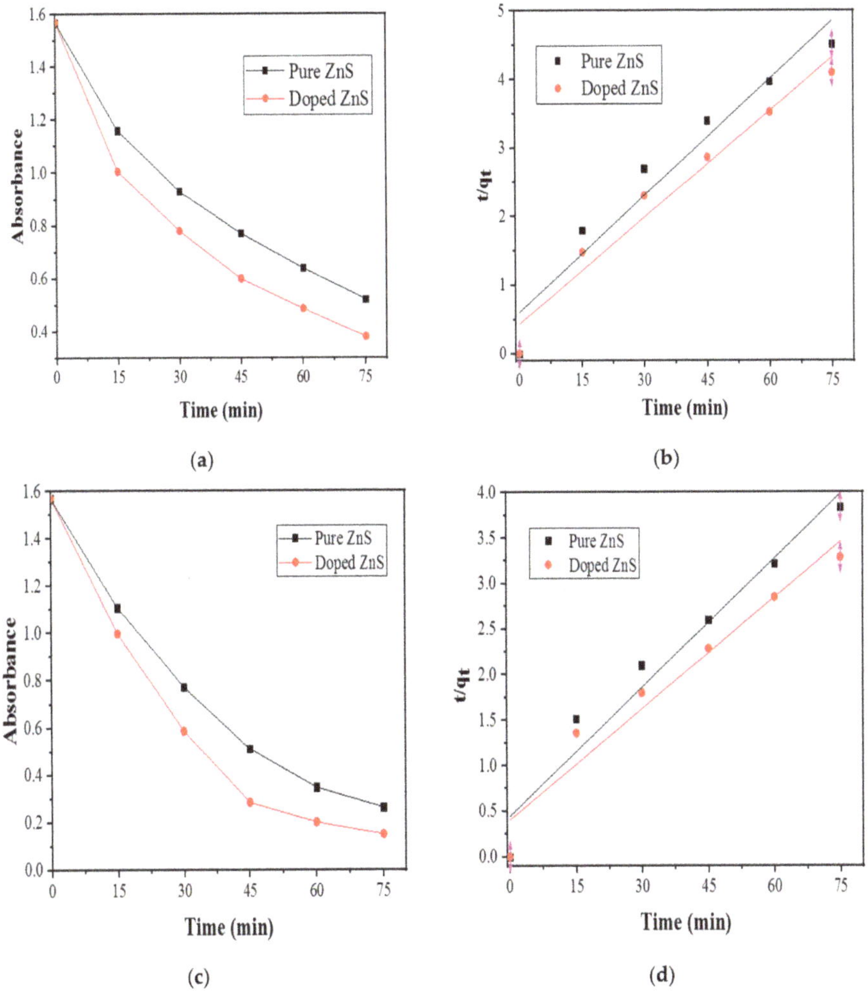


**Figure 10.** (**a**) SDB absorbance changes at maximum wavelength ($\lambda_{max}$ = 560 nm), (**b**) kinetic model for the ultrasound assisted degradation in the presence of pure and $Mn^{2+}$:ZnS Qds, under optimum conditions (15 mL 70 ppm of SDB, pH 6, 75 min ultrasound irradiation, 40 mg Qds, 40 kHz), (**c**) SDB absorbance changes at absorption maximum ($\lambda_{max}$ = 560 nm) and (**d**) kinetic model for the ultrasound and UV light assisted degradation in the presence of pure and $Mn^{2+}$:ZnS Qds at optimum conditions (15 mL of 70 ppm SDB, pH 6, 75 min ultrasound with UV light irradiation, 40 mg Qds, 40 kHz).

To enhance the degradation efficiency of the system, UV radiation assisted photocatalysis was coupled with ultrasound; with the combined process, sonophotocatalysis significantly enhanced the degradation rate of the system. The UV light source enhanced the process of photoexcitation, consequently after 60 min ultrasonicating, 80% degradation occurred in the presence of UV light. On further irradiation, 89% degradation was achieved after 75 min, as shown in Figure 10c,d. Thus, the sonophotocatalysis experiments established the semiconducting $Mn^{2+}$:ZnS Qds to be a prominent sonophotocatalyst for the degradation of SDB through the combined effect of ultrasound and UV light. Therefore, it was used for further studies.

#### 2.2.1. Kinetic Study

Kinetics can provide information about the efficiency and mechanism of a photocatalytic process. These experiments were performed at optimum experimental conditions for the SDB dye solution. The pseudo second order kinetic model given in Equation (2) [27,61] was relevant, applying a linear fitting of $q_t$ versus $t$, where $k$ is the rate constant (g mg$^{-1}$ min$^{-1}$), and $q_e$ and $q_t$ are the equilibrium adsorption capacity and adsorption capacity at time $t$, respectively [96].

$$t/q_t = 1/\left(k\, q_e^{\,2}\right) + t/q_e \tag{2}$$

The SDB degradation rates under sonophotocatalysis, sonocatalysis and photocatalysis were calculated from the slopes of Figure 11a plots. Figure 11b shows UV–Vis absorption spectra of the SDB suspension under sonophotocatalytic conditions, with $Mn^{2+}$:ZnS Qds nanocatalyst, for 75 min. As the time for sonophotocatalysis progresses, the absorption band steadily decreased, showing the decomposition of the SDB chromophoric structure. The values of $R^2$ indicate that pseudo second order model fits the experimental data. The values for the kinetic data, rate constants ($k$), correlation coefficients ($R^2$) and decolorization efficiency (DE) for pure and $Mn^{2+}$:ZnS Qds under different conditions of sono and/or photocatalysis are depicted in Table S1.

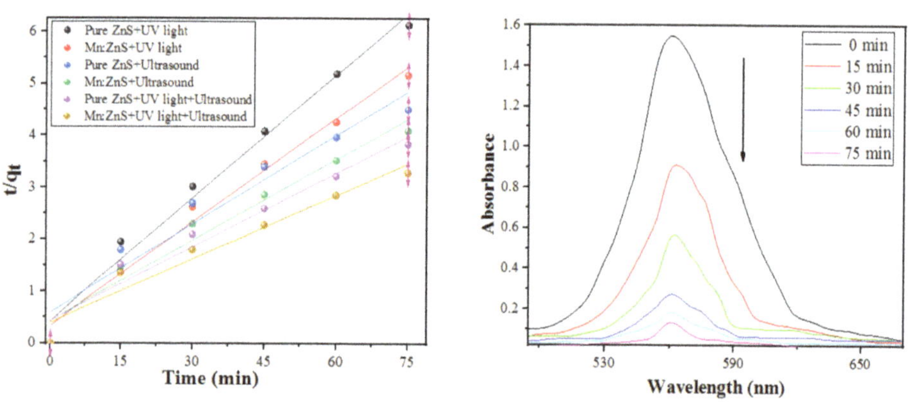


(a) (b)

**Figure 11.** (**a**) Second order kinetic model fitting to the Qds-based degradation of SDB data with pure and $Mn^{2+}$:ZnS Qds and (**b**) absorption spectra of the SDB aqueous solution during sonophotocatalysis in the presence of $Mn^{2+}$:ZnS Qds at optimum conditions (15 mL of SDB 70 ppm, pH 6, 75 min irradiation, 40 mg Qds).

#### 2.2.2. Effect of the Initial SDB Concentration

The influence of the initial SDB concentration was investigated with four different concentrations (30, 50, 70 and 90 ppm) as shown in Figure 12. The amount of dye degraded diminishes with an increase in SDB concentration, displaying a higher degradation rate for a low initial concentration. As the initial SDB concentration increases, the degradation rate also decreases, as more SDB molecules are adsorbed on the surface of Qds, but degradation is not successful. Since intensity and catalyst dose are constant, higher SDB concentration results in a smaller light penetration depth. Additionally, a maximum number of active sites may be available at that concentration. On further increasing the concentration, the active sites present get blocked-up, thus the dye degradation rate decreases [27,31].

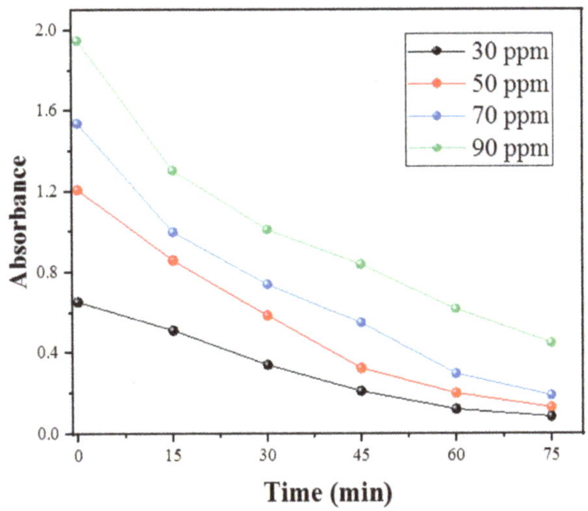


**Figure 12.** Effect of variation of the initial concentration of SDB on the sonophotocatalytic degradation under optimum conditions (15 mL of SDB, pH 6, 75 min irradiation, 40 mg Qds).

### 2.2.3. Effect of Catalyst Loading

The reactions were carried out by varying the quantity of catalyst (ranging from 10 to 50 mgs). The decomposition of SDB increased from 68% to 89% with the increase in the quantity of the nanocatalysts from 10 to 40 mg/15 mL and dropped to 87% for 50 mg catalyst loading. The initial enhancement in SDB decolorization might be due to an excess availability of active sites, as the amount of catalyst increases [97]. However, a further increase in the catalyst amount leads to an increase in the turbidity and reduces the light penetration depth, thus, the quantity of the photoactive suspension decreases [98,99]. Figure 13 shows that the nanocatalysts amount of 40 mg/15 mL is optimum for maximal SDB removal. Therefore, the remaining experiments were carried out with 40 mg of catalyst loading.

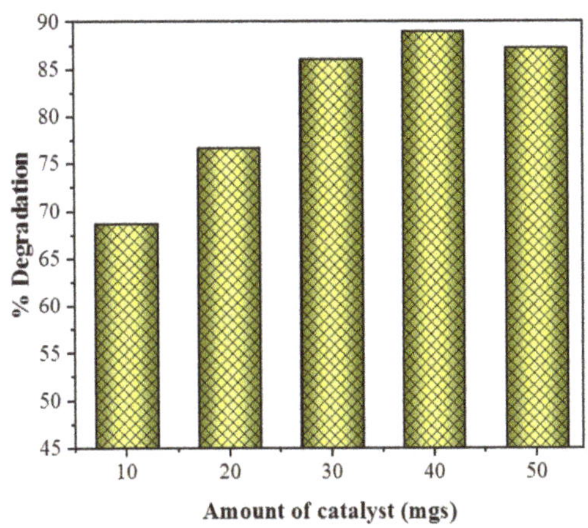


**Figure 13.** Effect of variation of the amount of $Mn^{2+}$:ZnS Qds catalyst on the sonophotocatalytic degradation of SDB under optimum conditions (15 mL of SDB 70 ppm, pH 6, 75 min irradiation, 40 kHz).

2.2.4. Effect of Initial pH

The pH of the solution is one of the significant parameters that affect the sonochemical and photocatalytic oxidation of various organic compounds. To study the pH effect on the SDB degradation, the experiments were done at various pH values (2.0–12.0), with 15 mL of 70 ppm initial SDB concentration and 40 mg Qds, for 75 min.

The sonocatalytic and photocatalytic degradation of SDB was higher at pH 6 (Figure 14) than at higher values of pH, as also observed in the previous results obtained with SDB [55,78]. The sonophotocatalytic degradation improved with pH increase, but maximum adsorption and degradation were obtained at pH 5–6. Higher values of pH caused lower levels of adsorption and decolorization of the dye.

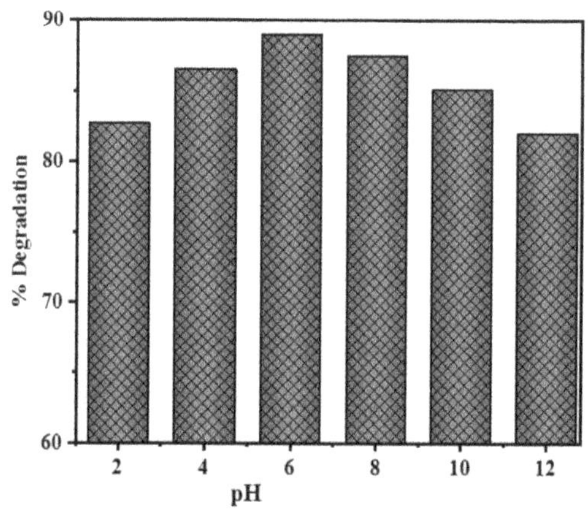


**Figure 14.** Influence of varying pH values on sonophotocatalytic degradation of SDB in the presence of $Mn^{2+}$:ZnS Qds at optimum conditions (15 mL of SDB 70 ppm, 75 min irradiation, 40 mg Qds).

The zero-point charge ($pH_{zpc}$) of ZnS is 7–7.5, therefore, the surface of Qds is negatively charged above a pH of 7, whereas it is positively charged below pH 7 [100]. As an anionic dye, SDB is absorbed by Qds without any trouble at acidic range. However, at pH values higher than the isoelectric point, the quantity of positively charged surface adsorbent sites decreases, while the negative site increases, which does not favor the adsorption of negatively charged dye anions [27]. Therefore, pH 6 was chosen as the optimum pH value for the subsequent experiments.

2.2.5. Effect of $Mn^{2+}$ Dopant Concentration

The doping of ZnS with Mn significantly influenced the photodegradation efficiency of the Qds. To study the doping effect, experiments were done under optimum conditions by varying the amount of dopant (1–5.0% *w*/*w*). On evaluating the SDB degradation efficiencies, with pure and doped ZnS Qds, it was observed that doping enhanced the efficiency of the catalyst through sonophotocatalysis, even though a negative effect was found at high concentrations. Figure 15 shows that the sonophotocatalytic degradation efficacies of the Qds increase from 0 to 3% and decrease above 3%, indicating potential activity at 3% $Mn^{2+}$ doping. The variations in the degradation efficiency may be attributed to differences in the size of the nanoparticles, as well as the recombination rates of the $e^-$ and $h^+$ upon doping, which affects their catalytic behavior [101]. Also, at high dopant concentration, $Mn^{2+}$ entraps both the charge carriers, which consequently recombine by quantum tunneling as the distance between the trapping sites reduces. At low dopant concentration, only the $h^+$ are trapped, which move to the surface and combine with the hydroxide ions present there, generating hydroxyl radicals ($HO^{\bullet}$), which are the primary oxidizing radicals for the dyes [102–104].

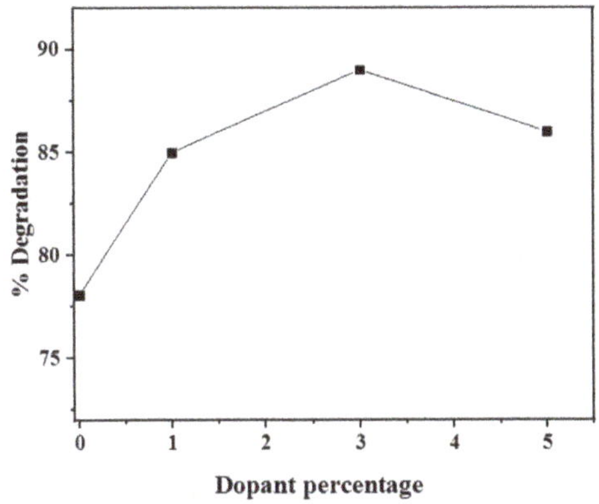


**Figure 15.** Effect of the variation of amount of $Mn^{2+}$ dopant ions on the sonophotocatalytic degradation of SDB at optimum conditions (15 mL of 70 ppm SDB, pH 6, 75 min irradiation, 40 mg Qds).

2.2.6. Effect of Ultrasonic Power on Degradation of SDB

To study the influence of power dissipation, experiments were carried out at two different power values (60 and 120 W). Figure 16 shows that an enhancement in the degradation rate was observed with an increase in power from 60 to 120 W. With the increase in power dissipation, the cavitational effects also increase, in turn producing additional turbulence and greater generation of free radicals, causing an enhancement in the degradation rate of SDB from 79 to 88%. Though the difference is minor, this may possibly be due to the cushioning effects, resulting from combination of a huge quantity of bubbles ensuing unproductive collapse action [105]. Thus, smaller energy gets used up for the free radical generation despite higher power dissipation [106,107]. Thus, 120 W was selected as the optimum power supply. The cavitational yield obtained for the system is $3.74 \times 10^{-12}$ mol/J.

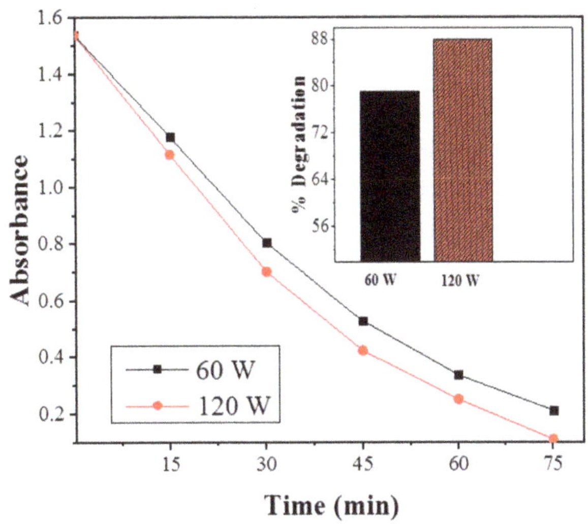


**Figure 16.** Effect of power on degradation of SDB at optimum conditions (15 mL of SDB, pH 6, 75 min irradiation, 40 mg Qds).

### 2.3. Re-Usability and Stability of Photocatalysts

Sustainability and reusability are important parameters of photocatalysts. In order to study the stability and durability of the as-prepared ZnS Qds, recycling experiments were performed for the removal of SDB. In the end of each cycle, the photocatalyst was removed, washed, dried and re-used [100]. During the washing process, loss of some catalyst amounts occurred, causing reduction in activity after the consecutive cycles [5,108]. Figure 17 shows that there is no significant loss of degradation efficiency after 5 consecutive cycles. Therefore, the doped ZnS Qds may be considered as a re-usable and photostable nanocatalyst during the degradation process.

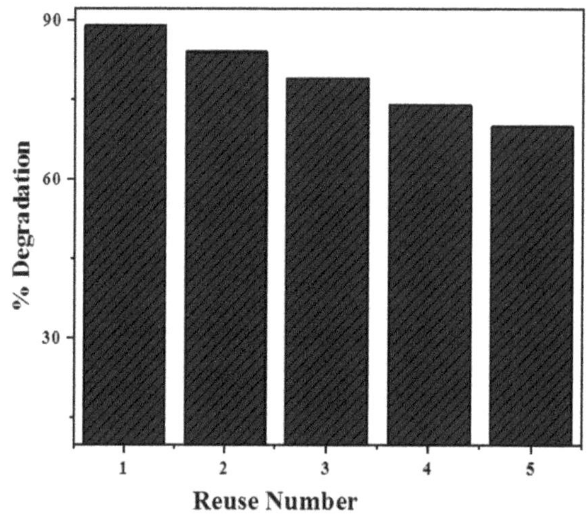


**Figure 17.** Recycling performance of $Mn^{2+}$:ZnS Qds over multiple cycles on sonophotocatalytic degradation of SDB.

The stability of the photocatalyst was also ascertained by XRD analysis. The XRD analysis of $Mn^{2+}$:ZnS Qds was performed before and after degradation of the SDB dye molecule (Figure 17). After sonophotocatalytic degradation, small transformations occurred and the peaks shifted to lower 2θ values, as observed in Figure 18. This is attributed to the ultrahigh strain rates generated by sonication [109].

### 2.4. Mechanism of the Sonophotocatalytic Degradation

Sonication of water is known to generate active radicals such as $OH^•$ and $H^•$ by cavitation, which degrades the organic compounds present in water [110]. The presence of the nanocatalyst augment this phenomenon since the small bubbles present in water have a tendency to break into smaller ones, causing an increase in the total area of high pressure and temperature [111]. The oxygen molecules present in water act as a source for nucleus cavitation, while the $HO^•$ radicals degrade the SDB dye species [112].

Moreover, the agglomerated molecules get dispersed by sonication. This deagglomeration enhances the surface area of the nanocatalyst, increasing the active sites for adsorption of the dye molecules as well as for the absorption of light producing more reactive species. Sonication also avoids catalyst deactivation, attributed to the upsurge of microstreaming and microbubbles which eliminates the molecules adsorbed at the surface of the nanocatalyst [113]. This causes cleaning of the catalyst surface, further enhancing the reaction.

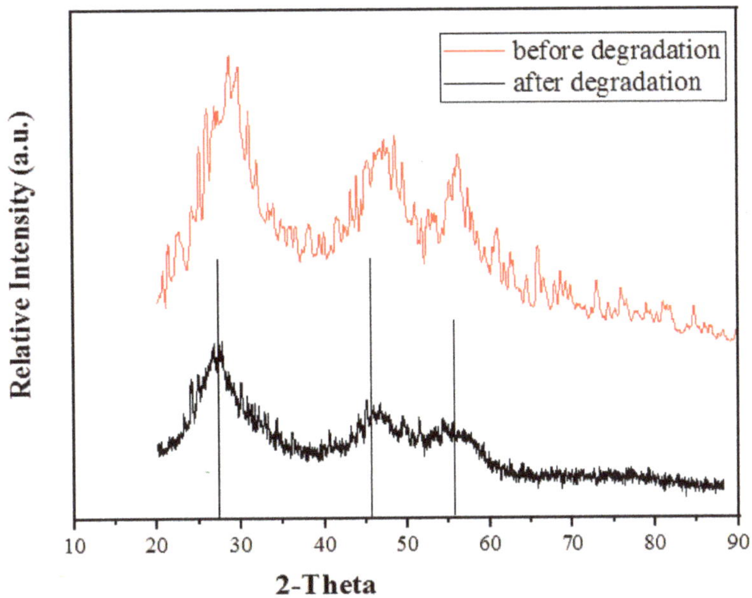


**Figure 18.** XRD patterns of $Mn^{2+}$:ZnS Qds before and after sonophotocatalytic degradation of SDB.

During photocatalysis, the irradiation of ZnS with UV light photons, results in $HO^•$ radical generation, due to oxidation of water by valence band holes. The active species electrons ($e^-$), holes ($h^+$), hydroxyl radicals ($HO^•$) and superoxide radicals ($O_2^{•-}$) are generally produced subsequent to UV irradiation. The $h^+$ with high oxidative potential allow direct oxidation of pollutants to highly reactive intermediates; also they could react with chemisorbed $H_2O$, generating reactive species, such as the hydroxyl ($HO^•$) radicals [114].

$$\text{Qds} + h\nu \rightarrow \text{Qds}\ (e^- + h^+) \quad (3)$$

$$h^+ + \text{SDB} \rightarrow \text{oxidation of SDB} \quad (4)$$

$$h^+ + OH^- \rightarrow HO^• \quad (5)$$

$$h^+ + H_2O \rightarrow H^+ + HO^• \quad (6)$$

The generated electrons react with the dissolved oxygen molecules, originating several radicals [101]. Transition metals at the surface of ZnS and oxygen atoms work as an electron sink and increase the electron hole separation. The electrons of the conduction band reduce the molecular oxygen, originating a superoxide anion at the catalyst surface, which then reacts with $H_2O$, forming $H_2O_2$, which originates $HO^•$ radicals [115]. The dye molecules can be degraded or oxidized by the hydroxyl radicals ($HO^•$), causing the dye to dissociate into smaller and not so toxic species.

$$e^- + O_2 \rightarrow O_2^{•-} \quad (7)$$

$$O_2^{•-} + H_2O \rightarrow H_2O_2 \quad (8)$$

$$H_2O_2 + e^- \rightarrow 2HO^• \quad (9)$$

The application of ultrasound in water causes acoustic cavitation. This comprises the formation, growth and collapse of cavity bubbles, entrapped gases or vapors surrounding water. During the sonolysis of water, it is well known that acoustic cavitation generates highly reactive primary radicals such as $OH^•$ and $H^•$, due to the thermal decomposition of water, as shown in reaction (10) [116,117]. A number of recombinations and other reactions (namely, reactions (11)–(14)) occur within the bubble following primary radical generation [118].

From a thermodynamic view, bubble collapse is significant, as it causes a large change in bubble volume. As the bubble collapse occurs quickly (<1 μs), the associated "work done" (PdV) leads to "near" adiabatic heating of the bubble contents, which results in the generation of very high temperatures and pressures within the bubble. As a result, numerous local hot spots with extremely high temperature and pressure are generated, consequently inducing the dissociation of water [116,118].

Thus, primary radical generation takes place due to various recombination and other reactions within the bubble. Among these radicals, $HO^\bullet$ is a powerful nonselective oxidant that has a high redox potential value (2.8 V) and can oxidize most of the organic pollutants.

$$H_2O \overset{))))))}{\rightarrow} HO^\bullet + H^\bullet \tag{10}$$

$$H^\bullet + H_2O \rightarrow HO^\bullet + H_2 \tag{11}$$

$$H^\bullet + O_2 \rightarrow HO_2{}^\bullet \tag{12}$$

$$HO^\bullet + H^\bullet \rightarrow H_2O \tag{13}$$

$$HO^\bullet + HO^\bullet \rightarrow H_2O_2 \tag{14}$$

$$2HO_2{}^\bullet \rightarrow H_2O_2 + O_2 \tag{15}$$

$$H_2O_2 + e^- \rightarrow OH^- + HO^\bullet \tag{16}$$

where )))))) refers to sonication.

Active $HO^\bullet$ radicals+ SDB molecules → Degradation of SDB

Thus, in both methods, the $HO^\bullet$ acts as a primary oxidizing radical, but other degradation processes can occur, since solutes have varying capacities to adsorb on the catalyst surface, as compared to the bubbles surface. Volatile solutes can be thermally degraded by entering the core of a collapsing bubble, but direct oxidation by the hole is also possible on the photocatalyst surface. Those extra processes can have improved effects when combined treatments are used, especially when the intermediates of the degradation process have different chemical properties than the parent molecules.

### 2.5. Role of Radical Trapping Agents

To elucidate the main contributors in the photodegradation reaction, the degradation rates of SDB in the presence of different scavengers were obtained. For this, the reactive species capture studies were carried out similar to the approach used for photocatalytic experiments. The experiments were performed by adding 0.01 M of different scavengers, for example, sodium azide ($NaN_3$), potassium iodide (KI), sodium chloride (NaCl) and formic acid (HCOOH). The scavengers were added prior to the addition of the photocatalyst into the dye solution.

Figure 19 shows various control experiments for the photodegradation of SDB. HCOOH was added as the $HO^\bullet$ scavenger, NaCl as the $h^+$ scavenger and $NaN_3$ for scavenging $^1O_2$ and $HO^\bullet$ [119]. Additionally, KI works for scavenging $h^+$ and $HO^\bullet$s at the catalyst surface [119,120]. The experiments were carried out by adding 0.01 M of different scavengers prior to the addition of photocatalyst into the dye solution. The maximum degradation (89%) of SDB was found without any scavenger. A small change in SDB photodegradation was found with the addition of NaCl, indicating that the photoexcited $h^+$ also contributes in photodegradation as a minor factor. With the addition of NaCl the rate of photodegradation of SDB slightly declined to 66%, signifying that $h^+$ were not the main active species. Further, the inhibition effect in photocatalytic efficiency was observed with a degradation of 19%, when HCOOH was added as the quencher, confirming the role of $HO^\bullet$ and $H^\bullet$ in the photocatalytic process [121–123]. Meanwhile, the addition of $NaN_3$ resulted in a significant decrease, with a degradation of 33%, indicating the important roles of $^1O_2$ and

$HO^\bullet$ in the photocatalytic process. The formation of $O_2^{\bullet -}$ is directly influenced by the reduction of $O_2$, as it determines the production of $HO^\bullet$ by its multistep reduction. Also, the photodegradation activity of SDB declined to 45% after the addition of KI, indicating the important roles of $h^+$ and $HO^\bullet$s in the photodegradation process.

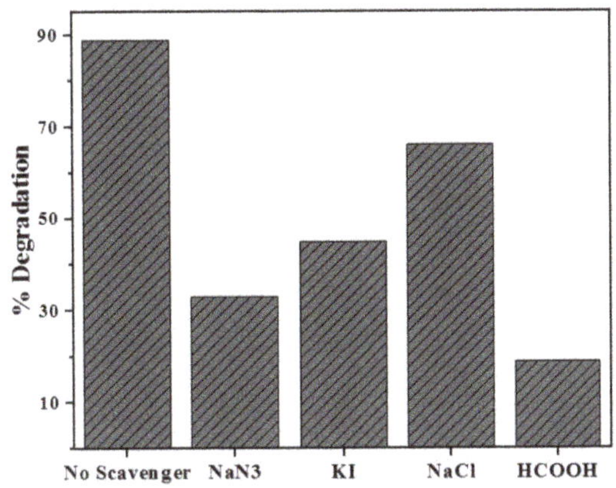


**Figure 19.** $Mn^{2+}$:ZnS Qds based sonophotocatalytic degradation of SDB with different scavengers.

Furthermore, to confirm the formation of $HO^\bullet$ radicals, a terephthalic acid test was conducted. The highest intensity peak in the fluorescence spectra of the terephthalic acid test represents the larger generation of $HO^\bullet$ radicals. Figure 20 confirms the generation of a higher amount of $HO^\bullet$ radicals during the sonophotocatalytic degradation of SDB with highest fluorescence intensity, as compared to the sonocatalytic and photocatalytic processes [124,125].

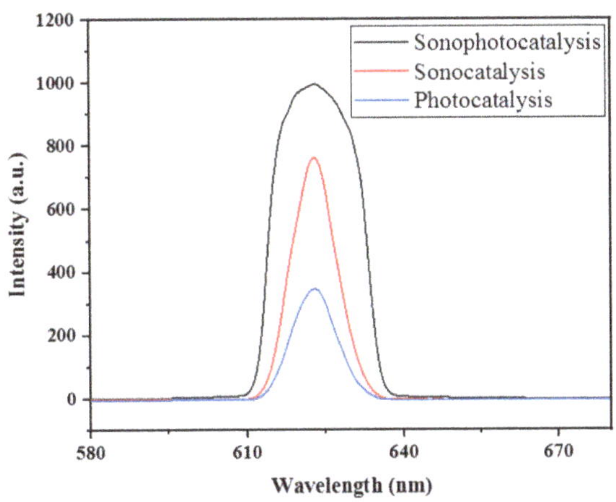


**Figure 20.** Trapping experiment: terephthalic acid tests for sonophotocatalytic, sonocatalytic and photocatalytic degradations of SDB.

### *2.6. Degradation Products of $Mn^{2+}$:ZnS Assisted Photodegradation of SDB*

The mechanism for the $Mn^{2+}$:ZnS assisted photodegradation of SDB was established using LC–MS. The results obtained for entirely degraded SDB are given in Figure S3. The SDB solution before photodegradation is shown in the inset. The main possible products from the SDB transformation are shown in Figure S4. The structure of those probable transformation products was proposed based only on LC-MS fragmentation. The peaks of SDB were observed at 461 m/z. Only the parent dye was present before irradiation, as expected [126]. Afterwards, quite a few fragments were obtained at m/z 124, 182, 240, 307.8 and 558, indicating subsequent photodegradation of SDB. Possible structures with those m/z values are depicted in Figure S4.

## 3. Experimental

### 3.1. Materials

Analytical grade chemicals, which did not require further purification, were used for the syntheses. Manganese carbonate, zinc acetate, sodium sulfide and nicotinic acid were purchased from Merck India Ltd., and used to prepare $Mn^{2+}$:ZnS Qds. The SDB dye (laboratory reagent grade) was also supplied by Merck India Ltd. Table S2 shows the formula and other data of this dye. In order to adjust the pH values of the suspensions, NaOH and HCl solutions were used (Merck India Ltd., Mumbai). Using double distilled water, varying concentrations of SDB solutions were made by diluting the prepared stock solution.

### 3.2. Apparatus

The surface morphology of the nanoparticles was studied using a scanning electron microscope JEOL JSM—6390LV, Tokyo, Japan. A small piece of extrudate of 10 mm diameter was mounted on specimen stubs using carbon tape and was over coated with gold using the JFC 1600. This ion sputtering device performs rapid and efficient gold coating on microscopic specimens, allowing surface visualization. The SEM measurements were performed at 15 kV accelerating voltage. Different voltages and magnifications were used as indicated in the SEM images.

The transmission electron microscopy was performed using a JEOL Model JM 2100 TEM device. An extremely small amount of material was suspended in water/ethanol (just enough to obtain slightly turbid solution). The solution was ultrasonicated to disperse the particles. A drop of the solution was then pipetted out and the drop was placed on a carbon-coated grid of 200 mesh. The measurements of particles observed in the TEM images were carried out using ImageJ software.

The absorption spectrum was used to find the optical properties of the as-prepared material, adopting a Cary Win UV spectrophotometer. The Belsorp mini II (BEL Japan Inc., Osaka, Japan) was used for $N_2$ gas adsorption–desorption analysis at −196 °C (with liquid nitrogen as coolant). The Brunauer–Emmett–Teller (BET) method was used to measure the surface areas, whereas the pore size distribution and volume were determined by the Barrett–Joyner–Halenda (BJH) model. The thermo gravimetric analysis (TGA) was conducted with a Perkin Elmer STA 6000 TG/DT model at 10 °C/min, from 30 to 650 °C, in $N_2$ atmosphere. For tuning the pH, a Systronic pH-meter was used. The ultrasound assistance in the dye degradation experiment was accomplished in an Aczet ultrasonic bath reactor with operating frequency of 40 kHz, a power rating of 120 W, and a 2.5 L capacity with dimensions of 235 × 135 × 100 mm. The UV irradiations were provided using a 40 W mercury lamp (Osram), at 254 nm emission, with incident light intensity of 221.23 $W/m^2$. Finally, product analysis was done by mass spectrometry by the Agilent 1290 Infinity UHPLC system (Agilent Technologies, Santa Clara, CA, USA).

### 3.3. Preparation of Quantum Dots

The synthesis of the $Mn^{2+}$:ZnS Qds sample was done in our lab by the chemical precipitation method following the procedure reported earlier [5,59,71]. In brief, all the aqueous solutions of precursors were prepared first, i.e., manganese carbonate, sodium sulfide and zinc acetate. For 1% manganese doping, we added 29.75 mL of 0.01 mol/L of manganese carbonate solution to 49.5 mL of a solution of 0.5 mol/L zinc acetate. Then, nicotinic acid (1.0 At. wt%) was mixed for the capping action. The pH was adjusted to 10 using a solution of NaOH 1 M. Then a $Na_2S$ solution was added, dropwise, with continuous stirring to obtain a white precipitate. The solution was then refluxed at 60 °C and centrifuged to obtain the precipitate. Finally, the precipitate was filtered, washed 3–4 times to eliminate impurities and dried.

*3.4. Quantum Dot Based Catalytic Experiments*

After characterization, the synthesized Qds were used in the sonophotocatalytic degradation experiments of SDB dye. A 100-ppm stock solution of SDB was prepared for the studies. In order to optimize the amount of catalyst, various concentrations of dye solutions were prepared, then the desired pH value was adjusted initially, and the appropriate amount of Qds catalyst was added to each suspension. The mixtures were stirred for 20 min in the dark before irradiation/ultrasound to ascertain an adsorption–desorption state of equilibrium. Then the sonicator and/or lamp were connected and the reaction started.

The solutions were irradiated by adopting a UV-C ($\lambda$ = 254 nm) light mounted above as the driving energy source, as shown in Figure S5. A distance of 12 cm between the light source and the solution was set for overall degradation experiments. An ultrasonic bath (40 kHz, 120 W) was used as the ultrasound source for the sonocatalytic and sonophotocatalytic experiments. The sonicator water was continuously replaced by circulation, in order to maintain the desired temperature (29 $\pm$ 1 °C, unless otherwise mentioned). The reactor position in the ultrasonic bath was the same.

The experimental conditions employed for photocatalysis, sonocatalysis and sonophotocatalysis were similar. After certain time intervals, sample aliquots were removed, centrifuged and analyzed for the remaining dye by recording the absorbance in spectrophotometer. The absorbance decrease (at $\lambda_{max}$ = 560 nm) for SDB samples after irradiation and/or sonication for a certain time showed the decolorization rate, that is, efficiency of decolorization of SDB, and therefore sonophotocatalytic activity of Qds.

## 4. Conclusions

In the present study, $Mn^{2+}$:ZnS Qds were synthesized by a simple and fast precipitation method. The photocatalytic, sonocatalytic and sonophotocatalytic degradation of SDB was investigated. Optimal parameters involved in the degradation such as the initial dye concentration, catalyst loading, pH, power dissipation and ultrasonic frequency were determined. The prepared nanocatalysts showed an improved efficiency for the sonophotocatalytic degradation of SDB in comparison to sonocatalysis or photocatalysis alone. An explanation can be found in the enhanced electron-holes separation at the heterointerface, by generation of highly reactive radicals and improved active surface area. The sonophotocatalytic dye removal process follows pseudo second-order kinetics. Therefore, $Mn^{2+}$:ZnS Qds provided effective removal of SDB, the process being straightforward and potentially useful for removal of organic pollutants from wastewater.

**Supplementary Materials:** The following are available online at https://www.mdpi.com/article/10.3390/catal11091025/s1, Figure S1: (a) $N_2$ adsorption-desorption isotherms of $Mn^{2+}$:ZnS Qds. (b) Pore size distribution obtained by BJH-adsorption, Figure S2: SDB absorbance changes at absorption maximum ($\lambda_{max}$ = 560 nm) for the ultrasound and UV light assisted degradation in the absence of pure and doped $Mn^{2+}$:ZnS Qds at optimum conditions (15 mL of 70 ppm SDB, pH 6, 75 min ultrasound and/or UV light irradiation), Figure S3: LC–MS of $Mn^{2+}$ doped ZnS assisted photodegraded SDB solution. Inset: SDB solution before and after degradation (after 75 mins), Figure S4: Degradation products of SDB photodegradation catalyzed by$Mn^{2+}$:ZnS Qds, Figure S5: Schematic diagram of sonophotocatalytic experimental setup, Table S1: Kinetic parameters: rate constants ($k$), correlation coefficients ($R^2$) and decolorization efficiency (DE) for the removal of Solochrome dark blue dye by $Mn^{2+}$:ZnS Qds, Table S2: Data for Solochrome dark blue dye.

**Author Contributions:** Conceptualization: J.P. and B.J.; methodology: J.P., B.J. and S.Y.; software: J.P., A.K.S., S.Y. and S.A.C.C.; validation: S.A.C.C. and A.K.S.; formal analysis: J.P.; investigation: J.P., B.J., S.Y. and A.K.S.; resources: A.K.S. and M.A.B.H.S.; data curation: J.P. and B.J.; writing—original draft preparation: J.P.; writing—review and editing: M.A.B.H.S. and S.A.C.C.; visualization: A.K.S.; supervision: A.K.S.; project administration: J.P., A.K.S., M.A.B.H.S. and S.A.C.C.; funding acquisition:

J.P., A.K.S., M.A.B.H.S. and S.A.C.C. All authors have read and agreed to the published version of the manuscript.

**Funding:** J.P. is thankful to DST, New Delhi, India for Research fellowship under Women Scientist Scheme (SR/WOS-A/CS-82/2018). This work has also been supported by FCT—Fundação para a Ciência e a Tecnologia, I.P., under the Scientific Employment Stimulus-Institutional Call (CEECINST/00102/2018) and the Associate Laboratory for Green Chemistry-LAQV which is financed by national funds from FCT/MCTES (UIDB/50006/2020 and UIDP/50006/2020).

**Institutional Review Board Statement:** Not applicable.

**Informed Consent Statement:** Not applicable.

**Data Availability Statement:** Data will be made available upon request.

**Acknowledgments:** Authors are also thankful to SAIF centers: STIC Kochi, IIT Madras, MNIT Jaipur, NIT Raipur and Center for Basic Sciences, Pt. RSSU Raipur for providing instrumental analysis facilities, and to the Department of Chemistry, Govt. V.Y.T.PG. Autonomous College, Durg (C.G.) for providing basic instruments for the kinetic study.

**Conflicts of Interest:** Authors declare no conflict of interest.

## References

1. Sugumar, S.; Thangam, B. BiodEnz: A database of biodegrading enzymes. *Biomed. Inform.* **2012**, *8*, 40–42.
2. Munter, R. Advanced oxidation processes—Current status and prospects. *Proc. Estonian Acad. Sci. Chem.* **2001**, *50*, 59–80.
3. Chen, D.; Chen, X. Luminescent perovskite quantum dots: Synthesis, microstructures, optical properties and applications. *J. Mater. Chem. C* **2019**, *7*, 1413–1416. [CrossRef]
4. Kumar, H.; Kumari, A.; Singh, R.R. Tunable narrow emission in ZnS/CdS/ZnS quantum well structures prepared by aqueous route. *Opt. Mater.* **2017**, *69*, 23–29. [CrossRef]
5. Patel, J.; Singh, A.K.; Carabineiro, S. Carabineiro, Assessing the photocatalytic degradation of fluoroquinolone norfloxacin by Mn:ZnS quantum dots: Kinetic study, degradation pathway and influencing factors. *Nanomaterials* **2020**, *10*, 964. [CrossRef] [PubMed]
6. Xiong, C.R.; Balkus, K.J. Mesoporous molecular sieve derived $TiO_2$ Nanofibers doped with $SnO_2$. *J. Phys. Chem. C* **2007**, *111*, 10359. [CrossRef]
7. Ji, P.; Takeuchi, M.; Cuong, T.M.; Zhang, J.; Matsuoka, M.; Anpo, M. Recent advances in visible light responsive Titanium Oxide-based photocatalysis. *Res. Chem. Intermed.* **2010**, *36*, 327–347. [CrossRef]
8. Wu, C.G.; Chao, C.C.; Kuo, F.T. Enhancement of the photocatalytic performance of $TiO_2$ catalysts via transition metal modification. *Catal. Today* **2004**, *97*, 103–112. [CrossRef]
9. Panda, D.; Manickam, S. Recent advancements in the sonophotocatalysis (SPC) and doped-sonophotocatalysis (DSPC) for the treatment of recalcitrant hazardous organic water pollutants. *Ultrason. Sonochem.* **2017**, *36*, 481–496. [CrossRef]
10. Neppolian, B.; Park, J.S.; Choi, H. Effect of fenton-like oxidation on enhanced oxidative degradation of para-chlorobenzoic acid by ultrasonic Irradiation. *Ultrason. Sonochem.* **2004**, *11*, 273. [CrossRef]
11. Guillard, C.; Theron, P.; Pichat, P.; Petrier, C. Evaluation of 1-Octanol degradation by photocatalysis and ultrasound using SPME. *Water Res.* **2002**, *36*, 4263. [CrossRef]
12. Brennen, C.E. *Cavitation and Bubble Dynamics*; New York Oxford Univ. Press: New York, NY, USA, 1995.
13. Suslick, K.S.; Flannigan, D.J. Inside a collapsing bubble: Sonoluminescence and the conditions during cavitation. *Ann. Rev. Phys. Chem.* **2008**, *59*, 659–683. [CrossRef]
14. Ciawi, E.; Rae, J.; Kumar, M.A.; Grieser, F. Determination of temperatures within acoustically generated bubbles in aqueous solutions at different ultrasound frequencies. *J. Phys. Chem. B* **2006**, *110*, 13656–13660. [CrossRef]
15. Jagannathan, M.; Grieser, F.; Kumar, M.A. Sonophotocatalytic degradation of Paracetamol using $TiO_2$ and $Fe^{3+}$. *Sep. Purif. Technol.* **2013**, *103*, 114–118. [CrossRef]
16. Selli, E.; Bianchi, C.L.; Pirola, C.; Bertelli, M. Degradation of Methyl tert-butyl ether in water: Effects of the combined use of sonolysis and photocatalysis. *Ultrason. Sonochem.* **2005**, *12*, 395–400. [CrossRef] [PubMed]
17. Sathishkumar, P.; Mangalaraja, R.V.; Rozas, O.; Vergara, C.; Mansilla, H.D.; Gracia-Pinilla, M.A.; Anandan, S. Sonophotocatalytic mineralization of norflurazon in aqueous environment. *Chemosphere* **2016**, *146*, 216–225. [CrossRef] [PubMed]
18. Babu, S.G.; Vinoth, R.; Neppoliana, B.; Dionysiou, D.D.; Kumar, M.A. Diffused sunlight driven highly synergistic pathway for complete mineralization of organic contaminants using reduced graphene oxide supported photocatalyst. *J. Hazard. Mater.* **2015**, *291*, 83–92. [CrossRef] [PubMed]
19. Babu, S.G.; Karthik, P.; John, M.C.; Lakhera, S.K.; Kumar, M.A.; Khim, J. Neppolian, synergistic effect of sono-photocatalytic process for the degradation of organic pollutants using CuO-$TiO_2$/rGO. *Ultrason. Sonochem.* **2019**, *50*, 218–223. [CrossRef]

20. Sathishkumar, P.; Mangalaraja, R.V.; Rozas, O.; Mansilla, H.D.; Gracia-Pinilla, M.A.; Anandan, S. Low frequency ultrasound (42 kHz) assisted degradation of Acid Blue 113 in the presence of visible light driven rare earth nanoclusters loaded $TiO_2$ nanophotocatalysts. *Ultrason. Sonochem.* **2014**, *21*, 1675–1681. [CrossRef]
21. Kaur, S.; Singh, V. Visible light induced sonophotocatalytic degradation of reactive red dye 198 using dye sensitized $TiO_2$. *Ultrason. Sonochem.* **2007**, *14*, 531. [CrossRef]
22. Neppolian, B.; Doronila, A.; Grieser, F.; Kumar, M.A. Simple and efficient sonochemical method for the oxidation of Arsenic (III) to Arsenic (V). *Environ. Sci. Technol.* **2009**, *43*, 6793–6798. [CrossRef] [PubMed]
23. Fouad, O.A.; Ismail, A.A.; Zaki, Z.I.; Mohamed, R.M. Zinc oxide thin films prepared by thermal evaporation deposition and its photocatalytic activity. *Appl. Catal. B Environ.* **2006**, *62*, 144–149. [CrossRef]
24. Abhilash, M.R.; Akshatha, G.; Srikantaswamy, S. Photocatalytic dye degradation and biological activities of $Fe_2O_3/Cu_2O$ nanocomposite. *RSC Adv.* **2019**, *9*, 8557–8568. [CrossRef]
25. Samadi, M.; Zirak, M.; Naseri, A.; Kheirabadi, M.; Ebrahimi, M.; Moshfegh, A.Z. Design and tailoring of one-dimensional ZnO nanomaterials for photocatalytic degradation of organic dyes: A review. *Res. Chem. Intermediat.* **2019**, *45*, 2197–2254. [CrossRef]
26. Shamsipur, M.; Rajabi, H.R. Study of photocatalytic activity of ZnS quantum dots as efficient nanoparticles for removal of methyl violet: Effect of ferric ion doping. *Spectrochim. Acta Mol. Biomol. Spectrosc.* **2014**, *122*, 260–267. [CrossRef]
27. Rajabi, H.R.; Farsi, M. Effect of transition metal ion doping on the photocatalytic activity of ZnS quantum dots: Synthesis, characterization and application for dye decolorization. *J. Mol. Catal. A Chem.* **2015**, *399*, 53–61. [CrossRef]
28. Chen, Z.; Li, D.; Zhang, W.; Shao, Y.; Chen, T.; Sun, M.; Fu, X. Photocatalytic degradation of dyes by $ZnIn_2S_4$ microspheres under visible light irradiation. *J. Phys. Chem. C* **2009**, *113*, 4433–4440. [CrossRef]
29. Wang, C.; Gao, X.; Ma, Q.; Su, X. Aqueous synthesis of mercaptopropionic acid capped $Mn^{2+}$-doped ZnSe quantum dots. *J. Mater. Chem.* **2009**, *19*, 7016–7022. [CrossRef]
30. Jia, Y.; Shen, S.; Wang, D.; Wang, X.; Shi, J.; Zhang, F.; Han, H.; Li, C. Composite$Sr_2TiO_4/SrTiO_3$(La,Cr) heterojunction based photocatalyst for hydrogen production under visible light irradiation. *J. Mater. Chem. A* **2013**, *1*, 7905–7912. [CrossRef]
31. Bhargava, R.N.; Gallagher, D.; Hong, X.; Nurmikko, A. Optical properties of manganese-doped nanocrystals of ZnS. *Phys. Rev. Lett.* **1994**, *72*, 416–419. [CrossRef]
32. Colvin, V.L.; Schlamp, M.C.; Alivisatos, A.P. Light-emitting diodes made from cadmium selenide nanocrystals and a semiconducting polymer. *Nature* **1994**, *370*, 354–357. [CrossRef]
33. Klimov, V.I.; Ivanov, S.A.; Nanda, J.; Achermann, M.; Bezel, I.; McGuire, J.A. Piryatinski, single-exciton optical gain in semiconductor nanocrystals. *Nature* **2007**, *447*, 441–446. [CrossRef]
34. Radovanovic, P.V.; Gamelin, D.R. High-temperature ferromagnetism in $Ni^{2+}$-doped ZnO aggregates prepared from colloidal diluted magnetic semiconductor quantum dots. *Phys. Rev. Lett.* **2003**, *91*, 1572021–1572024. [CrossRef]
35. Hanif, K.M.; Meulenberg, R.W.; Strouse, G.F. Magnetic ordering in doped Cd $(1-x)$ Co(x)Se diluted magnetic quantum dots. *J. Am. Chem.Soc.* **2002**, *124*, 11495–11502. [CrossRef]
36. Greenham, N.C.; Peng, X.; Alivisatos, A.P. Charge separation and transport in conjugated-polymer/semiconductor-nanocrystal composites studied by photoluminescence quenching and photoconductivity. *Phys. Rev. B Condens. Matter* **1996**, *54*, 17628–17637. [CrossRef]
37. Rao, S.S.; Durga, I.K.; Varma, C.V.T.; Punnoose, D.; Kim, S.K.; Kim, H.J. Enhance the performance of quantum dot-sensitized solar cell by manganese-doped ZnS films as a passivation layer. *Org. Electron.* **2015**, *26*, 200–207.
38. Chang, L.; Wu, H.; He, X.; Chen, L.; Zhang, Y. A highly sensitive fluorescent turn-on biosensor for glycoproteins based on boronic acid functional polymer capped Mn-doped ZnS quantum dots. *Anal. Chim. Acta* **2017**, *995*, 91–98. [CrossRef] [PubMed]
39. Diestra, D.D.; Huarac, J.B.; Rincon, D.P.; Gonzalez-Feliciano, J.A.; Gonzalez, C.I.; Weiner, B.R.; Morell, G. Biocompatible ZnS:Mn quantum dots for reactive oxygen generation and detection in aqueous media. *J. Nanopart. Res.* **2015**, *17*, 46101–46114. [CrossRef] [PubMed]
40. Chan, W.C.; Nie, S. Quantum dot bioconjugates for ultrasensitive non isotopic detection. *Science* **1998**, *281*, 2016–2018. [CrossRef]
41. Patel, J.; Jain, B.; Singh, A.K.; Susan, M.A.B.H.; Paul, L.J. Mn-doped ZnS quantum dots–an effective nanoscale sensor. *Microchem. J.* **2020**, *155*, 104755. [CrossRef]
42. Geng, Y.; Guo, M.; Tan, J.; Huang, S.; Tang, Y.; Tan, L.; Liang, Y. The fabrication of highly ordered fluorescent molecularly imprinted mesoporous microspheres for the selective sensing of sparfloxacinin biological samples. *Sens. Actuators B Chem.* **2019**, *281*, 821–829. [CrossRef]
43. Pradhan, N.; Goorskey, D.; Thessing, J.; Peng, X.G. An alternative of CdSe nanocrystal emitters: Pure and tunable impurity emissions in ZnSe nanocrystals. *J. Am. Chem. Soc.* **2005**, *127*, 17586–17587. [CrossRef] [PubMed]
44. Gurung, G.; Ekanayaka, T.K.; Yost, A.J.; Paudel, T.R. Absorption enhancement by transition metal doping in ZnS. *Mater. Res. Express* **2019**, *6*, 126550. [CrossRef]
45. Ramasamy, V.; Praba, K.; Murugadoss, G. Synthesis and study of optical properties of transition metals doped ZnS nanoparticles. *Spectrochim. Acta A Mol. Biomol. Spectrosc.* **2012**, *96*, 963–971. [CrossRef]
46. Sridevi, D.; Rajendran, K.V. Enhanced photoluminescence of ZnS nanoparticles doped with transition and rare earth metallic ions. *Chalcogenide Lett.* **2010**, *7*, 397–401.
47. Chen, H.; Shi, D.; Qi, J. Comparative studies on the magnetic properties of ZnS nanowires doped with transition metal atoms. *J. Appl. Phys.* **2011**, *109*, 084338. [CrossRef]

48. Chen, H.; Shi, D.; Qi, J.; Wang, B. First-principles study on the magnetic properties of transition-metal atoms doped $(ZnS)_{12}$ cluster. *J. Magn. Magn. Mater.* **2011**, *323*, 781–788. [CrossRef]
49. Akhtar, M.S.; Malik, M.A.; Alghamdi, Y.G.; Ahmad, K.S.; Riaz, S.; Naseem, S. Chemical bath deposition of Fe-doped ZnS thin films: Investigations of their ferromagnetic and half-metallic properties. *Mater. Sci. Semicond. Process.* **2015**, *39*, 283–291. [CrossRef]
50. Lovric, J.; Bazzi, H.S.; Cuie, Y.; Fortin, G.R.A.; Winnik, F.M. Maysinger, differences in subcellular distribution and toxicity of green and red emitting CdTe quantum dots. *J. Mol. Med.* **2005**, *83*, 377–385. [CrossRef] [PubMed]
51. Kirchner, C.; Liedl, T.; Kudera, S.; Pellegrino, T.; Javier, A.M.; Gaub, H.E.; Stolzle, S.; Fertig, N.; Parak, W.J. Cytotoxicity of colloidal CdSe and CdSe/ZnS nanoparticles. *Nano Lett.* **2005**, *5*, 331–338. [CrossRef]
52. Gobara, H.M.; Elsalamony, R.A.; Hassan, S.A. Sonophotocatalytic degradation of eriochrome black-T dye in water using Ti grafted SBA-15. *J. Porous Mater.* **2016**, *23*, 1311–1318. [CrossRef]
53. Gautam, A.; Kshirsagar, A.; Biswas, R.; Banerjee, S.; Khanna, P.K. Photodegradation of organic dyes based on anatase and rutile TiO2 nano-particles. *RSC Adv.* **2016**, *6*, 2746–2759. [CrossRef]
54. Mamba, G.; Mbianda, X.Y.; Mishra, A.K. Enhanced visible light photocatalytic degradation of eriochrome black T and eosin blue shade in water using tridoped titania decorated on SWCNTs and MWCNTs: Effect of the type of carbon nanotube incorporated. *Mater. Chem. Phys.* **2015**, *149–150*, 734–742. [CrossRef]
55. Jassal, V.; Shanker, U.; Kaith, B.S.; Shankar, S. Green synthesis of potassium zinc hexacyanoferrate nanocubes and their potential application in photocatalytic degradation of organic dyes. *RSC Adv.* **2015**, *5*, 26141–26149. [CrossRef]
56. Sharma, S.; Chaturvedi, N.; Chaturvedi, R.K.; Sharma, M.K. Photocatalytic degradation of eriochrome black t using ammonium phosphomolybdate semiconductor. *Int. J. Chem. Sci.* **2010**, *8*, 1580–1590.
57. Kazeminezhad, I.; Sadollahkhani, A. Photocatalytic degradation of Eriochrome black-T dye using ZnO nanoparticles. *Mater. Lett.* **2014**, *120*, 267–270. [CrossRef]
58. La Porta, F.A.; Ferrer, M.M.; Santana, Y.V.B.; Raubach, C.W.; Longo, V.M.; Sambrano, J.R.; Longo, E.; Andres, J.; Li, M.S.; Varela, J.A. Synthesis of wurtzite ZnS nanoparticles using the microwave assisted solvothermal method. *J. Alloys. Compds.* **2013**, *556*, 153–159. [CrossRef]
59. Zhou, C.; Song, J.; Zhou, L.; Zhong, L.; Liu, J.; Qi, Y. Greener synthesis and optimization of highly photoluminescence $Mn^{2+}$-Doped ZnS quantum dots. *J. Lumin.* **2015**, *158*, 176–180. [CrossRef]
60. Yu, F.P.; Ou, S.L.; Yao, P.C.; Wu, B.R.; Wu, D.S. Structural, surface morphology and optical properties of ZnS Films by chemical bath deposition at various Zn/S molar ratios. *J. Nanomater.* **2014**, *2014*, 594952. [CrossRef]
61. Rajabi, H.R.; Farsi, M. Quantum dot based photocatalytic decolorization as an efficient and green strategy for the removal of anionic dye. *Mater. Sci. Semicond. Process.* **2015**, *31*, 478–486. [CrossRef]
62. Shah, S.I.; Li, W.; Huang, C.P.; Jung, O.; Ni, C. Study of $Nd^{3+}$, $Pd^{2+}$, $Pt^{4+}$, and $Fe^{3+}$ dopant effect on photoreactivity of $TiO_2$ nanoparticle. *Proc. Natl. Acad. Sci. USA* **2002**, *99*, 6482–6486. [CrossRef]
63. Tauc, J. *Optical Properties of Solids*; Academic Press Inc.: New York, NY, USA, 1966.
64. Pouretedal, H.R.; Keshavarz, M.H.; Yosefi, M.H.; Shokrollahi, A.; Zali, A. Photodegradation of HMX and RDX in the presence of nanocatalyst of zinc sulfide doped with copper. *Iran. J. Chem. Chem. Eng.* **2009**, *28*, 13–19.
65. Mall, M.; Kumar, L. Optical studies of $Cd^{2+}$ and $Mn^{2+}$ Co-doped ZnS nanocrystals. *J. Lumin.* **2010**, *130*, 660–665. [CrossRef]
66. Lippens, P.E.; Lannoo, M. Calculation of the band gap for small CdS and ZnS crystallites. *Phys. Rev. B* **1989**, *39*, 10935–10942. [CrossRef] [PubMed]
67. Brus, L. Electronic wave functions in semiconductor clusters: Experiment and theory. *J. Phys. Chem.* **1986**, *90*, 2555–2560. [CrossRef]
68. Dong, B.; Cao, L.; Sua, G.; Liu, W. Synthesis and characterization of Mn doped ZnS d-dots with controllable dual-color emissions. *J. Coll. Interf. Sci.* **2012**, *367*, 178–182. [CrossRef] [PubMed]
69. Gong, Y.; Fan, Z. Melamine modulated mercaptopropionic acid capped manganese doped zinc sulphide quantum dots as a room-temperature phosphorescence sensor for detecting clenbuterol in biological fluids. *Sens. Actuators B* **2014**, *202*, 638–644. [CrossRef]
70. Wang, Y.; Liang, X.; Ma, X.; Hu, Y.; Hu, X.; Li, X.; Fan, J. Simple and greener synthesis of highly photoluminescence $Mn^{2+}$-doped ZnS quantum dots and its surface passivation mechanism. *Appl. Surf. Sci.* **2014**, *316*, 54–61. [CrossRef]
71. Singhal, M.; Sharma, J.K.; Jeon, H.C.; Kang, T.W.; Kumar, S. Effect of pyridine capping on morphological and optical properties of Zns:$Mn^{2+}$ core–shell quantum dots. *J. Mater. Sci. Mater. Electron.* **2016**, *27*, 3003–3010. [CrossRef]
72. Singhal, M.; Sharma, J.K.; Kumar, S. Morphological and optical behaviour of pyridine capped bio-compatible ZnS quantum dots. *Adv. Sci. Eng. Med.* **2012**, *4*, 133–139. [CrossRef]
73. Taylor, L.D. The infrared spectrum of nicotinic acid. *J. Org. Chem.* **1962**, *27*, 4064–4065. [CrossRef]
74. Trivedi, M.K.; Branton, A.; Trivedi, D.; Nayak, G.; Bairwa, K.; Jana, S. Spectroscopic characterization of disulfiram and nicotinic acid after biofield treatment. *J. Anal. Bioanal. Tech.* **2015**, *6*, 1000265.
75. Basha, S.J.; Khidhirbrahmendra, V.; Madhavi, J.; Thampy, U.U.; Reddy, C.V.; Ravikumar, R.V.S.S.N. Structural, optical, magnetic and thermal investigations on $Cr^{3+}$ ions doped ZnS nanocrystals by co-precipitation method. *J. Sci. Adv. Mater. Devices* **2019**, *4*, 260–266. [CrossRef]

76. Basha, S.J.; Khidhirbrahmendra, V.; Avinash, M.; Reddy, N.B.; Zyryanov, G.V.; Ravikumar, R.V.S.S.N. Structural, magnetic and thermal propertiesof $Mn^{2+}$ doped ZnS nanocrystals for device applications. *AIP Conf. Proc.* **2019**, *2063*, 040006.
77. Shanmugam, N.; Cholan, S.; Viruthagiri, G.; Gobi, R.; Kannadasan, N. Synthesis and characterization of $Ce^{3+}$-doped flowerlike ZnS nanorods. *Appl. Nanosci.* **2014**, *4*, 359–365. [CrossRef]
78. Saha, B.; Das, S.; Saikia, J.; Das, G. Preferential and enhanced adsorption of different dyes on iron oxide nanoparticles: A comparative study. *J. Phys. Chem. C* **2011**, *115*, 8024–8033. [CrossRef]
79. Kansal, S.K.; Sood, S.; Umar, A.; Mehta, S.K. Photocatalytic degradation of Eriochrome Black T dye using well-crystalline anatase $TiO_2$ nanoparticles. *J. Alloys Compd.* **2013**, *581*, 392–397. [CrossRef]
80. Singla, P.; Sharma, M.; Singh, K.; Pandey, O.P. Synthesis and characterization of Zinc doped nano $TiO_2$ for efficient photocatalytic degradation of Eriochrome Black T. *AIP Conf. Proc.* **2013**, *1536*, 103–104.
81. Derazkola, S.M.; Ajabshir, S.Z.; Niasari, M.S. Facile hydrothermal and novel preparation of nanostructured $Ho_2O_3$ for photodegradation of eriochrome black T dye as water pollutant. *Adv. Powder Technol.* **2017**, *28*, 747–754. [CrossRef]
82. Ajabshir, S.Z.; Derazkola, S.M.; Niasari, M.S. Schiff-base hydrothermal synthesis and characterization of $Nd_2O_3$ nanostructures for effective photocatalytic degradation of eriochrome black T dye as water contaminant. *J. Mater. Sci. Mater. Electron.* **2017**, *28*, 17849–17859. [CrossRef]
83. Kaur, J.; Singhal, S. Highly robust light driven ZnO catalyst for the degradation of eriochrome black T at room temperature. *Superlattice Microst.* **2015**, *83*, 9–21. [CrossRef]
84. Saritha, B.; Chockalingam, M.P. Photodegradation of eriochrome black-t dye from aqueous medium by photocatalysis. *IJPAMS* **2017**, *116*, 183–187.
85. Preethi, M.E.L.; Priya, A.; Thiriveni, S. Solar light driven degradation of Eriochrome Black T by photocatalysis. *IOSR-JAC* **2015**, *8*, 55–62.
86. Siddhi, I.; Disha, D. Bioremediation of azo dye: Eriochrome black t by the novel organism Bacillus lentus. *Bio. Disc.* **2017**, *8*, 771.
87. Deepa, K.P.; Panneerselvam, A.; Thajuddin, N. A Study on the waning effect of oil spill isolated miocroalga *Coelastrella* Sp. on a synthetic dye, eriochrome black t. *Asian J. Microbiol. Biotechnol. Environ. Sci.* **2019**, *21*, 205.
88. Abdelmalek, F.; Ghezzar, M.R.; Belhadj, M.; Addou, A. Bleaching and degradation of textile dyes by nonthermal plasma process at atmospheric pressure jean-louis brisset. *Ind. Eng. Chem. Res.* **2006**, *45*, 23–29. [CrossRef]
89. Dave, P.N.; Kaur, S.; Khosla, E. Removal of erichrome black t by adsorption on to eucalyptus bark using green technology. *Ind. J. Chem. Technol.* **2011**, *18*, 53–60.
90. Wen, H.; Zhang, D.; Gu, L.; Yu, H.; Pan, M.; Huang, Y. Preparation of sludge-derived activated carbon by fenton activation and the adsorption of eriochrome black t. *Materials* **2019**, *12*, 882. [CrossRef]
91. Almeida, J.M.F.; Oliveira, E.S.; Silva, I.N.; de Souza, S.P.M.C.; Fernandes, N.S. Adsorption of Erichrome Black T from aqueous solution onto expanded perlite modified with orthophenanthroline. *Rev. Virtual Quim.* **2017**, *9*, 502–513. [CrossRef]
92. Rao, Y.; Yang, H.; Xue, D.; Guo, Y.; Qi, F.; Ma, J. Sonolytic and sonophotolytic degradation of carbamazepine: Kinetic and mechanism. *Ultrason. Sonochem.* **2016**, *32*, 371–379. [CrossRef]
93. Geng, N.; Chen, W.; Xu, H.; Ding, M.; Liu, Z.; Shen, Z. A sono-photocatalyst for humic acid removal from water: Operational parameters, kinetics and mechanism. *Ultrason. Sonochem.* **2019**, *57*, 242–252. [CrossRef]
94. Hartmann, J.; Bartels, P.; Mau, U.; Witter, M.; Tumpling, W.V.; Hofmann, J.; Nietzschmann, E. Degradation of the drug diclofenac in water by sonolysisin presence of catalysts. *Chemosphere* **2008**, *70*, 453–461. [CrossRef]
95. Ince, N.H.; Tezcanli-Guyer, G.; Belen, R.K.; Apikyan, I.G. Ultrasound as a catalyzer of aqueous reaction systems: The state of the art and environmental applications. *Appl. Catal. B Environ.* **2001**, *29*, 167–176. [CrossRef]
96. Yadav, S.; Asthana, A.; Chakraborty, R.; Jain, B.; Singh, A.K.; Carabineiro, S.A.C.; Susan, M.A.B.H. Cationic dye removal using novel magnetic/activated charcoal/β-cyclodextrin/alginate polymer nanocomposite. *Nanomaterials* **2020**, *10*, 170. [CrossRef] [PubMed]
97. Shrivastava, V.S. Photocatalytic degradation of methylene blue dye and chromium metal from wastewater using nanocrystalline $TiO_2$ semiconductor. *Arch. Appl. Mech.* **2012**, *4*, 1244–1254.
98. Daneshvar, N.; Khataee, A.R. Removal of azo dye C.I. acid red from contaminated water using Fenton, UV/$H_2O_2$, UV/$H_2O_2$/Fe(II), UV/$H_2O_2$/Fe(III) and UV/$H_2O_2$/Fe(III)/oxalate processes: A comparative study. *J. Environ. Sci. Health A* **2006**, *41*, 315–328. [CrossRef] [PubMed]
99. So, C.M.; Cheng, M.Y.; Yu, J.C.; Wong, P.K. Degradation of azo dye procion red MX-5B by photocatalytic oxidation. *Chemosphere* **2002**, *46*, 905–912. [CrossRef]
100. Rajabi, H.R.; Karimi, F.; Kazemdehdashti, H.; Kavoshic, L. Fast sonochemically-assisted synthesis of pure and doped zinc sulfide quantum dots and their applicability in organic dye removal from aqueous media. *J. Photochem. Photobiol. B Biol.* **2018**, *181*, 98–105. [CrossRef]
101. Rajabi, H.R.; Khani, O.; Shamsipur, M.; Vatanpour, V. High-performance pure and $Fe^{3+}$-ion Doped ZnS quantum dots as green nanophotocatalysts for the removal of malachite green under UV-light irradiation. *J. Hazard. Mater.* **2013**, *250*, 370–378. [CrossRef]
102. Yang, Q.; Liao, Y.; Mao, L. Kinetics of photocatalytic degradation of gaseous organic compounds on modified TiO2/AC composite photocatalyst. *Chin. J. Chem. Eng.* **2012**, *20*, 572–576. [CrossRef]
103. Ullah, R.; Dutta, J. Photocatalytic degradation of organic dyes with manganese doped ZnO nanoparticles. *J. Hazard. Mater.* **2008**, *156*, 194–200. [CrossRef]

104. Herrmann, J.M. Heterogeneous photocatalysis: Fundamentals and applications to the removal of various types of aqueous pollutants. *Catal. Today* **1999**, *53*, 115–129. [CrossRef]
105. Daware, G.B.; Gogate, P.R. Sonochemical degradation of 3-methylpyridine (3MP) intensified using combination with various oxidants. *Ultrason. Sonochem.* **2020**, *67*, 105120. [CrossRef] [PubMed]
106. Sivkumar, M.; Pandit, A.B. Ultrasound enhanced degradation of Rhodamine B: Optimization with power density. *Ultrason. Sonochem.* **2001**, *8*, 223–240. [CrossRef]
107. Wang, S.; Wu, X.; Wang, Y.; Li, Q.; Tao, M. Removal of organic matter and ammonia nitrogen from landfill leachate by ultrasound. *Ultrason. Sonochem.* **2008**, *15*, 933–937. [CrossRef] [PubMed]
108. Karimi, H.; Rajabi, H.R.; Kavoshi, L. Application of decorated magnetic nanophotocatalysts for efficient photodegradation of organic dye: A comparison study on photocatalytic activity of magnetic zinc sulfide and graphene quantum dots. *J. Photochem. Photobiol. A Chem.* **2020**, *397*, 112534. [CrossRef]
109. Dai, H.; Li, H.; Li, Z.; Zhao, J.; Yu, X.; Sun, J.; An, Q. Sonication induced amorphisation in Ag nanowires. *Sci. Rep.* **2019**, *9*, 2114. [CrossRef]
110. Kotronatou, A.; Mills, G.; Hoffmann, M.R. Ultrasonic irradiation of p-nitrophenol in aqueous solution. *J. Phys. Chem.* **1991**, *95*, 3630–3638. [CrossRef]
111. Suslick, K.S.; Crum, L.A. *Encyclopedia of Acoustics*; Crocker, M.J., Ed.; Wiley Interscience: New York, NY, USA, 1997; Volume 1, pp. 271–282.
112. Jyothi, K.P.; Yesodharan, S.; Yesodharan, E.P. Yesodharan ultrasound (US), ultraviolet light (UV) and combination (US + UV) assisted semiconductor catalysed degradation of organic pollutants in water: Oscillation in the concentration of hydrogen peroxide formed in situ. *Ultrason. Sonochem.* **2014**, *21*, 1787–1796. [CrossRef]
113. Chen, Y.C.; Smirniotis, P. Enhancement of photocatalytic degradation of phenol and chlorophenols by ultrasound. *Ind. Eng. Chem. Res.* **2002**, *41*, 5958–5965. [CrossRef]
114. Montazerozohori, M.; Nasr-Esfahani, M.; Joohari, S. Photocatalytic degradation of an organic dye in some aqueous buffer solutions using nano titanium dioxide: A kinetic study. *Environ. Prot. Eng.* **2012**, *38*, 45–55. [CrossRef]
115. Kumar, K.; Chitkara, M.; Sandhu, I.S.; Mehta, D.; Kumar, S. Photocatalytic and magnetic properties of $Zn_{1-x}Cr_xO$ nanocomposites prepared by a co-precipitation method. *Mater. Sci. Semicond. Process.* **2015**, *30*, 142–151. [CrossRef]
116. Kumar, M.A.; Grieser, F. Single bubble sonoluminescence- a chemist's overview. *Chem. Phys. Chem.* **2004**, *5*, 439–448.
117. Kumar, M.A.; Cozzoli, P.D. *Advanced Wet-Chemical Synthetic Approaches to Inorganic Nanostructures*; Transworld Research Network: Kerala, India, 2008; pp. 107–131.
118. Neppolian, B.; Ciceri, L.; Bianchi, C.L.; Grieser, F.; Kumar, M.A. Sonophotocatalytic degradation of 4-chlorophenol using $Bi_2O_3/TiZrO_4$ as a visible light responsive photocatalyst. *Ultrason. Sonochem.* **2011**, *18*, 135–139. [CrossRef]
119. Zhang, L.S.; Wong, K.H.; Zhang, D.Q.; Hu, C.; Yu, J.C.; Chan, C.Y.; Wong, P.K. Zn:In(OH)ySz solid solution nanoplates: Synthesis, characterization, and photocatalytic mechanism. *Environ. Sci. Technol.* **2009**, *43*, 7883–7888. [CrossRef]
120. Kaur, A.; Kansal, S.K. $Bi_2WO_6$ nanocuboids: An efficient visible light active photocatalyst for the degradation of levofloxacin drug in aqueous phase. *Chem. Eng. J.* **2016**, *302*, 194–203. [CrossRef]
121. Leitner, N.K.V.; Dore, M. Hydroxyl radical induced decomposition of aliphatic acids in oxygenated and deoxygenated aqueous solutions. *J. Photochem. Photobiol. A* **1996**, *99*, 137–143. [CrossRef]
122. Buxton, G.V.; Greenstock, C.L.; Helman, W.P.; Ross, A.B. Critical-review of rate constants for reactions of hydrated electrons, hydrogen-atoms and hydroxyl radicals ($^{\bullet}OH/^{\bullet}O$) in aqueous-solution. *J. Phys. Chem. Ref. Data* **1988**, *17*, 513–886. [CrossRef]
123. Von Sonntag, C.; Schuchmann, H.P. *Peroxyl Radicals in Aqueous Solutions*; John Wiley & Sons: Chichester, UK, 1997.
124. Acharya, L.; Nayak, S.; Pattnaik, S.P.; Acharya, R.; Parida, K. Resurrection of boron nitride in p-n type-II boron nitride/B-doped-g-$C_3N_4$ nanocomposite during solid-state Z-scheme charge transfer path for the degradation of tetracycline hydrochloride. *J. Colloid Interface Sci.* **2020**, *566*, 211–223. [CrossRef] [PubMed]
125. Balu, S.; Chen, Y.L.; Juang, R.C.; Yang, T.C.K.; Juan, J.C. Morphology-controlled synthesis of ae$Fe_2O_3$ nanocrystals impregnated on g-$C_3N_4$e$SO_3$H with ultrafast charge separation for photoreduction of Cr (VI) under visible light. *Environ. Pollut.* **2020**, *267*, 115491. [CrossRef]
126. Borhade, A.; Tope, D.; Kushare, S. Mercenaria shell powder as a cost-effective and eco-friendly photocatalyst for the degradation of eriochrome black t dye. *Iran. J. Sci. Technol. Trans. Sci.* **2020**, *44*, 75–83. [CrossRef]

*catalysts*

*Article*

# Identification of Active Species in Photodegradation of Aqueous Imidacloprid over g-$C_3N_4$/$TiO_2$ Nanocomposites

**Thawanrat Kobkeatthawin [1], Jirawat Trakulmututa [1], Taweechai Amornsakchai [1], Puangrat Kajitvichyanukul [2,*] and Siwaporn Meejoo Smith [1,*]**

1 Center of Sustainable Energy and Green Materials and Department of Chemistry, Faculty of Science, Mahidol University, 999 Phuttamonthon Sai 4 Rd, Salaya 73170, Thailand; kunthidakob@gmail.com (T.K.); jirawat.trk@student.mahidol.edu (J.T.); taweechai.amo@mahidol.edu (T.A.)

2 Department of Environmental Engineering, Faculty of Engineering, Chiang Mai University 239, Huay Kaew Road, Muang District, Chiang Mai 50200, Thailand

* Correspondence: puangrat.k@cmu.ac.th (P.K.); siwaporn.smi@mahidol.edu (S.M.S.)

**Abstract:** In this work, g-$C_3N_4$/$TiO_2$ composites were fabricated through a hydrothermal method for the efficient photocatalytic degradation of imidacloprid (IMI) pesticide. The composites were fabricated at varying loading of sonochemically exfoliated g-$C_3N_4$ (denoted as CNS). Complementary characterization results indicate that the heterojunction between the CNS and $TiO_2$ formed. Among the composites, the 0.5CNS/$TiO_2$ material gave the highest photocatalytic activity (93% IMI removal efficiency) under UV-Vis light irradiation, which was 2.2 times over the pristine g-$C_3N_4$. The high photocatalytic activity of the g-$C_3N_4$/$TiO_2$ composites could be ascribed to the band gap energy reduction and suppression of photo-induced charge carrier recombination on both $TiO_2$ and CNS surfaces. In addition, it was found that the active species involved in the photodegradation process are OH• and holes, and a possible mechanism was proposed. The g-$C_3N_4$/$TiO_2$ photocatalysts exhibited stable photocatalytic performance after regeneration, which shows that g-$C_3N_4$/$TiO_2$ is a promising material for the photodegradation of imidacloprid pesticide in wastewater.

**Keywords:** imidacloprid (IMI); graphitic carbon nitride (g-$C_3N_4$); titanium dioxide ($TiO_2$); g-$C_3N_4$/$TiO_2$ composite; photocatalytic activity

**Citation:** Kobkeatthawin, T.; Trakulmututa, J.; Amornsakchai, T.; Kajitvichyanukul, P.; Smith, S.M. Identification of Active Species in Photodegradation of Aqueous Imidacloprid over g-$C_3N_4$/$TiO_2$ Nanocomposites. *Catalysts* **2022**, *12*, 120. https://doi.org/10.3390/catal12020120

Academic Editors: Ioan Balint and Monica Pavel

Received: 23 December 2021
Accepted: 16 January 2022
Published: 19 January 2022



## 1. Introduction

Imidacloprid (IMI), which is the most widely used pesticide in the group of neonicotinoids, is a pesticide that is used in agriculture such as in crop protection against aphids, leafhoppers, psyllids beetles, etc. [1], and parasite management [2]. The use of neonicotinoids has been registered in approximately 120 countries worldwide [3], and IMI is one of the top ten global agrochemicals used as a pesticide worldwide [4]. It acts as a nicotinic acetylcholine receptor (nAChR) agonist that interferes with the transmission in the central nervous system of insects and results in paralysis and death [5]. With their widespread use, persistent nature, and high solubility (610 mg/L in 20 °C $H_2O$; log $K_{ow}$ = 0.57), IMI can cause damage to the environment via transportation in water, soil, and air [6]. Furthermore, the use of IMI can affect human health which includes neurological effects [7,8], in addition to gastrointestinal symptoms, lethargy [9], emaciation thyroid lesions, and cardiorespiratory failure [10]. Thus, the removal of these pollutants from water is essential due to their harmful influence on human health and aquatic ecosystems. Various methods can be applied for the degradation of IMI from aqueous solutions such as microfiltration membrane [11], biological degradation [12], adsorption [13,14], and advanced oxidation processes (AOPs) [15,16]. Among the AOP methods, photocatalytic activity has been used effectively in wastewater treatment for the removal of organic pollutants due to its simplicity, high activity, low cost, and ability to reduce $CO_2$ [17,18].

Graphitic carbon nitride (g-$C_3N_4$) has attracted significant attention as a visible photocatalyst for water purification due to its stability, high surface area, eco-friendliness, and facile synthesis [19,20]. However, the disadvantage of pure g-$C_3N_4$ is the fast recombination of photogenerated electron–hole pairs which lead to low photocatalytic efficiency [21]. Many strategies have been tried to improve the photocatalytic performance such as nanostructure design [22,23], metal and non-metal doping [24], and composite photocatalysts [25–28]. Among the various strategies, photocatalysis by coupling with other semiconductor materials is a beneficial method to improve the electron recombination process and extend the visible light absorption, which can enhance the photocatalytic performance. $TiO_2$ is an n-type semiconductor that has been widely used owing to the high efficiency, low cost, non-toxicity, and long-term stability of this compound. However, because of the large band gap energy of 3.2 eV of $TiO_2$, this results in the ineffective utilization of visible light, low quantum efficiency, and fast recombination [29]. It is expected that coupling $TiO_2$ with g-$C_3N_4$ can improve electron–hole pair recombination, broaden the photo-response range, and promote oxidation and reduction processes.

Herein, g-$C_3N_4$/$TiO_2$ photocatalysts were synthesized by a simple hydrothermal method. The phase structure, chemical composition, morphology, and scavenger trapping were investigated in detail. The g-$C_3N_4$/$TiO_2$ photocatalysts were used to degrade imidacloprid pesticide in wastewater under UV-Vis light irradiation. The recyclability of the composite was studied. In addition, the possible photodegradation mechanism was also proposed in this study.

## 2. Results and Discussion

### 2.1. Characterization

The XRD patterns of bulk-CN, CNS, $TiO_2$, and 0.5$TiO_2$/g-$C_3N_4$ are shown in Figure 1a. The g-$C_3N_4$ has two main diffraction peaks at 13.1° and 27.5°, which corresponds to the (001) plane caused by the arrangement of the tri-s-triazine units and the (002) plane caused by the interlayer stacking of the conjugated aromatic ring (JCPDS 87-1526) [29]. After the exfoliation of the bulk-CN, the decrease of CNS intensity (002) peak indicated that the interlayer structure was partially destroyed [30,31], and the slight shift of the (002) peak is attributed to the decreased distance of the basic sheets in the nanosheets [32]. The peaks of pure $TiO_2$ at 25.3, 37.8, 48.0, 53.9, 62.7, 68.8, 70.3, 75.0, and 82.6° correspond to the (101), (004), (200), (211), (204), (116), (220), (215), and (224) crystal planes of anatase $TiO_2$ (JCPDS 21-1272) [33]. Hydrothermally synthesized g-$C_3N_4$/$TiO_2$ photocatalysts showed the patterns related to both pure g-$C_3N_4$ and $TiO_2$. In addition, there is no obvious change in the peaks of $TiO_2$ in the composites, indicating that coupling with g-$C_3N_4$ did not influence the phase structure of $TiO_2$.

The Raman spectra of g-$C_3N_4$, $TiO_2$, and g-$C_3N_4$/$TiO_2$ composites are shown in Figure 1b. The characteristic peaks of g-$C_3N_4$ appeared at 707 $cm^{-1}$ and 1230 $cm^{-1}$ which were assigned to the breathing modes of the tri-s-triazine ring and C-N heterocycles, respectively [34]. Moreover, all the Raman bands observed for bulk-CN can be found in the CNS. The Raman spectrum of pure $TiO_2$ exhibited peaks at 148, 395, 510, and 640 $cm^{-1}$ corresponding to anatase-phase $TiO_2$ [35]. 15CNS/$TiO_2$ showed a combination peak of g-$C_3N_4$ and $TiO_2$ which confirms the formation of composites. No peak shifts were observed, which means no structural changes occurred during the preparation of the composites with pure $TiO_2$ and g-$C_3N_4$.

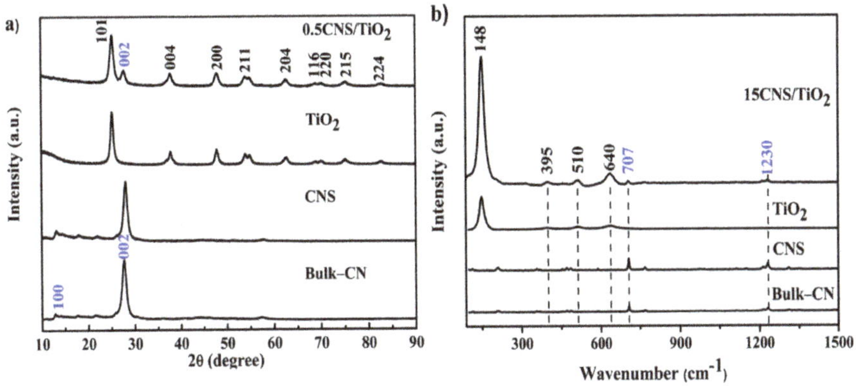


**Figure 1.** (**a**) Powder XRD patterns of bulk-CN, CNS, $TiO_2$, and 0.5CNS/$TiO_2$. (**b**) Raman spectra of bulk-CN, CNS, $TiO_2$, and 15CNS/$TiO_2$. Figure S1 shows the PXRD of the rest of nanocomposites.

The chemical binding states of g-$C_3N_4$, $TiO_2$, and composites were studied through XPS analysis. Figure 2a displays the survey scan of bulk-CN, CNS, $TiO_2$, and g-$C_3N_4$/$TiO_2$ in various weight ratios which confirmed the presence of C, N, Ti, and O atoms in the composites. Figure 2b shows three high-resolution C 1s spectrums at binding energies of 285.0, 288.3, and 289.2 eV, assigned as C-C, N–C=N, and $sp^2$ hybridized carbon in the tri-s-triazine ring ($N_2$-C=N) for g-$C_3N_4$.

Four binding energies in N 1s spectra (Figure 2c) can be observed, which can be classified into to $sp^2$ hybridized nitrogen C-N=C (398.8 eV), tertiary nitrogen N-$(C)_3$ (399.2 eV), amino functional groups N–H (400.3 eV), and $\pi$-excitation (401.2 eV), respectively [36–38]. The C 1s and N 1s spectra are slightly shifted from primitive g-$C_3N_4$ which suggests that there is a chemical bond connection between g-$C_3N_4$ and $TiO_2$ [39]. The C/N ratio of g-$C_3N_4$ is 0.90, indicating the presence of nitrogen vacancies that probably occurred during the thermal reduction process [40]. EPR spectra can provide evidence for probing the surface vacancies in photocatalysts. As shown in Figure 2f, the EPR intensity signal of CNS is significantly enhanced, revealing the increase of nitrogen vacancies generated in g$C_3N_4$ [41]. Figure 2d shows the high-resolution Ti 2p spectrum. The binding energy peaks of Ti $2p^{3/2}$ and Ti $2p^{1/2}$ appeared at 459.3 and 465.0 eV, which represent $Ti^{4+}$ species in the form of $TiO_2$ clusters [42]. In addition, there might be another Ti species in the material due to the poor XPS peak fitting for the $Ti^{4+}$ alone. A better XPS profile fitting was later obtained by including a peak at 460.2 eV, being assigned as the $Ti^{3+}$ defects on the composite surface [43,44]. The O 1s spectrum in Figure 2e can be devised into three peaks in $TiO_2$ with the binding energy of 530.5, 531.9, and 533.2 eV which can be assigned to (Ti-O), oxygen vacancy ($V_o$), and water molecules adsorbed on the surface of $TiO_2$, respectively [45]. Figure 2f shows the result of the solid ESR measurement which was used to confirm the presence of $Ti^{3+}$. A strong EPR signal of $TiO_2$ and the composites was observed with g of 1.997, which corresponds with $Ti^{3+}$ defect ($3d^1$, S = 1/2) and oxygen vacancy (Vo) [46]. It is possible that $Ti^{4+}$ was reduced to $Ti^{3+}$ by the loss of oxygen from the surface of $TiO_2$ because of the hydrothermal treatment at a high temperature [47].

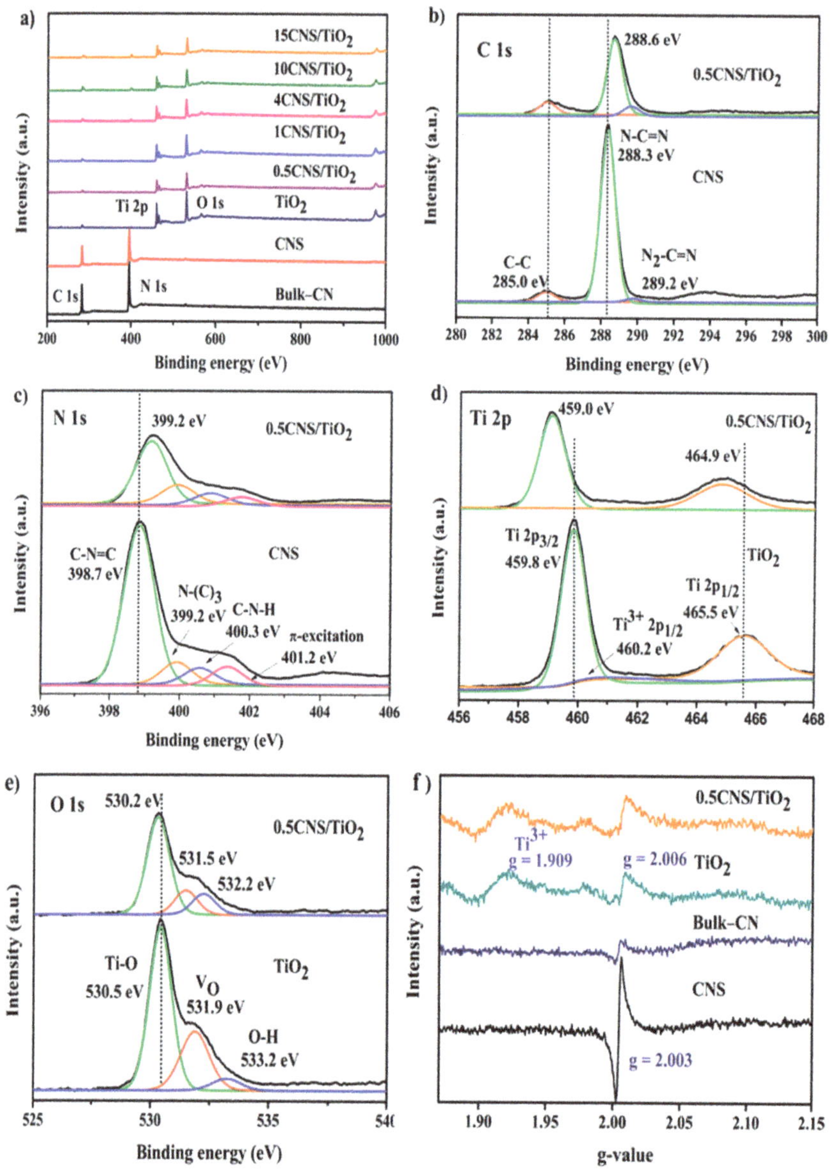


**Figure 2.** (**a**) The survey scan of all samples. (**b**) The C1s spectra. (**c**) The N1s spectra. (**d**) The Ti 2p spectra. (**e**) The O1s spectra. (**f**) Solid EPR spectra of g-$C_3N_4$, $TiO_2$, and g-$C_3N_4$/$TiO_2$.

## 2.2. Morphology Study

The morphologies of bulk-CN, CNS, $TiO_2$, and 0.5CNS/$TiO_2$ composites were examined by SEM as shown in Figure 3. We can see that bulk-CN presents in the form of bulk morphology with a layered structure (Figure 3a) [48]. CNS revealed smaller particles like the nanosheet structure after thermal exfoliation with $HNO_3^-$ (Figure 3b). The BET surface area of bulk-CN and CNS were calculated to be 68.26 and 91.61 $m^2g^{-1}$, respectively. It is possible that the van der Waals forces and hydrogen bonds of g-$C_3N_4$ might be destroyed with thermal treatment which results in the separation of g-$C_3N_4$ into small layers [49,50]. In addition, nitric acid might be intercalated into interlayers of bulk-CN which caused the expansion of their interlayer space and reduction of the layer thickness and hence increased its surface area [51,52]. The obtained $TiO_2$ showed spherical-like morphology with a particle size around 10 nm. From Figure 3d, it can be seen that the CNS particles having thin-layered structures are well distributed on the surface of $TiO_2$ particles, which is consistent with the presence of the peaks of g-$C_3N_4$ in XRD, XPS, and Raman spectra.

**Figure 3.** (**a**) SEM images of (**a**) bulk-CN, (**b**) CNS, (**c**) $TiO_2$, and (**d**) 0.5CNS/$TiO_2$.

## 2.3. *Optical Study*

Figure 4a shows the UV-Vis diffuse reflectance spectra of g-$C_3N_4$, $TiO_2$, and g-$C_3N_4$/$TiO_2$ composites. The exfoliated g-$C_3N_4$ nanosheets show an absorption edge at 470 nm with a band gap of 2.93 eV, which was in agreement with previous reports [53]. The absorption spectra of the $TiO_2$ shows an absorption edge at around 400 nm with a band gap of 3.20 eV. The presence of g-$C_3N_4$ resulted in the red shift of the absorption edge in all composites, revealing that the composites can be applied to visible-light photocatalysis. In addition, the presence of $Ti^{3+}$-$TiO_2$ can narrow the wide band gap of $TiO_2$ for harvesting visible light and can provide an increase in electronic conductivity [54].

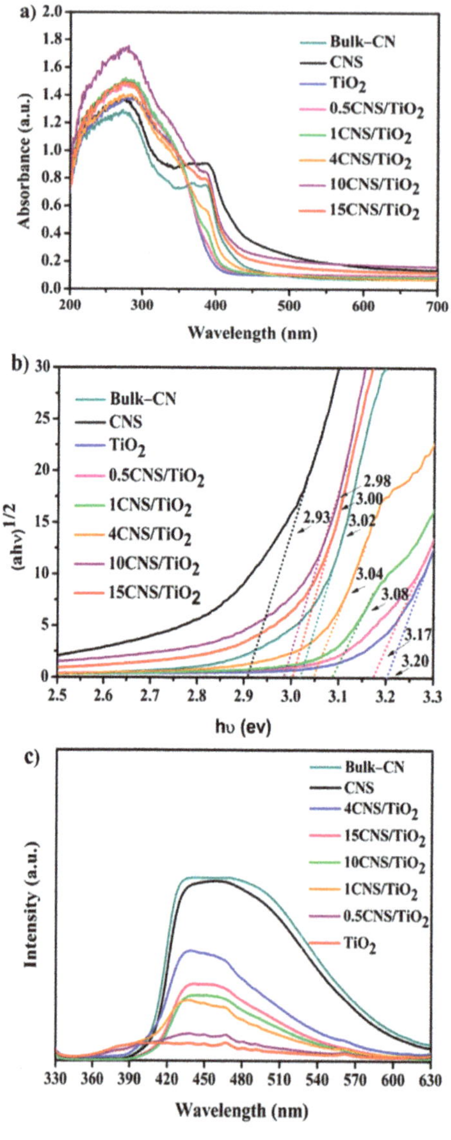


**Figure 4.** (**a**,**b**) UV-Vis DRS spectra and Tauc plot of g-$C_3N_4$, $TiO_2$, and g-$C_3N_4/TiO_2$ photocatalysts, (**c**) photoluminescence spectra of the as-synthesized g-$C_3N_4$, $TiO_2$, and g-$C_3N_4/TiO_2$.

The band gap energy was calculated using the Tauc plot in Equation (1) and is shown in Figure 4b [55].

$$\alpha h\nu = A(h\nu - Eg)^{1/2} \tag{1}$$

where $\alpha$ is the optical absorption coefficient, h is Planck's constant, $\nu$ is photon frequency, A is constant, and Eg is band gap.

The band gaps of pure g-$C_3N_4$, $TiO_2$, and 0.5CNS/$TiO_2$ were calculated to be 2.93, 3.20, and 3.17 eV, respectively.

Photoluminescence analysis was performed in order to determine the electron–hole recombination which is shown in Figure 4c. Under excitation at 320 nm, the emission peak of g-$C_3N_4$ appears at around 457 nm. The bulk-CN and CNS showed high PL intensity because of the fast recombination of electron–hole pairs, whereas $TiO_2$ showed a broad emission peak at 410 nm and a lower maximum peak than that of the g-$C_3N_4$ system. After the hybridization of g-$C_3N_4$ and $TiO_2$, the composite showed a much weaker emission peak, implying that the recombination of charge carriers may be effectively inhibited.

### 2.4. Photocatalytic Study

The photocatalytic performance of g-$C_3N_4$, $TiO_2$, and composites was evaluated for IMI degradation under UV-Vis light irradiation in Figure 5. Photolysis of IMI degradation was carried out under the same conditions, as can be seen from Figure S2. It was found that the photolysis is not the main cause of effective degradation of IMI. On the other hand, the treatments of IMI with catalysts are less effective in dark conditions. From this result, the g-$C_3N_4$ system exhibited low photocatalytic efficiency in the degradation of IMI. This could be because of the fast recombination of electron–hole pairs, as evidenced by PL spectra (Figure 4c). However, it was found that exfoliated g-$C_3N_4$ showed higher photocatalytic activity than bulk-g-$C_3N_4$. It could be explained as the effects of a larger specific surface area, narrow band gap, and nitrogen defects which improved photogenerated charge separation and transfer [56]. In addition, the incorporation of $TiO_2$ clearly enhances the photocatalytic activity of g-$C_3N_4$. Specifically, 0.5CNS/$TiO_2$ and 1CNS/$TiO_2$ exhibited excellent photocatalytic activity, and photocatalysts were able to degrade 93.1% and 88.3% of IMI within 150 min, while pure $TiO_2$ and g-$C_3N_4$ only degraded IMI by 79.7% and 51.8%, respectively. It is reasonable that there might have been a synergetic effect between $TiO_2$ and g-$C_3N_4$. g-$C_3N_4$ can narrow the band gap energy and increase solar absorption efficiency. In addition, $Ti^{3+}$ and oxygen vacancies (Ov) in $TiO_2$ can suppress the recombination of photogenerated electron–hole pairs and promote charge separation [57] which led to high photocatalytic degradation of IMI. As seen in Figure 4c, CNS gave a very broad PL spectrum having a very high intensity. In addition, the PL intensity from 0.5CNS/$TiO_2$ was found just slightly higher than that from the $TiO_2$ material (much weaker than that of CNS). Several works [58,59] related the intensity of PL spectra to the oxidation–reduction potential between the conduction band and the valence band. PL spectra with lower intensity described a low probability of photogenerated electron–hole recombination. Although PL results suggest a slightly faster recombination rate on the 0.5CNS/$TiO_2$, its relatively narrow band gap (compared with $TiO_2$) promoted superior IMI removal efficiencies (Figure 5). It should be noted that the loading level of g-$C_3N_4$ played an important role in improving the IMI photodegradation. It was found that the photocatalytic rate of activity slightly decreased after 90 min irradiation time when the loading of g-$C_3N_4$ was increased from 4% to 15%. This might be due to the fast recombination of the electron–hole pair in g-$C_3N_4$. Furthermore, from Figure 5, 0.5CNS/$TiO_2$ gave a higher IMI degradation rate than that of 1CNS/$TiO_2$, as seen from the slope. However, the IMI removal efficiencies obtained from the 4CNS/$TiO_2$ treatment were higher during 0–120 min. Effective photodegradation of organic compounds requires a suitable amount of stable radical species in the aqueous media. Too-high concentrations of radical species may cause termination of the radical reaction pathway, while insufficient radical concentrations resulted in slow degradation rates and low removal efficiencies. We could explain the removal efficiencies by the varied concentrations of radicals over time. Hence, after 150 min, three samples (i.e., 0.5CNS/$TiO_2$, 1CNS/$TiO_2$, and $TiO_2$) gave % IMI removal efficiencies of 80% and above, likely due to the suitable amount of stable radical species in the aqueous media through the prolonged degradation process.

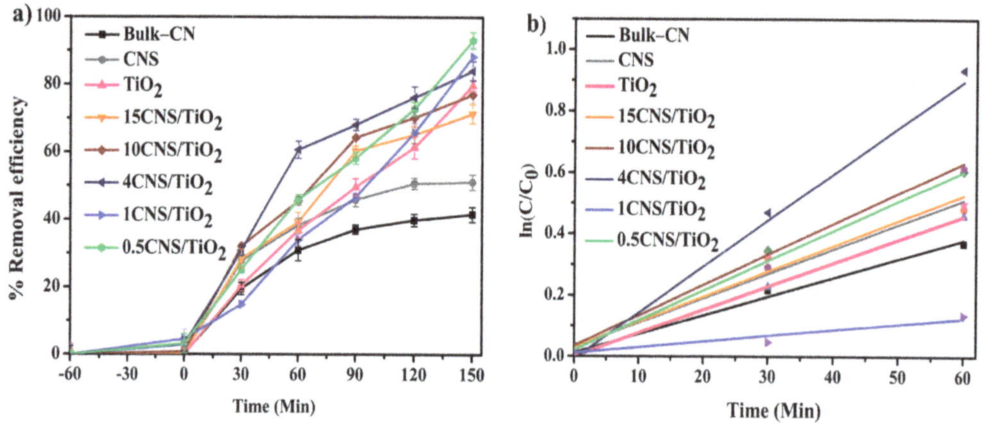


**Figure 5.** (**a**) Photocatalytic degradation of imidacloprid when treated with g-$C_3N_4$, $TiO_2$, and g-$C_3N_4$/$TiO_2$ composites under UV-Vis light irradiation (10 ppm of pesticide and 1g/L of catalyst loading), and (**b**) the first-order kinetic fitting curve of the photocatalytic IMI degradation during 30 min irradiation time.

The initial rate constants, derived from the first-order kinetic fitting curve (Figure 5b), for the photodegradation of IMI from the highest to the lowest, are given in the order of 4CN/$TiO_2$ ($1.50 \times 10^{-2}$ $min^{-1}$), 10CNS/$TiO_2$ ($9.96 \times 10^{-3}$ $min^{-1}$), 0.5CNS/$TiO_2$ ($9.70 \times 10^{-3}$ $min^{-1}$), 15CNS/$TiO_2$ ($8.26 \times 10^{-3}$), CNS ($8.00 \times 10^{-3}$ $min^{-1}$), $TiO_2$ ($7.60 \times 10^{-3}$ $min^{-1}$), bulk-CN ($6.13 \times 10^{-3}$ $min^{-1}$), and 1CNS/$TiO_2$ ($1.88 \times 10^{-3}$ $min^{-1}$) catalysts. As a result, the initial rate constants are poorly correlated with the IMP removal efficiencies after 180 min of irradiation time, possibly due to the stability of radical species as a function of time discussed earlier.

### *2.5. Reusability and Regeneration*

The stability of the photocatalysts was evaluated over multiple cycles of IMI degradation. As shown in Figure 6a, the IMI removal efficiency of 0.5CN/$TiO_2$ decreased significantly in the fourth cycle. The SEM image (Figure 6b,c) shows that the sheet-like morphology of the photocatalyst remained. However, the surface of the catalysts could be covered either by reactants or products that hindered photocatalytic performance. A regeneration experiment was carried out. After the photocatalysis experiment, the catalyst was separated from the reaction mixture by centrifugation. The used photocatalyst was regenerated by stirring in water (dark) for 1 h and irradiated for 2.5 h before using it in the next cycle. It was found that the 0.5CN/$TiO_2$ composite still kept ~91% regeneration efficiency at the end of the fourth cycle, indicating a relatively high regeneration potential of the nanocomposite.

From this work, the bulk carbon nitride is less suitable than the exfoliated material to be incorporated with $TiO_2$ for photocatalytic applications. The IMI removal efficiencies obtained from the 4CNS/$TiO_2$ treatment are significantly higher (ca. 30%) than those obtained from 4CN/$TiO_2$ (Figure S1, Supplementary Data). The photocatalytic performance of several carbon nitride based composites in the degradation of imidacloprid is given in Table 1.

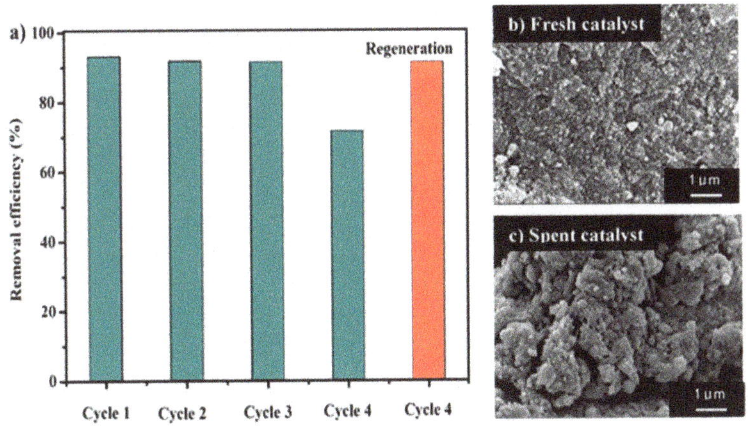


**Figure 6.** (**a**) Reusability and regeneration performance test of $0.5CN/TiO_2$ for imidacloprid degradation, SEM images of (**b**) fresh and (**c**) spent $0.5CN/TiO_2$ photocatalyst

**Table 1.** Comparative photocatalytic degradation of imidacloprid pesticide at varying conditions, over various carbon nitride based materials.

| Photocatalyst | Light Source | Cat. Loading (g/L) | Initial (IMI) (ppm) | Irradiation Time (h) | Best Removal Eff. (%) | Ref. |
|---|---|---|---|---|---|---|
| $g\text{-}C_3N_4$ (urea)<br>$g\text{-}C_3N_4$ (melamine) | λ > 400 nm (8 W) | 0.5<br>1.0 | 20 | 5.0 | 90<br>43 | [60] |
| $Ag_2O/g\text{-}C_3N_4$ | Infrared lamp (250 W) | 1.0 | 10 | 2.0 | 80 | [61] |
| $g\text{-}C_3N_4$<br>P doped $g\text{-}C_3N_4$ (PCN)<br>$0.04C_{60}/PCN$ | LED lamp (35 W) | 0.6 | 26 | 9.0 | 60<br>72<br>95 | [62] |
| $g\text{-}C_3N_4$<br>$Ag\text{-}Bi_2O_3/g\text{-}C_3N_4$ | LED lamp (35 W) | 0.5 | 26 | 8.0 | 65<br>98 | [63] |
| $Bi_2WO_6$: $NH_2$-MOF | Xe lamp | 0.4 | 10 | 3.0 | 84 | [64] |
| $Ag_4V_2O_7/g\text{-}C_3N_4$ | Xe lamp (300 W) | 1.0 | 10 | 4.0 | 38 | [65] |
| $g\text{-}C_3N_4$ (urea)<br>CNS<br>$0.5CNS/TiO_2$ | W lamp (300 W) | 1.0 | 10 | 2.5 | 42<br>51<br>93 | This work |

As seen in Table 1, a quite prolonged reaction time (5 h) was required in order to achieve high IMI removal efficiencies in the photocatalytic treatments of IMI (aq) over the $g\text{-}C_3N_4$ materials, and the photocatalytic performance of $g\text{-}C_3N_4$ is precursor-dependent. Direct comparison of the catalytic performance of the reported photocatalysts and those developed in this work could not be entirely appropriate as each report utilized specific performance testing setups and conditions (initial concentration, catalyst loading, and reaction time). Nevertheless, a greater number of steps and expensive chemicals would be required to prepare several functional photocatalysts ($0.04C_{60}/PCN$, $Ag\text{-}Bi_2O_3/g\text{-}C_3N_4$, $Bi_2WO_6$: $NH_2$-MOF, $Ag_4V_2O_7/g\text{-}C_3N_4$), compared to this work.

### *2.6. Photocatalytic Mechanism*

To find out the major active species for the photocatalytic oxidation, several scavengers were added to the photocatalytic system individually to trap and remove active species

(Figure 7). Ammonium oxalate (AO), isopropanol (IPA), and benzoquinone (BQ) act as scavengers to holes ($h^+$), hydroxyl radical (•OH), and superoxide radical ($•O_2^-$), respectively. The addition of p-benzoquinone had a little effect on the photocatalytic degradation of IMI, implying that $•O_2^-$ has a minor role in the reaction as an oxidative species. In contrast, the photodegradation activity of the 0.5CNS/$TiO_2$ had a dramatic decrease with the addition of IPA and AO, suggesting that both $OH^-$ and holes are the main oxidative species in this system.

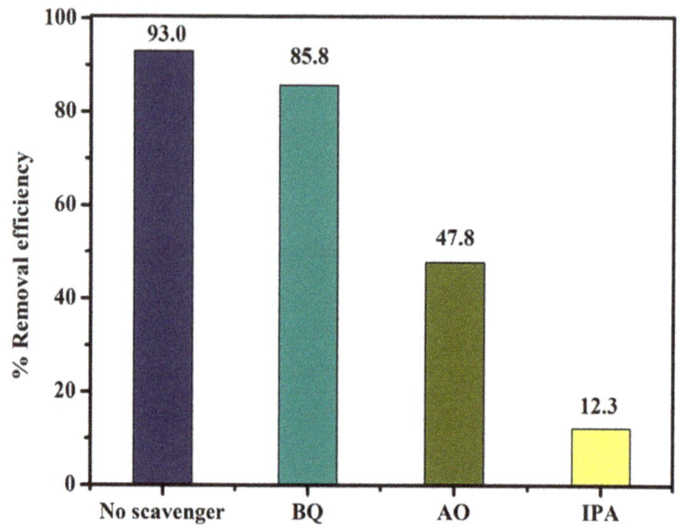


**Figure 7.** Effects of different scavengers on the photocatalytic degradation of IMI.

In order to describe the photocatalytic mechanism of 0.5CNS/$TiO_2$ for the degradation of IMI, the CB and VB edge potentials of g-$C_3N_4$ and $TiO_2$ were calculated from Equations (2) and (3) [66].

$$E_{CB} = X - E_c - 1/2E_g \tag{2}$$

$$E_{VB} = E_{CB} + E_g \tag{3}$$

where X is the absolute electronegativity of the atom semiconductor, and the X values of $TiO_2$ and g-$C_3N_4$ are 5.8 eV and 4.73 eV, respectively [66]. $E_c$ is the energy of free electrons of the hydrogen scale (4.5 Ev). $E_g$ is the band gap of the semiconductor which is 2.93 and 3.20 eV for g-$C_3N_4$ and $TiO_2$, respectively. Therefore, the reductive potentials of the conduction band (CB) are −0.30 and −1.23 V for $TiO_2$ and g-$C_3N_4$, and the oxidizing potentials of the valence band (VB) of $TiO_2$ and g-$C_3N_4$ are +2.90 and +1.70 V, respectively.

Based on the above results, the possible Z-scheme photocatalytic mechanism of g-$C_3N_4$/$TiO_2$ was proposed as shown in Figure 8. Under UV-Vis irradiation, $TiO_2$ absorbed photon energy, and then electrons were excited from the VB to the CB. The photogenerated holes tended to stay in the VB of $TiO_2$, whereas photogenerated electrons on the CB of $TiO_2$ can be directly transferred into the VB of g-$C_3N_4$ due to their proximity to each other. Then, the electrons in the VB of g-$C_3N_4$ are further excited into the CB. This resulted in an efficient charge separation of the photo-induced electron–hole pair and an enhancement in their oxidation–reduction ability. Specifically, the presence of $Ti^{3+}$ and oxygen vacancy could be an important reason for the hindrance of the electron–hole recombination. It was found that the photogenerated holes ($h^+$) in the VB of $TiO_2$ ($E_{VB}$ = 2.90 V vs. NHE) have the ability to oxidize $H_2O$ or hydroxyl ions ($OH^-$) to hydroxyl radicals (•OH), while the photogenerated $h^+$ in the VB of g-$C_3N_4$ ($E_{VB}$ = 1.70 V vs. NHE) is not sufficient for the oxidation of $H_2O$ to hydroxyl radicals. In addition, the photogenerated electron in the CB of g-$C_3N_4$ was trapped on the surface to form reactive superoxide radical ions ($\bullet O_2^-$). The photocatalytic mechanism was consistent with the scavenger experiments in which the hydroxyl radical and holes were the principal reactive species for the IMI degradation, whereas the superoxide radical had a minor role. The Z-scheme photocatalyst was suggested since the photogenerated h+ on the $TiO_2/g\text{-}C_3N_4$ composite has a sufficient oxidation potential for producing $\bullet OH$ radicals [67]. Evaluated by using Equations (2) and (3), the reduction potential of $g\text{-}C_3N_4$ (+1.70 V) is less positive to oxidize $H_2O$ to $\bullet OH$ (+1.99 V). Thus, the holes in the VB of $g\text{-}C_3N_4$ cannot adsorb water molecules near the surface of $g\text{-}C_3N_4$ to generate hydroxyl radicals ($\bullet OH$). Note that $\bullet OH$ radicals can be produced on semiconductors with an oxidation potential of 2.4 V (and above) versus NHE. The scavenging testing indicated that $\bullet OH$ radicals are the key radicals promoting effective IMI degradation. The Z-scheme $g\text{-}C_3N_4/TiO_2$ composites showed better photocatalytic performance than $TiO_2$ or $g\text{-}C_3N_4$ alone. However, with the content of $g\text{-}C_3N_4$ in $g\text{-}C_3N_4/TiO_2$ being in excess, numerous photo-induced electrons and holes would recombine easily. Therefore, the 0.5CNS/$TiO_2$ sample displayed the best photocatalytic performance among these different $g\text{-}C_3N_4/TiO_2$ photocatalysts.

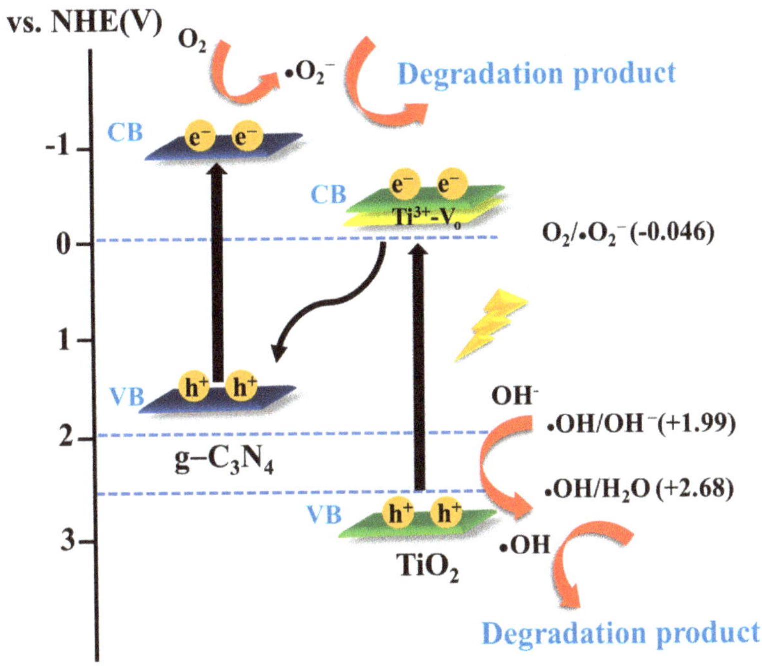


**Figure 8.** Photocatalytic mechanism of 0.5CNS/$TiO_2$ for degradation of imidacloprid.

## 3. Materials and Methods

### *3.1. Chemicals*

Urea ($CH_4N_2O$) was obtained from Kemaus, Australia. Ammonium oxalate $(NH_4)_2C_2O_4$, nitric acid ($HNO_3$), and methanol ($CH_3OH$) were purchased from Merck, Darmstadt, Germany. Benzoquinone, isopropyl alcohol, titanium (IV) oxysulfate ($TiOSO_4$) and imidacloprid were obtained from Sigma-Aldrich, USA. All reagents were of analytical grade and were used without further purification. Deionized water was used in preparing all aqueous solutions.

### 3.2. Characterization

The sample was determined using powder X-ray diffraction (Bruker AXS, D8 advance, Germany) with CuKα radiation (λ = 1.54 Å) and was collected in 2θ range from 10° to 90°. The X-ray photoelectron spectroscopy (XPS) was carried out to determine the surface electronic state of the samples with a monochromatized Al Kα radiation source (AXIS Ultra DLD, Japan). The Raman spectra were recorded using Raman microscope at laser wavelength of 785 nm (Horiba, XploRA Plus, France). The electron paramagnetic resonance (EPR) signals of free radicals were recorded at ambient temperature (Bruker; Elexsys 500, Germany). The morphologies and elemental composites of the samples were examined by a scanning electron microscope (JEOL, JSM-IT500, Japan) and Field emission scanning electron microscope (FE-SEM, JEOL, JSM-7610FPlus, Japan). Photoluminescence (PL) spectrum was performed with excitation at 320 nm (Horiba, FluoroMax, France). The band gap energy of the prepared samples was carried out by UV-Vis NIR spectrophotometer (Shimadzu, UV3600 plus, Tokyo, Japan). The degradation of IMI was monitored by measuring the absorbance at 268 nm with a UV-Vis spectrophotometer (Perkin Elmer, Lambda 800, MA, USA).

### 3.3. Synthesis of g-$C_3N_4$

Bulk-g-$C_3N_4$ was prepared by direct pyrolysis of urea according to a reported procedure with modification [68]. In detail, 125 g of urea was put into an alumina crucible with a cover and heated with a heating rate of 10 °C/min to 600 °C for 4 h. After being cooled to room temperature, the pale yellow of bulk-g-$C_3N_4$ (CN) was obtained. The bulk-CN was exfoliated into a nanosheet structure by thermal exfoliation in the presence of acid, and 2.5 g of bulk-CN was stirred in 65 % of $HNO_3$ solution (100 mL) for 12 h. The dispersion was filtrated and washed several times with D.I. water, followed by annealing at 500 °C for 4 h to obtain g-$C_3N_4$ nanosheet (CNS).

### 3.4. Synthesis of g-$C_3N_4$/$TiO_2$ Composites

g-$C_3N_4$/$TiO_2$ composites were prepared by a hydrothermal method. Firstly, $TiOSO_4$ suspensions were obtained by sonication for 35 min in D.I. water. After dispersing, CNS was added into the solution and was sonicated continuously for 30 min. The solution was transferred to a Teflon-lined autoclave which was then further heated at 180 °C for 4 h. The obtained solution was centrifuged, washed with D.I. water, and dried at 65 °C for 24 h. According to the above method, different weight ratios of g-$C_3N_4$ to $TiO_2$ at 0.5%, 1%, 4%, 10%, and 15% were synthesized and labeled as 0.5CNS/$TiO_2$, 1CNS/$TiO_2$, 4CNS/$TiO_2$, 10CNS/$TiO_2$, and 15CNS/$TiO_2$, respectively. $TiO_2$ was also prepared by the same procedure without adding g-$C_3N_4$.

### 3.5. Photocatalytic Activity

The photocatalytic behavior of the catalyst was evaluated by the photodegradation of imidacloprid in an aqueous solution with an initial concentration of 10 mg/L under 300 watts of W lamp. First, 10 mg of photocatalyst was added to 10 mL of IMI solution. The suspension was stirred using a magnetic stirrer in the dark at room temperature for 1 h before irradiation. The solution was collected every 30 min to 150 min using a syringe with a microspore filter (0.45 μm). The concentration of IMI was analyzed by UV-Vis spectrophotometry at 268 nm, and the removal efficiency was calculated via Equation (4) [69].

Where $C_0$ is the initial concentration of IMI, and $C_t$ is the concentration of IMI after t minutes.

To study the reaction kinetic, the obtained data were fitted by a first-order kinetic model which is shown in Equation (2).

$$\ln(C_0/C) = kt \tag{4}$$

where k is the pseudo-first-order rate constant, $C_0$ is the initial IMI concentration, and C is imidacloprid equilibrium concentration in aqueous solution at time t.

### 3.6. Scavenger Activity

Active species capture experiments were used to study the photocatalysis mechanism. First, isopropyl alcohol (IPA; 0.5 mM) was used as the hydroxyl scavenger ($\bullet OH^{\cdot}$), benzoquinone (BQ; 0.5 mM) was employed as the superoxide scavenger ($\bullet O_2^-$), and ammonium oxalate (AO; 0.5 mM) was used as the hole scavenger ($h^+$) [70]. Different scavengers were used in the trapping experiments to check the inhibitory effect of scavengers during the photocatalytic reaction under analogous irradiation experimental conditions.

## 4. Conclusions

In this work, g-$C_3N_4$ nanosheets were exfoliated from bulk-g-$C_3N_4$ by thermal exfoliation in the presence of $HNO_3$. The exfoliated g-$C_3N_4$ resulted in nanosheets with a large specific surface area and N vacancy defects, which can be prepared and further used in the fabrication of $TiO_2$ base composites having superior photocatalytic activity, under UV-Vis light irradiation, to bulk the g-$C_3N_4$ and g-$C_3N_4$/$TiO_2$ composites. Results show that the g-$C_3N_4$/$TiO_2$ photocatalyst exhibited higher removal efficiency for IMI than g-$C_3N_4$ and $TiO_2$, indicating that a synergistic effect exists between $Ti^{3+}$-$TiO_2$ and g-$C_3N_4$. With the increase of g-$C_3N_4$ loading, the photocatalytic activity of g-$C_3N_4$/$TiO_2$ composites may also decrease the photocatalytic activity of g-$C_3N_4$. The sample of 0.5CNS/$TiO_2$ showed the highest photocatalytic activity with 93% removal efficiency within 150 min. The enhanced photocatalytic performance of the g-$C_3N_4$/$TiO_2$ composites could be due to the generation of reactive oxidation species induced by photogenerated electrons and the effective suppression of the recombination of the charge carriers. In addition, the g-$C_3N_4$/$TiO_2$ photocatalyst showed good stability for multiple recycling. Thus, the g-$C_3N_4$/$TiO_2$ could be effectively used as material for the photodegradation of imidacloprid pesticide in wastewater. Comprehensive photoelectrochemical analysis of the g-$C_3N_4$/$TiO_2$ materials should be further studied to obtain the photogenerated charge recombination rates in detail.

**Supplementary Materials:** The following supporting information can be downloaded at: https://www.mdpi.com/article/10.3390/catal12020120/s1, Figure S1: Powder XRD patterns of 1CNS/$TiO_2$, 4CNS/$TiO_2$, 10CNS/$TiO_2$, and 15CNS/$TiO_2$, Figure S2: Photolysis of imidacloprid.

**Author Contributions:** Conceptualization, methodology, formal analysis, investigation, visualization, T.K. and S.M.S.; writing—original draft preparation, T.K. and J.T.; writing— review and editing, S.M.S.; supervision, T.A. and S.M.S.; writing—review and editing, S.M.S.; resources, P.K. and S.M.S.; funding acquisition, P.K. and S.M.S. All authors have read and agreed to the published version of the manuscript.

**Funding:** National Research Council of Thailand (Grant No. IRN62W0005).

**Data Availability Statement:** The data presented in this study are openly available in Mendeley repository at doi:10.17632/3v2mgmgzpz.1.

**Acknowledgments:** This work was partially supported by the National Research Council Thailand (Grant No. IRN62W0005). We thank Mahidol University Frontier Research Facility (MU-FRF) for instrument support and the MU-FRF scientists, Nawapol Udpuay and Suwilai Chaveanghong, for their kind assistance in Raman and FE-SEM analyses.

**Conflicts of Interest:** The authors declare no conflict of interest.

## References


1. Wumuerhan, P.; Yuntao, J.; Deying, M. Effects of exposure to imidacloprid direct and poisoned cotton aphids Aphis gossypii on ladybird Hippodamia variegata feeding behavior. *J. Pestic. Sci.* **2020**, *45*, 24–28. [CrossRef]
2. Stanneck, D.; Ebbinghaus-Kintscher, U.; Schoenhense, E.; Kruedewagen, E.M.; Turberg, A.; Leisewitz, A.; Jiritschka, W.; Krieger, K.J. The synergistic action of imidacloprid and flumethrin and their release kinetics from collars applied for ectoparasite control in dogs and cats. *Parasites Vectors* **2012**, *5*, 73. [CrossRef]
3. Borsuah, J.F.; Messer, T.L.; Snow, D.D.; Comfort, S.D.; Mittelstet, A.R. Literature review: Global neonicotinoid insecticide occurrence in aquatic environments. *Water* **2020**, *12*, 3388. [CrossRef]
4. Berheim, E.H.; Jenk, J.A.; Lundgren, J.G.; Michel, E.S.; Grove, D.; Jensen, W.F. Effects of neonicotinoid insecticides on physiology and reproductive characteristics of captive female and fawn white-tailed deer. *Sci. Rep.* **2019**, *9*, 4534. [CrossRef]
5. Yang, E.C.; Chuang, Y.C.; Chen, Y.L.; Chang, L.H. Abnormal foraging behavior induced by sublethal dosage of imidacloprid in the honey bee (*Hymenoptera: Apidae*). *J. Econ. Entomol.* **2008**, *101*, 1743–1748. [CrossRef]
6. Nyman, A.-M.; Hintermeister, A.; Schirmer, K.; Ashauer, R. The insecticide imidacloprid causes mortality of the freshwater amphipod *gammarus pulex* by interfering with feeding behavior. *PLoS ONE* **2013**, *8*, e62472. [CrossRef]
7. Sriapha, C.; Trakulsrichai, S.; Tongpool, A.; Pradoo, A.; Rittilert, P.; Wananukul, W. Acute imidacloprid poisoning in Thailand. *Ther. Clin. Risk Manag.* **2020**, *16*, 1081–1088. [CrossRef]
8. Mundhe, S.A.; Birajdar, S.V.; Chavan, S.S.; Pawar, N.R. Imidacloprid poisoning: An emerging cause of potentially fatal poisoning. *Indian J. Crit Care Med.* **2017**, *21*, 786–788. [CrossRef]
9. Crall, J.D.; Switzer, C.M.; Oppenheime, R.L.; Versypt, A.N.F.; Dey, B.; Brown, A.; Eyster, M.; Guérin, C.; Pierce, N.E.; Combes, S.A.; et al. Neonicotinoid exposure disrupts bumblebee nest behavior, social networks, and thermoregulation. *Science* **2018**, *362*, 683–686. [CrossRef]
10. Patil, A.L.; Patil, P.N.; Gogate, P.R. Degradation of imidacloprid containing wastewaters using ultrasound-based treatment strategies. *Ultrason. Sonochem.* **2014**, *21*, 1778–1786. [CrossRef]
11. Zhang, Y.; Lu, H.; Wang, B.; Zhang, Z.; Lin, X.; Chen, Z.; Li, B. Removal of imidacloprid and acetamiprid from tea infusions by microfiltration membrane. *Int. J. Food Sci.* **2015**, *50*, 1397–1404.
12. Sabourmoghaddam, N.; Zakaria, M.P.; Omar, D. Evidence for the microbial degradation of imidacloprid in soils of Cameron Highlands. *J. Saudi Soc. Agric. Sci.* **2015**, *14*, 182–188. [CrossRef]
13. Urbain, K.Y.; Fodjo, E.K.; Ardjouma, D.; Serge, B.Y.; Aimé, E.S.; Marc, G.B.I.; Albert, T. Removal of imidacloprid using activated carbon produced from ricinodendron heudelotii shells. *Bull. Chem. Soc. Ethiop.* **2017**, *31*, 397–409. [CrossRef]
14. Mohammad, S.G.; El-Sayed, M.M.H. Removal of imidacloprid pesticide using nanoporous activated carbons produced via pyrolysis of peach stone agricultural wastes. *Chem. Eng. Commun.* **2020**, *208*, 1069–1080. [CrossRef]
15. Soltani-nezhad, F.; Saljooqi, A.; Shamspur, T.; Mostafavi, A. Photocatalytic degradation of imidacloprid using $GO/Fe_3O_4/TiO_2$-NiO under visible radiation: Optimization by response level method. *Polyhedron* **2019**, *165*, 188–196. [CrossRef]
16. Lacson, C.F.Z.; de Luna, M.D.G.; Dong, C.; Garcia-Segura, S.; Lu, M.-C. Fluidized-bed fenton treatment of imidacloprid: Optimization and degradation pathway. *Sustain. Environ. Res.* **2018**, *28*, 309–314. [CrossRef]
17. Lee, Y.-J.; Ka, J.-K.; Park, S.-J.; Lee, C.-G.; Moon, J.-K.; Alvarez, P.J.J. Photocatalytic degradation of neonicotinoid insecticides using sulfate-doped $Ag_3PO_4$ with enhanced visible light activity. *Chem. Eng. Sci.* **2020**, *402*, 126183. [CrossRef]
18. Kumar, A.; Pandey, G. A review on the factors affecting the photocatalytic degradation of hazardous materials. *Mater. Sci. Eng. Int.* **2017**, *1*, 106–114. [CrossRef]
19. Wen, J.; Xie, J.; Chen, X.; Li, X. A review on g-$C_3N_4$-based photocatalysts. *Appl. Surf. Sci.* **2017**, *391*, 72–123. [CrossRef]
20. Liu, R.; Chen, Z.; Yao, Y.; Li, Y.; Cheema, W.A.; Wang, D.; Zhu, S. Recent advancements in g-$C_3N_4$-based photocatalysts for photocatalytic $CO_2$ reduction: A mini review. *RSC Adv.* **2020**, *10*, 29408. [CrossRef]
21. Zou, L.R.; Huang, G.-F.; Li, D.-F.; Liu, J.-H.; Pan, A.-L.; Huang, W.-Q. A facile and rapid route for synthesis of g-$C_3N_4$ nanosheets with high adsorption capacity and photocatalytic activity. *RSC Adv.* **2016**, *6*, 86688–86694. [CrossRef]
22. Yang, Y.; Chen, J.; Mao, Z.; An, N.; Wang, D.; Fahkman, B.D. Ultrathin g-$C_3N_4$ nanosheets with an extended visible-light-responsive range for significant enhancement of photocatalysis. *RSC Adv.* **2017**, *7*, 2333. [CrossRef]
23. Wang, P.; Yang, M.; Tang, S.; Chen, F.; Li, Y. Preparation of cellular $C_3N_4/CoSe_2/GA$ composite photocatalyst and its $CO_2$ reduction activity. *Chem. J. Chin. Univ.* **2021**, *42*, 6.
24. Jiang, L.; Yuan, X.; Pan, Y.; Liang, J.; Zeng, G.; Wu, Z.; Wang, H. Doping of graphitic carbon nitride for photocatalysis: A review. *Appl. Catal. B* **2017**, *217*, 388–406. [CrossRef]
25. Zhang, H.; Liu, F.; Wu, H.; CaO, X.; Sun, J.; Lei, W. In situ synthesis of g-$C_3N_4/TiO_2$ heterostructures with enhanced photocatalytic hydrogen evolution under visible light. *RSC Adv.* **2017**, *7*, 40327. [CrossRef]
26. Kanwal, M.; Tariq, S.R.; Chotana, G.A. Photocatalytic degradation of imidacloprid by Ag-ZnO composite. *Environ. Sci. Pollut. Res.* **2018**, *25*, 27307–27320. [CrossRef]
27. Lin, X.; Du, S.; Li, C.; Li, G.; Li, Y.; Chen, F.; Fang, P. Consciously Constructing the Robust NiS/g-$C_3N_4$ Hybrids for Enhanced Photocatalytic Hydrogen Evolution. *Catal. Lett.* **2020**, *150*, 1898–1908. [CrossRef]
28. Chen, P.; Liu, F.; Ding, H.; Chen, S.; Chen, L.; Li, Y.-J.; Au, C.-T.; Yin, S.-F. Porous double-shell CdS@$C_3N_4$ octahedron derived by in situ supramolecular self-assembly for enhanced photocatalytic activity. *Appl. Catal. B Environ.* **2019**, *252*, 33–40. [CrossRef]

29. Schneider, J.; Matsuoka, M.; Takeuchi, M.; Zhang, J.; Horiuchi, Y.; Anpo, M.; Bahnemann, D.W. Understanding $TiO_2$ photocatalysis: Mechanisms and materials. *Chem. Rev.* **2014**, *114*, 9919–9986. [CrossRef]
30. Ma, Y.; Liu, E.; Hu, X.; Tang, C.; Wan, J.; Li, J.; Fan, J. A simple process to prepare few-layer g-$C_3N_4$ nanosheets with enhanced photocatalytic activities. *Appl. Surf. Sci.* **2015**, *358*, 246–251. [CrossRef]
31. Niu, P.; Zhang, L.; Liu, G.; Cheng, H.-M. Graphene-like carbon nitride nanosheets for improved photocatalytic activities. *Adv. Funct. Mater.* **2012**, *22*, 4763–4770. [CrossRef]
32. Li, Y.; Yang, M.; Xing, Y.; Liu, X.; Yang, Y.; Wang, X.; Song, S. Preparation of carbon-rich g-$C_3N_4$ nanosheets with enhanced visible light utilization for efficient photocatalytic hydrogen production. *Small* **2017**, *13*, 1701552. [CrossRef]
33. Robotti, M.; Dosta, S.; Cana, I.G.; Concustell, A.; Cinca, N.; Guilemany, J.M. Attrition and cryogenic milling powder production for low pressure cold gas spray and composite coatings characterization. *Adv. Powder Technol.* **2016**, *27*, 1257–1264. [CrossRef]
34. Wang, M.; Ma, F.; Wang, Z.; Hu, D.; Xu, X.; HaO, X. Graphitic carbon nitride, a saturable absorber material for the visible waveband. *Photonics Res.* **2018**, *6*, 307–313. [CrossRef]
35. El-Deen, S.S.; Hashem, A.M.; Abdel Ghany, A.E.; Indris, S.; Ehrenberg, H.; Mauger, A.; Julien, C.M. Anatase $TiO_2$ nanoparticles for lithium-ion batteries. *Ionics* **2018**, *24*, 2925–2934. [CrossRef]
36. Zhang, J.-R.; Ma, Y.; Wang, S.-Y.; Ding, J.; Gao, B.; Kan, E.; Hua, W. Accurate K-edge X-ray photoelectron and absorption spectra of g-$C_3N_4$ nanosheets by first-principles simulations and reinterpretations. *Phys. Chem. Chem. Phys.* **2019**, *21*, 22819. [CrossRef]
37. Che, W.; Cheng, W.; Yao, T.; Tang, F.; Liu, W.; Su, H.; Huang, Y.; Liu, O.; Hu, F.; Pan, Z.; et al. Fast photoelectron transfer in (C ring)-$C_3N_4$ plane heterostructural nanosheets for overall water splitting. *J. Am. Chem. Soc.* **2017**, *139*, 3021–3026. [CrossRef] [PubMed]
38. Bi, G.; Wen, J.; Li, X.; Liu, W.; Xie, J.; Fang, Y.; Zhang, W. Efficient visible-light photocatalytic $H_2$ evolution over metal-free g-$C_3N_4$ co-modified with robust acetylene black and $Ni(OH)_2$ as dual co-catalysts. *RSC Adv.* **2016**, *6*, 31497. [CrossRef]
39. Du, X.; Bai, X.; Xu, L.; Yang, L.; Jin, P. Visible-light activation of persulfate by $TiO_2$/g-$C_3N_4$ photocatalyst toward efficient degradation of micropollutants. *Chem. Eng. Sci.* **2020**, *384*, 123245. [CrossRef]
40. Cao, J.; Pan, C.; Ding, Y.; Li, W.; Lv, K.; Tang, H. Constructing nitrogen vacancy introduced g-$C_3N_4$ p-n homojunction for enhanced photocatalytic activity. *J. Environ. Chem. Eng.* **2019**, *7*, 102984. [CrossRef]
41. Wang, J.; Gao, B.; Dou, M.; Huang, X.; Ma, Z. A porous g-$C_3N_4$ nanosheets containing nitrogen defects for enhanced photocatalytic removal meropenem: Mechanism, degradation pathway and DFT calculation. *Environ. Res.* **2020**, *184*, 109339. [CrossRef]
42. Xi, J.; Zhang, Y.; Chen, X.; Hu, Y. A simple sol-gel hydrothermal method for the synthesis of defective $TiO_2$ nanocrystals with excellent visible-light photocatalytic activity. *Res. Chem. Intermed.* **2020**, *46*, 2205–2214. [CrossRef]
43. Bharti, B.; Kumar, S.; Lee, H.-N.; Kumar, R. Formation of oxygen vacancies and $Ti^{3+}$ state in $TiO_2$ thin film and enhanced optical properties by air plasma treatment. *Sci. Rep.* **2016**, *6*, 32355. [CrossRef] [PubMed]
44. HaO, J.; Hu, Y.; Jiang, H.; Li, C. In situ surface hydrogenation synthesis of Ti3+self-doped TiO2 with enhanced visible light photoactivity. *Nanoscale* **2014**, *6*, 9078. [CrossRef] [PubMed]
45. Cheng, D.; Li, Y.; Yang, L.; Luo, S.; Yang, L.; Luo, X.; Luo, Y.; Li, T.; Gao, J.; Dionysiou, D.D. One step reductive synthesis of $Ti^{3+}$ self–doped elongated anatase $TiO_2$ nanowires combined with reduced graphene oxide for adsorbing and degrading waste engine oil. *J. Hazard. Mater.* **2019**, *378*, 120752. [CrossRef]
46. Swaminathan, J.; Ravichandran, S. Insights into the defect-centered electrocatalytic behavior of reduced titania ($TiO_{1.23}$). *J. Phys. Chem. C* **2018**, *122*, 1670–1680. [CrossRef]
47. Mohajernia, S.; Andryskova, P.; Zoppellaro, G.; Hejazi, S.; Kment, S.; Zboril, R.; Schmidt, J.; Schmuki, P. Influence of $Ti^{3+}$ defect-type on heterogeneous photocatalytic $H_2$ evolution activity of $TiO_2$. *J. Mater. Chem. A* **2020**, *8*, 1432–1442. [CrossRef]
48. Wang, H.; Bu, Y.; Wu, G.; Zouc, X. The promotion of the photocatalytic nitrogen fixation ability of nitrogen vacancy-embedded graphitic carbon nitride by replacing the corner-site carbon atom with phosphorus. *Dalton Tran.* **2019**, *48*, 11724–11731. [CrossRef]
49. Yang, L.; Liu, X.; Liu, Z.; Wang, C.; Liu, G.; Li, Q.; Feng, X. Enhanced photocatalytic activity of g-$C_3N_4$ 2D nanosheets through thermal exfoliation using dicyandiamide as precursor. *Ceram. Int.* **2018**, *44*, 20613–20619. [CrossRef]
50. Che, H.; Liu, L.; Che, G.; Dong, H.; Liu, C.; Li, C. Control of energy band, layer structure and vacancy defect of graphitic carbon nitride by intercalated hydrogen bond effect of $NO_3^-$ toward improving photocatalytic performance. *Chem. Eng. J.* **2019**, *357*, 209–219. [CrossRef]
51. Dong, F.; Li, Y.H.; Wang, Z.Y. Enhanced visible light photocatalytic activity and oxidation ability of porous graphene-like g-$C_3N_4$ nanosheets via thermal exfoliation. *Appl. Surf. Sci.* **2015**, *358*, 393–403. [CrossRef]
52. Wu, X.; Gao, D.; Yua, H.; Yu, J. High-yield lactic acid-mediated route for g-$C_3N_4$ nanosheet photocatalyst with enhanced $H_2$-evolution performance. *Nanoscale* **2019**, *11*, 9608. [CrossRef] [PubMed]
53. Ye, C.; Li, J.-X.; Li, Z.-J.; Li, X.-B.; Fan, X.-B.; Zhang, L.-P.; Chen, B.; Tung, C.-H.; Wu, L.-Z. Enhanced driving force and charge separation efficiency of protonated g-$C_3N_4$ for photocatalytic $O_2$ evolution. *ACS Catal.* **2015**, *5*, 6973–6979. [CrossRef]
54. Li, K.; Huang, Z.; Zeng, X.; Huang, B.; Gao, S.; Lu, J. Synergetic effect of $Ti^{3+}$ and oxygen doping on enhancing photoelectrochemical and photocatalytic properties of $TiO_2$/g-$C_3N_4$ heterojunctions. *ACS Appl. Mater. Interfaces* **2017**, *9*, 11577–11586. [CrossRef] [PubMed]
55. Gao, L.; Gan, W.; Qiu, Z.; Zhan, X.; Qiang, T.; Li, J. Preparation of heterostructured $WO_3/TiO_2$ catalysts from wood fibers and its versatile photodegradation abilities. *Sci. Rep.* **2017**, *7*, 1102. [CrossRef] [PubMed]

56. Xu, J.; Zhang, L.; Shi, R.; Zhu, Y. Chemical exfoliation of graphitic carbon nitride for efficient heterogeneous photocatalysis. *J. Mater. Chem.* **2013**, *1*, 14766–14772. [CrossRef]
57. Na, S.; Sao, S.; Lee, H. Recent Developments of advanced $Ti^{3+}$-self-doped $TiO_2$ for efficient visible-light-driven photocatalysis. *Catalysts* **2020**, *10*, 679. [CrossRef]
58. Lu, D.; Zhang, G.; Wan, Z. Visible-light-driven g-$C_3N_4$/$Ti^{3+}$-$TiO_2$ photocatalyst co-exposed {0 0 1} and {1 0 1} facets and its enhanced photocatalytic activities for organic pollutant degradation and Cr(VI) reduction. *Appl Surf. Sci.* **2015**, *358*, 223–230. [CrossRef]
59. Lin, H.; Zhao, L. Novel g-$C_3N_4$/$TiO_2$ nanorods with enhanced photocatalytic activity for water treatment and $H_2$ production. *J. Mater. Sci. Mater. Electron.* **2019**, *30*, 18191–18199. [CrossRef]
60. Liu, X.; Wu, X.; Long, Z.; Zhang, C.; Ma, Y.; Hao, X.; Zhang, H.; Pan, C. Photodegradation of Imidacloprid in Aqueous Solution by Metal Free Catalyst Graphitic Carbon Nitride using An Energy-Saving Lamp. *J. Agric. Food Chem.* **2015**, *63*, 4754–4760. [CrossRef]
61. Liang, S.; Zhang, D.; Pu, X.; Yao, X.; Han, R.; Yin, J.; Ren, Z. A novel $Ag_2O$/g-$C_3N_4$ p-n heterojunction photocatalysts with enhanced visible and near-infrared light activity. *Sep. Purif. Technol.* **2019**, *210*, 786–797. [CrossRef]
62. Sudhaik, A.; Raizada, P.; Singh, P.; Hosseini-Bandegharaei, A.; Thakur, V.K.; Nguyen, V.-H. Highly effective degradation of imidacloprid by $H_2O_2$/ fullerene decorated P-doped g-$C_3N_4$ photocatalyst. *J. Environ. Chem. Eng.* **2020**, *8*, 104483. [CrossRef]
63. Raizada, P.; Sudhaik, A.; Singha, P.; Hosseini-Bandegharaei, A.; Gupta, V.K.; Agarwal, S. Silver-mediated $Bi_2O_3$ and graphitic carbon nitride nanocomposite as all solid-state Z scheme photocatalyst for imidacloprid pesticide abatement from water. *Desalin. Water Treat.* **2019**, *171*, 344–355. [CrossRef]
64. Chen, M.L.; Lu, T.H.; Li, S.S. Photocatalytic degradation of imidacloprid by optimized $Bi_2WO_6$/$NH_2$-MIL-88B(Fe) composite under visible light. *Environ. Sci Pollut Res.* **2021**. [CrossRef]
65. Zhang, T.; Zhao, D.; Wang, Y.; Chang, Y.; Zhang, D.; Tang, Y.; Pu, X.; Shao, X. Facial synthesis of a novel $Ag_4V_2O_7$/g-$C_3N_4$ heterostructure with highly efficient photoactivity. *J. Am. Ceram. Soc.* **2019**, *102*, 3897–3907. [CrossRef]
66. Bi, X.; Yu, S.; Liu, E.; Liu, L.; Zhang, K.; Zang, J.; Zhao, Y. Construction of g-$C_3N_4$/$TiO_2$ nanotube arrays Z-scheme heterojunction to improve visible light catalytic activity. *Colloids Surf. A* **2020**, *603*, 125193. [CrossRef]
67. Liao, G.; Li, C.; Li, X.; Fang, B. Emerging polymeric carbon nitride Z-scheme system for photocatalysis. *Cell Rep.* **2021**, *2*, 100355. [CrossRef]
68. Fang, H.-B.; Luo, Y.; Zheng, Y.-Z.; Ma, W.; TaO, X. Facile large-scale synthesis of urea-derived porous graphitic carbon nitride with extraordinary visible-light spectrum Photodegradation. *Ind. Eng. Chem. Res.* **2016**, *55*, 4506–4514. [CrossRef]
69. Rana, A.G.; Minceva, M. Analysis of photocatalytic degradation of phenol with exfoliated graphitic carbon nitride and light-emitting diodes using response surface methodology. *Catalysts* **2021**, *11*, 898. [CrossRef]
70. Wang, B.; Li, L.; Chen, J.; Duan, C.; Song, J.; Wang, R.; Zhang, B. Synthesis of $BiOCl_{0.5}I_{0.5}$/$TiO_2$ heterojunctions with enhanced visible-light photocatalytic properties. *J. Nanopart. Res.* **2018**, *20*, 175. [CrossRef]

catalysts

MDPI

*Article*

# Advancements on Basic Working Principles of Photo-Driven Oxidative Degradation of Organic Substrates over Pristine and Noble Metal-Modified $TiO_2$. Model Case of Phenol Photo Oxidation

**Alexandra Sandulescu [1], Crina Anastasescu [1], Florica Papa [1], Monica Raciulete [1], Anca Vasile [1], Tanta Spataru [1], Monica Scarisoreanu [2], Claudiu Fleaca [2], Cristian N. Mihailescu [2], Valentin S. Teodorescu [3], Nicolae Spataru [1,*], Maria Zaharescu [1] and Ioan Balint [1,*]**

[1] "Ilie Murgulescu" Institute of Physical Chemistry of the Romanian Academy, Spl. Independentei 202, 060021 Bucharest, Romania; asandulescu@icf.ro (A.S.); canastasescu@icf.ro (C.A.); frusu@icf.ro (F.P.); mpavel@icf.ro (M.R.); avasile@icf.ro (A.V.); tspataru@icf.ro (T.S.); mzaharescu@icf.ro (M.Z.)

[2] National Institute for Lasers, Plasma and Radiation Physics, Atomistilor Str., No. 409, 077125 Bucharest-Magurele, Romania; monica.scarisoreanu@inflpr.ro (M.S.); claudiu.fleaca@inflpr.ro (C.F.); cristi.mihailescu@inflpr.ro (C.N.M.)

[3] Valentin Teodorescu, National Institute of Materials Physics, 405A Atomistilor Street, 077125 Bucharest-Magurele, Romania; teoval@infim.ro

* Correspondence: nspataru@icf.ro (N.S.); ibalint@icf.ro (I.B.)

**Citation:** Sandulescu, A.; Anastasescu, C.; Papa, F.; Raciulete, M.; Vasile, A.; Spataru, T.; Scarisoreanu, M.; Fleaca, C.; Mihailescu, C.N.; Teodorescu, V.S.; et al. Advancements on Basic Working Principles of Photo-Driven Oxidative Degradation of Organic Substrates over Pristine and Noble Metal-Modified $TiO_2$. Model Case of Phenol Photo Oxidation. *Catalysts* **2021**, *11*, 487. https://doi.org/10.3390/catal11040487

Academic Editor: Ewa Kowalska

Received: 16 March 2021
Accepted: 7 April 2021
Published: 10 April 2021



**Abstract:** The specific roles played by both support and noble metals in light absorption, charge separation, and the formation of ·OH and $O_2^-$ (ROS) are analyzed for light-triggered oxidation of phenol (Ph) over pristine and over noble metal (Ag, Au, Pt) -loaded $TiO_2$. Experiments show that the supported noble metals act as a light visible absorber, assist the separation of photo-charges and reduction of $O_2$ to $O_2^-$. The $O_2^-$ oxidizes mildly Ph to oxygenated products (hydroquinone, benzoquinone, and 1,2-dihydroxibenzene). In a parallel process, ·OH radicals, yielded by $TiO_2$, mineralize Ph to $CO_2$ by fast reaction sequences. Radical quenching and photo electrochemical measurements (surface photovoltage) confirm independently that the production of ·OH and $O_2^-$ scale with oxidative conversion of Ph. The selectivity to $CO_2$ and mild oxidation products is the result of the interplay between catalyst activity for ·OH and for $O_2^-$ production.

**Keywords:** photocatalytic mild oxidation; photocatalytic mineralization; phenol oxidation; $TiO_2$ modified with noble metals; surface photovoltage; $O_2$ photo reduction; reactive oxygen species; charge recombination

## 1. Introduction

The sun light driven photo oxidation processes of organic matter are of great importance for several practical reasons: (i) imply low material and operational costs, (ii) are potentially able to clean water and air by mineralization of organic pollutants to $CO_2$ [1–4], and (iii) are attractive alternative routes for selective synthesis of high-added value oxygenated products [5–7].

The thermodynamic of organics oxidation is downhill process ($\Delta G < 0$), the light being used to speed up chemical reaction via generation of charge carriers. The general accepted steps in photocatalytic processes are: (i) light absorption by photocatalysts followed by generation of $e^-$ and $h^+$ charges; (ii) charge transfer to reactant substrate intermediated by reactive oxygen species, ROS; (iii) development of redox processes with participation of $e^-$, $h^+$, and ROS on surface and vicinity of photocatalysts. Metals are involved actively in all essential reaction steps, determining the final overall photocatalytic efficiency [8,9].

The prevalent reaction mechanism in liquid phase depends on a series of factors including nature of photocatalyst, reaction media, and reacting organic substrate. The reaction mechanism, described well by the Langmuir–Hinshelwood adsorption equation [10], implies the interaction of photogenerated charges with the adsorbed species, at a time scale of $10^{-10}$ to $10^{-5}$ s [11]. The photo generated charges ($e^-$ and $h^+$) react first with hydroxyl groups of adsorbed $H_2O$ and $O_2$ to yield reactive $\cdot OH$ ($H_2O + h^+ \rightarrow \cdot OH + H^+$) and $\cdot O_2^-$ ($O_2 + e^- \rightarrow O_2^-$) ROS [9]. The oxidative conversion of organic compounds is intermediated by formation and diffusion of ROS to reaction scene [12], which can be remote from the illuminated surface [13].

Metals dispersed on surface of active materials (i) help separation of photo generated charges, (ii) work as cocatalyst by mediating the charge transfer to reacting substrates, (iii) favor the formation of $O_2^-$, (iv) control the selectivity of oxidation process, (v) bend the energy bands of photocatalysts at solid-liquid interfaces, (vi) modify the light absorbing property of materials, (vii) contribute to enhancement of photocharge production in visible wavelength domain by surface plasmon resonance (SPR) phenomenon [9]. The Schottky regions built at metal-oxide interfaces contribute to electron and hole separation, leading to increased efficiency of photo-driven redox processes. The charge separation efficiency is validated experimentally by the comparing the PL (Photoluminiscence) emission intensity of metal-loaded photocatalysts with pristine oxide [14,15]. The bending of valence band (VB) and conduction band (CB) depends on nature of metal and pH of solution. Metals shift the light absorption edge, in many cases with beneficial effects on efficiency. Noble metals exhibit visible light absorption peaks due SPR phenomenon, which is a collective electron oscillation in metal nanoparticles induced by visible light absorption. The SPR was reported to have in some cases favorable effects on photo-driven redox processes performed in visible light [16–18].

In spite of a large number of published researches, essential issues in photocatalysis remain to be elucidated. For example, the specific contribution of metal and support to formation of $O_2^-$ and $\cdot OH$ is, in many cases, controversial, although it is of crucial importance. The predominance of one reaction pathway over the other affects the selectivity of photocatalytic oxidation reaction. The reaction of an organic matter with $\cdot OH$ is unselective, yielding $CO_2$, whereas oxidation pathway with $\cdot O_2^-$ on surface of solid or on its premises proceed apparently with high selectivity to oxygenated products.

The general aim of this research is to gain a deeper insight on particularities showed by metal-modified oxide photocatalysts compared to pristine semiconductor oxides in what concerns ROS generation, charge separation, reaction mechanism of organic compounds oxidative conversion. The role of active participants in the oxidative degradation pathways are analyzed in detail. The light-driven (sun or visible light) oxidation of phenol over pristine and metal (Ag, Au, Pt)-modified $TiO_2$ was chosen as model reaction. To uncover the complex reaction network associated to photo-driven oxidation of organic compounds, we analyzed comparatively: (i) the relative efficiency of supported noble metals in separation of photo charges and implicitly the impact on photocatalytic activity, (ii) the specific contribution of metal and of oxide support to ROS ($\cdot OH$ and $O_2^-$) formation, (iii) the relationship between $O_2^-$ formation and reaction selectivity to oxygenated products, (iv) the correlation between the activity of catalyst for mineralization of organic substrate to $CO_2$ and the amount of $\cdot OH$ generated under light irradiation, and (v) the relationship between nature of supported metals and surface photovoltage (SPV) generated under light irradiation in connection with material capacity to generate $O_2^-$, with obvious implications in reaction mechanism.

## 2. Results and Discussion

### 2.1. Characterization Data

The TEM (Transmission Electron Microscopy) analysis of noble metals dispersed on $TiO_2$ prepared by laser pyrolysis shows well dispersed Pt on $TiO_2$ support (Figure 1A). The Ag and Au particles are significantly larger than of Pt (see TEM images in Figure 1B,C). Individual spherical Au nanoparticle of around 5 nm can be observed in Figure 1C.

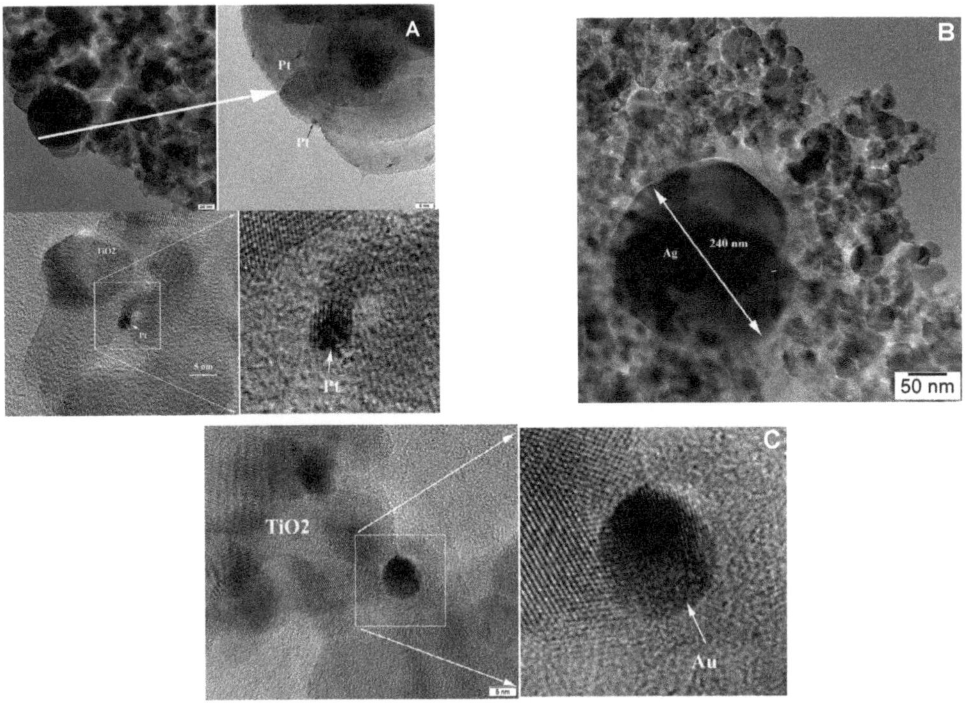


**Figure 1.** Suggestive TEM images of metal-loaded $TiO_2$ samples: Pt/$TiO_2$ (**A**), Ag/$TiO_2$ (**B**) and Au/$TiO_2$ (**C**).

The most relevant characterization data obtained by various experimental methods are presented comparatively in Table 1. More details are given in Supplementary Information.

**Table 1.** Comparative characterization data of investigated materials collected from EDAX (Energy Dispersive X-ray Analysis), XPS (X-ray Photoelectron Spectroscopy), and TEM measurements.

| Samples | Cryst. Size/nm [a] | Metal Content/at.% (wt%) | | Metal Particle Size/nm | | $Ti^{3+}$/at. % [f] | Oxid. State [g] |
|---|---|---|---|---|---|---|---|
| | | EDAX [b] | XPS [c] | TEM [d] | Chem. [e] | | |
| $TiO_2$ | 18.8 | - | - | - | - | 0 | - |
| Ag/$TiO_2$ | 19.3 | 0.04 (2.7) | 0.04 (2.7) | >50 | - | 22.9 | $Ag^0$ |
| Au/$TiO_2$ | 19.8 | 0.01 (2.5) | 0.02 (5.0) | 6 | - | 10.9 | $Au^0$ |
| Pt/$TiO_2$ | 20.6 | 0.01 (2.4) | 0.002 (4.8) | 2 | 1 | 8.8 | $Pt^0$ |

[a]—$TiO_2$ crystallite size determined by Scherrer method from XRD diffractograms; [b]—metal loading evaluated by EDAX; [c]—metal content estimated from XPS data; [d]—metal particle size determined from TEM micrographs; [e]—metal particle size determined by CO chemisorption measurements; [f]—atom percentage of $Ti^{3+}$ in $TiO_2$ matrix derived from XPS measurements; [g]—metal oxidation state analysis performed XPS measurements.

The XPS analysis revealed that the supported noble metals on $TiO_2$ were in metallic state (see the XPS data presented in Supplementary Information). Titanium in $TiO_2$ was in the form of $Ti^{4+}$ whereas the metal-loaded $TiO_2$ contained variable amounts of $Ti^{3+}$ (see Table 1 and Supplementary Information).

The average size of Pt particles supported on $TiO_2$ estimated from CO chemisorption measurements is $\approx$1 nm, corresponding to metal dispersion of around 40%. This result is in fair agreement with TEM result evidencing supported Pt nanoparticles of 1–2 nm (see Figure 1A).

The light absorption features of all investigated materials exhibit the characteristic band edge energies of $TiO_2$ at $\approx$ 400 nm (see Figure 2). The plots of Kubeka–Munk function

(F(R)) in Figure 2 describe light absorbance of solid samples. The SPR maxima of Ag and Au nanoparticles are clearly visible at 460 and 544 nm, respectively. Assuming the indirect allowed transitions, the optical band gaps of all investigated materials, obtained by extrapolation of linear part of $([F(R)]\ h\nu)^{1/2}$ versus hν plots, are around 3.1 eV (see the inset of Figure 2). The close values of optical band gaps, makes difficult to predict the order of photocatalytic activity, based only on light absorption data.

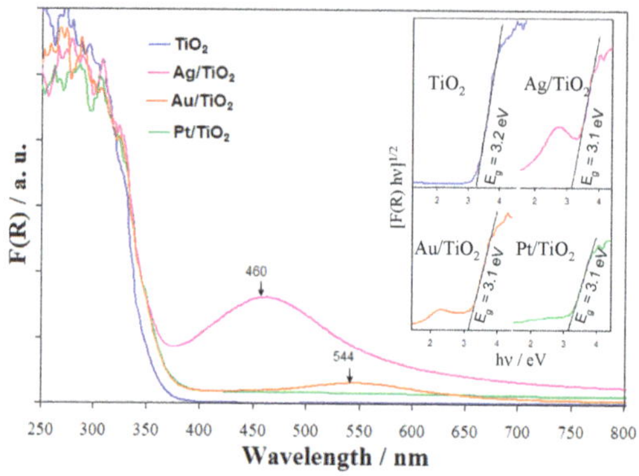


**Figure 2.** Comparative F(R) spectra and indirect optical band gap derived from Tauc representation (inset) for investigated photocatalysts.

### 2.2. Photocatalytic Test Results

The experimental data concerning reactant and product distribution, carbon balance, and conversion of phenol (Ph) over bare and metal-modified $TiO_2$ in aqueous media after 6 h of reaction time are presented in Table 2.

**Table 2.** Phenol conversion to gaseous and liquid products and carbon balance measured after 6 h of reaction time.

| Photocatalysts | C Inlet [a] | C Outlet [a] | | | | | | |
|---|---|---|---|---|---|---|---|---|
| | Ph/μmoles | Ph/μmoles | $CO_2$/μmoles | HQ/μmoles | BQ/μmoles | 1, 2 DHBz/μmoles | Carbon Balance/% | Phenol Conversion/% [b] |
| $TiO_2$ | 362.9 | 331.5 | 15.6 | 0 | 0 | 0 | 95.6 | 8.7 |
| Ag/$TiO_2$ | 369.8 | 324.5 | 6.5 | 7.9 | 0 | 1.3 | 92.0 | 12.2 |
| Au/$TiO_2$ | 356.2 | 324.6 | 13.8 | 11.1 | 0 | 0 | 98.1 | 8.9 |
| Pt/$TiO_2$ | 367.9 | 309.7 | 21.6 | 13.3 | 0.4 | 1.5 | 94.2 | 15.8 |

[a]—μmoles of carbon contained by inlet phenol, non-reacted outlet phenol (Ph) and formed reaction products (hydroquinone (HQ), benzoquinone (BQ), 1,2-dihydroxibenzene (1, 2 DHBz)), and $CO_2$; [b]—Phenol conversion after 6 h of reaction time (reaction conditions: 110 mL of 50 mg·L$^{-1}$ phenol aqueous solution.0.05 g photocatalyst, T = 18 °C, light source AM 1.5).

The carbon balance ((C(outlet)/C(inlet))x100) in our experiments was better than 92%. The Ph conversion ranged between 8.7% (for bare $TiO_2$) and 15.8% (for Pt/$TiO_2$). From the Ph conversion point of view, metal deposition enhances the activity of $TiO_2$ (Pt/$TiO_2$ > Ag/$TiO_2$ > Au/$TiO_2$ > $TiO_2$). The formation of Ph mild oxidation products at end of reaction time, hydroquinone (HQ), benzoquinone (BQ) and 1,2-dihydroxibenzene (1,2 DHBz), could be observed only on metal-loaded $TiO_2$ (Me = Ag, Au, Pt). In contrast, over bare $TiO_2$, Ph was mineralized directly to $CO_2$. The brief analysis of our results suggests that metal deposition on $TiO_2$ favor the formation of oxygenated products, whereas over pristine $TiO_2$, Ph is mineralized directly to $CO_2$, without intermediate formation. As we shall show in this article, the choice of metal is crucial in controlling the selectivity of oxidation reaction.

The time course of products formation during photocatalytic oxidative degradation of phenol over bare and metal-modified $TiO_2$ exposed to simulated solar light is presented in Figure 3. It can be observed that the formation of oxygenated products takes place only on metal-loaded $TiO_2$. The amount of HQ increases rapidly in the first hour of the reaction, then the formation rate is stabilized at ≈ 0.1 µmoles $h^{-1}$. The activity order for HQ formation over metal-loaded $TiO_2$ is $Ag/TiO_2 < Au/TiO_2 < Pt/TiO_2$. Transient formation of BQ was observed only over $Pt/TiO_2$ and $Au/TiO_2$. The amount of BQ peaked at ≈ 2.3 and 0.4 µmoles for the former and second photocatalysts, respectively, after 2 h of reaction. For a longer reaction time, the amount of BQ decreases progressively, vanishing completely for $Au/TiO_2$ and remaining at low concentration (≈ 0.5 µmoles) in the case of $Pt/TiO_2$.

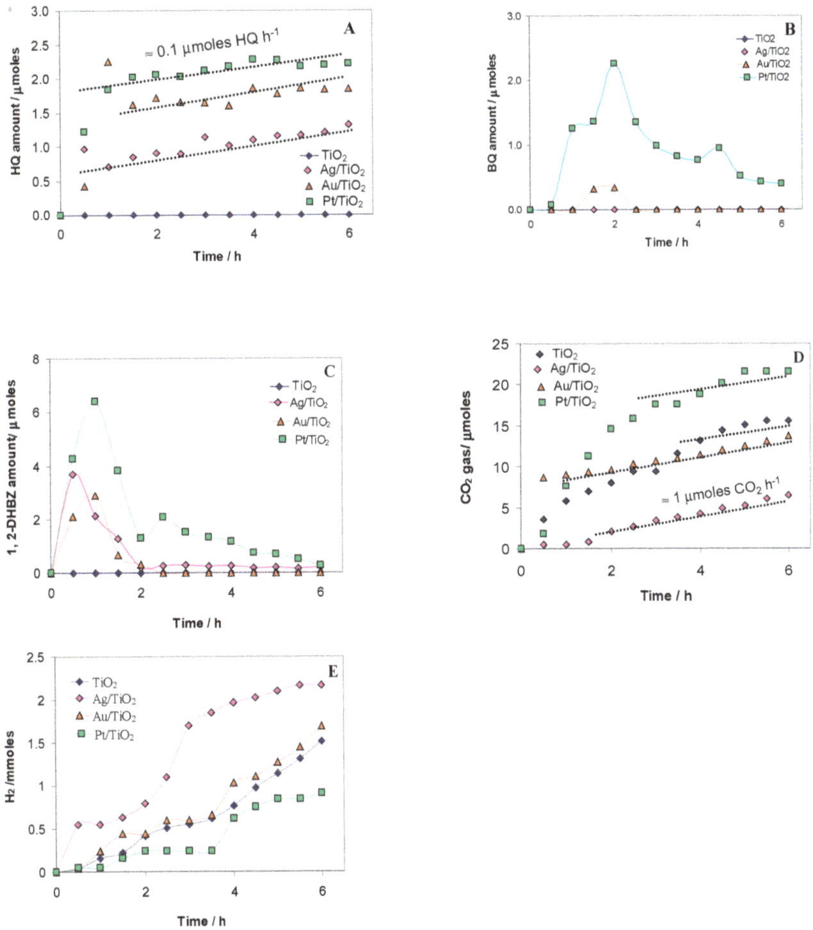


**Figure 3.** Solar light-driven phenol oxidative conversion to oxygenated compounds (HQ (**A**), BQ (**B**), and 1,2 DHBz (**C**)), $CO_2$ (**D**) and to $H_2$ (**E**) over bare and metal modified $TiO_2$ synthesized by laser pyrolysis. Experimental conditions: 0.05 g suspended in 110 mL of 50 mg·$L^{-1}$ phenol aqueous solution, reaction temperature 18 °C, simulated solar light AM 1.5.

The evolution of reaction selectivity to oxygenated compounds and $CO_2$ is presented in Figure 4A–E. Selectivity to 1,2 DHBz reaches a maximum at 30 min for all metal-loaded catalysts (78% for $Ag/TiO_2$, 73% for $Pt/TiO_2$, 53% for $Au/TiO_2$) (Figure 4C). Highest selectivity to oxygenated products (1,2 DHBz + HQ + BQ) of ≈95% was measured for $Pt/TiO_2$ after 0.5 h of reaction time, followed by $Au/TiO_2$ (77% at 1 h), $Ag/TiO_2$ (61% at 2 h), and $TiO_2$ (0%) (see Figure 4E).

### 2.3. *Noble Metals Role in Charge Separation and ROS Generation*

The next step of our investigation was to elucidate in more details the role played by metals in photocatalytic oxidation processes, specifically in (i) charge separation and in (ii) ROS generation.

Electron-hole recombination is one of the main energy loss routes through radiative and nonradiative processes [19]. Photoluminescence (PL) experiments were designed to observe whether, in our case, metal deposition is effective to decrease charge recombination by PL emission.

It is documented that, PL emission intensity depends on photogenerated charge concentration [20]. The PL spectra in Figure 5 show that the energy loss by radiative recombination decreases because of metal deposition on $TiO_2$, due to a better separation of photocharges at the metal–oxide interfaces [14,15]. In light of experimental results, the most efficient charge separation takes place on $Au/TiO_2$, followed, in order, by $Ag/TiO_2$ and $Pt/TiO_2$. Improvement in charge separation is expected to enhance photocatalytic activity because a greater number of electrons and holes become available for redox processes associated with photocatalytic reactions. Our results confirm that a higher conversion of Ph is observed over metal-loaded $TiO_2$ compared to bare $TiO_2$ (see Table 2). However, based only on PL emission intensity results, it is difficult to predict the precise order of activity because, beside the important role played by metals in charge separation, metals work as co-catalysts, mediating charge transfer to reacting substrates.

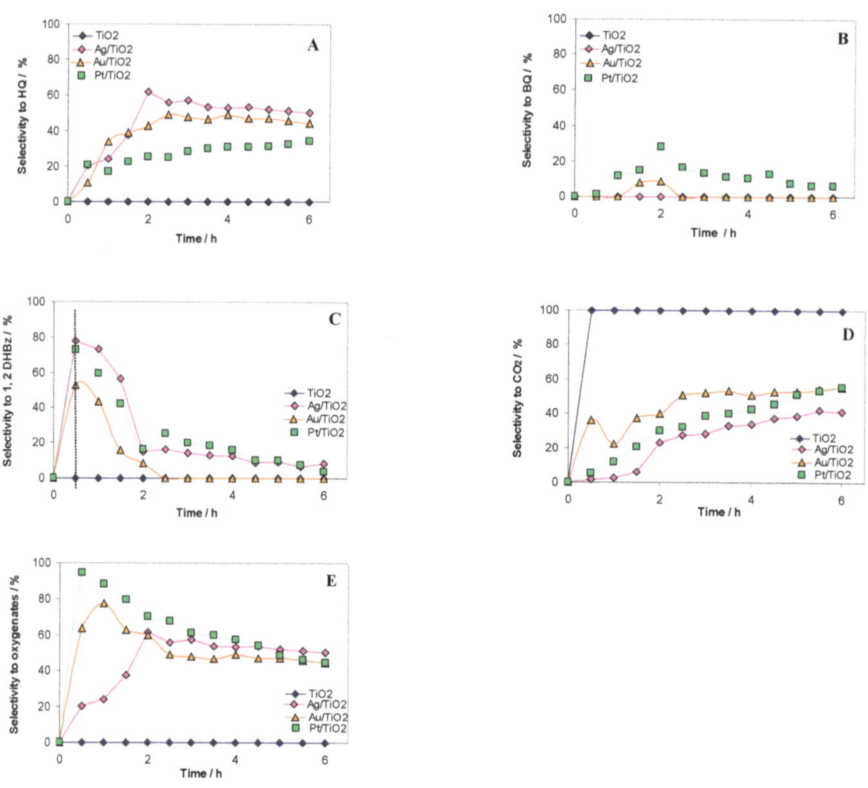


**Figure 4.** Time course of selectivity for phenol photocatalytic oxidative conversion to HQ (**A**), BQ (**B**), 1,2 DHBz (**C**), $CO_2$ (**D**) as well as overal selectivity to oxygenates (**E**) over bare and metal-modified $TiO_2$ exposed to simulated solar light.

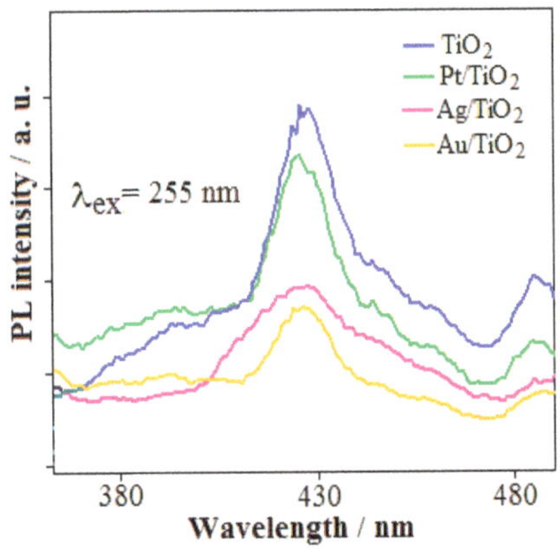


**Figure 5.** Comparative PL emission spectra of bare and metal-loaded $TiO_2$. Experimental conditions: 0.5 mg catalysts suspended by ultrasonication in 3 mL of water.

*2.4. ROS Formation on Bare and Metal-Loaded $TiO_2$*

We have considered three main reaction pathways for the oxidative conversion of Ph: (i) straight charge injection to adsorbed organic substrate on catalyst surface, (ii) reaction of organic substrate with ·OH or with $O_2^-$. In case of oxidative degradation reaction mechanism, the photogenerated charges are shuttled to Ph by intermediation of ROS (·OH and $O_2^-$). It is well documented that ·OH is a powerful, non-selective, oxidant, whereas $O_2^-$ is a weak oxidant [21].

To get further information on the relationship between ROS formation and the photocatalytic behavior of our materials, we have assessed the formation of ·OH and $O_2^-$ under light irradiation by using selective radical quenchers.

The formation of free ·OH radicals was probed by monitoring the development of fluorescent umbelliferone resulted in the reaction between non-fluorescent coumarin and ·OH radicals. The amount of ·OH raises gradually in time, for all photocatalyst exposed to solar light (see Figure 6A,B). From data presented in Figure 6B, the estimated amounts of ·OH formed in 6 h of irradiation time in 110 mL of solution of reactor are: Pt

$TiO_2$ (168 µmoles $g^{-1}{}_{cat}$) > $TiO_2$ (156 µmoles $g^{-1}{}_{cat}$) ≈ Au/$TiO_2$ (155 µmoles $g^{-1}{}_{cat}$) > Ag/$TiO_2$ (131 µmoles $g^{-1}{}_{cat}$). From comparison with photocatalytic data, it comes out that the ·OH quantity is proportional with that of $CO_2$. The experimentally measured $CO_2$, after of 6 h of reaction time, over 0.05 g of photocatalysts, was: Pt/$TiO_2$ (430 µmoles $g^{-1}{}_{cat}$) > $TiO_2$ (310 µmoles $g^{-1}{}_{cat}$) ≈ Au/$TiO_2$ (276 µmoles $g^{-1}{}_{cat}$) > Ag/$TiO_2$ (128 µmoles $g^{-1}{}_{cat}$) (see comparatively Figures 6B and 4D). Taking into account that the probability of ·OH trapping by coumarin or by Ph vary as a function of experimental conditions [22], it can be observed, based on the good matching between radical quenching and photocatalytic results, that $CO_2$ formation relates to ·OH production. The plots in Figure 6C show a clear correlation between relative amounts of ·OH and $CO_2$ formed over the investigated materials. Therefore, we assume that ·OH radicals are responsible for the mineralization of Ph to $CO_2$ (non-selective oxidation route). The formation of $CO_2$ cannot be prevented over $TiO_2$-based materials dispersed in aqueous media because the formation of ·OH radicals is unavoidable.

The formation of $O_2^-$ over metal loaded $TiO_2$ was evidenced indirectly by detection of formazan, which is the product of reaction between XTT and $O_2^-$ The specific absorbance peak of formazan is at 485 nm (Figure 7A). The reduction efficiency of $O_2$ to $O_2^-$, estimated from the amount of formazan, decreases in the order Pt/$TiO_2$ > Au/$TiO_2$ > Ag/$TiO_2$ (Figure 7B). The formation of $O_2^-$ could not be evidenced via formation of formazan on the bare $TiO_2$ sample, prepared by laser pyrolysis. There are, however, reports claiming that $O_2^-$ is formed on $TiO_2$. For example, Goto et al. [23] detected the formation of $O_2^-$

on rutile particles, suggesting that electron transfer takes place from an organic moiety (2-propanol) to $O_2$.

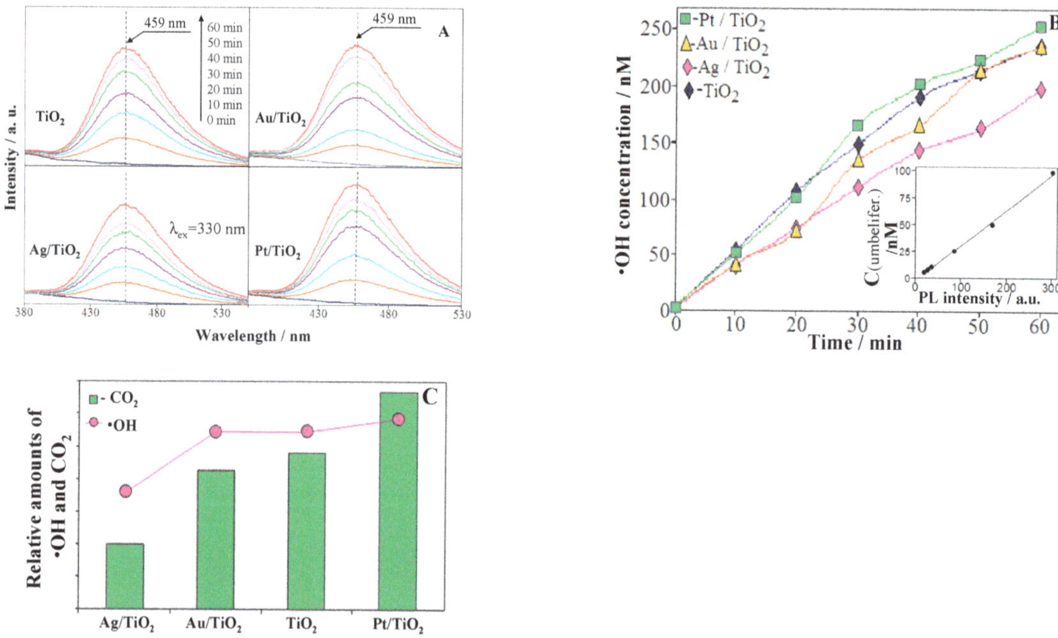


**Figure 6.** Time course of umbelliferone PL (**A**) and evolution in time of ·OH concentration (**B**) over the investigated photocatalysts exposed to solar light, as well as the relative amounts of $CO_2$ and ·OH formed over the photocatalysts exposed to simulated solar light for 6 h (**C**). The formation of ·OH radical was evidenced by observing PL peak of umbelliferone at ≈ 450 nm for $\lambda_{exc}$ = 330 nm (coumarin traps selectively ·OH to form umbelliferone). Inset of figure B represents the calibration curve obtained by plotting the PL response against umbelliferone concentration. Experimental conditions: 1 mg catalyst was dispersed by ultrasonication in 40 mL of 11 mM coumarin solution and then exposed to simulated solar light AM 1.5.

The quantity of oxygenated products resulted by photocatalytic oxidation of Ph scale with the relative amounts of $O_2^-$ (see Figure 7C). From here, it comes that, $O_2^-$ is the ROS responsible for Ph mild oxidation. The main outcomes from $O_2^-$ quenching experiments are: (i) supported metals catalyze $O_2^-$ formation and (ii) $O_2^-$ is the main player in Ph mild oxidative route. The eventual role played by $O_2^-$ for degradation of oxygenated compounds to $CO_2$ should not be completely disregarded, although $O_2^-$ is a significantly weaker oxidant compared to ·OH. The BQ is indicated as an effective $O_2^-$ quencher [1]. We have observed indeed the rapid degradation of BQ, formed only over Pt/$TiO_2$ and Au/$TiO_2$ (see Figure 3B).

The formation of ROS was checked also in visible light domain (λ > 420 nm).

The results of Figure 8 show that, the formation of ·OH radical does not proceed under visible light for any of the investigated materials (Figure 8), which is in line with the absence of $CO_2$ formation during photocatalytic tests conducted in visible light. Our selective radical quenching results demonstrate that $CO_2$ formation is due to ·OH appearance.

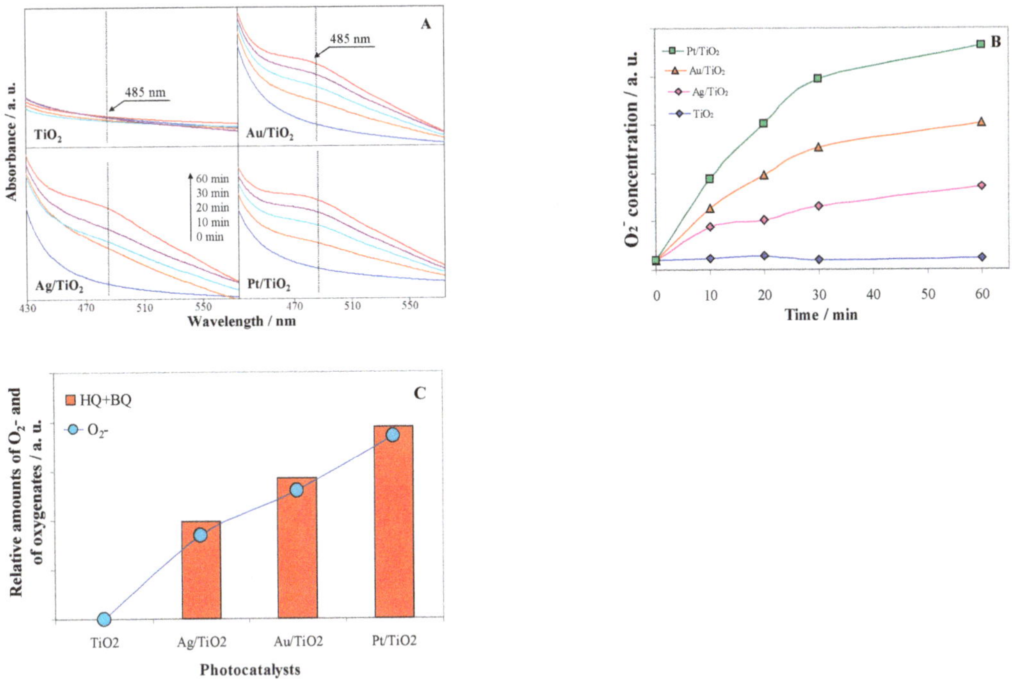


**Figure 7.** Time course of formazan absorbance (**A**), formed in the reaction between $O_2^-$ and XTT probe molecule, and evolution of $O_2^-$ relative concentration over the catalysts exposed to solar light (**B**). Relative amounts of oxygenates (HQ + BQ) measured at end of reaction in comparison with that of $O_2^-$ (**C**). Experimental conditions: 4 mg of catalysts, dispersed into 3 mL of XTT sodium salt solution, were exposed to simulated solar light to induce the formation formazan, which was put in evidence by the UV-VIS absorption peak at ≈ 470 nm.

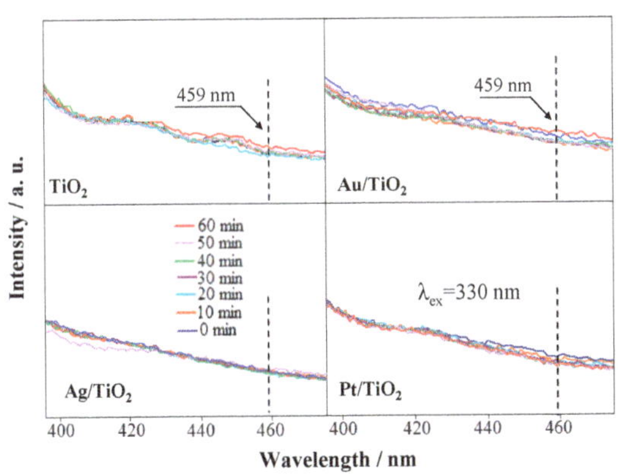


**Figure 8.** Coumarin formation survey, indicative of ·OH radical formation, upon exposure to visible light (λ > 420 nm) of catalysts dispersed in aqueous media.

The survey conducted in visible light (λ > 420 nm) reveal that $O_2^-$ formation does not take place over the scrutinized materials with exception of $Au/TiO_2$ (Figure 9). The formation of $O_2^-$ takes place by reaction between hot electrons of Au plasmon and adsorbed $O_2$ in vicinity of Au nanoparticles [14]. Participation of $TiO_2$ in $O_2^-$ formation, via Au plasmon electron injection in $TiO_2$ conduction followed by $O_2$ reduction on $TiO_2$, was also suggested [24]. However, the very short lifetime of plasmons of 2–10 fs associated with the low energy of electrons [25,26] decrease the probability of $O_2$ reduction. We have observed experimentally only tinny amounts of $O_2^-$ formed under visible light exposure of $Au/TiO_2$, which are not enough to react with Ph at rates high enough to make possible the identification of mild oxidation reaction products by HPLC. The catalytic test results, carried out over all photocatalysts at $\lambda > 420$ nm, evidenced the formation of small amounts of $H_2$ only over $Au/TiO_2$ ($\approx 2.5$ μmoles in 5 h of reaction). The experiments performed in visible light show that ROS are not produced, because visible light ($\lambda > 400$) nm is not absorbed by $TiO_2$.

In the UV region, both ·OH radicals and $O_2^-$ are produced, the former on $TiO_2$ and the second on metals. The ·OH oxidizes non-selectively the organic substrate(s) to $CO_2$ and $H_2O$. The reactions implying ·OH participation are important for environmental applications, where the scope is to mineralize rapidly the organic pollutant to non-harmful $CO_2$. In our case, the mild oxidant $O_2^-$, is produced only on supported noble metal particles, only under exposure to UV light. In addition to radical trapping experiments, the results of our photocatalytic tests evidence that, $CO_2$ is the only reaction product of phenol oxidation over bare $TiO_2$. The metals mediate the transfer of the photogenerated electrons from $TiO_2$ to adsorbed $O_2$. The activity order for $O_2^-$ formation is $Pt/TiO_2 > Au/TiO_2 > Ag/TiO_2$ ($TiO_2$ shows no activity). The interplay between material activity for ·OH and $O_2^-$ production determines the catalyst selectivity to oxygenated products and $CO_2$. $Pt/TiO_2$ is the most active to produce both ·OH and $O_2^-$, thus it will give finally the best phenol conversion. This result proves that $Pt/TiO_2$ generates the highest amount of photogenerated charges ready to participate in redox processes. The $Ag/TiO_2$ is the less active generator of ·OH and $O_2^-$ and consequently shows the smallest phenol conversion among metal-loaded photocatalysts. The supported metals have certain influence on activity of $TiO_2$ support to produce ·OH: $Pt/TiO_2 > TiO_2 \approx Au/TiO_2 > Ag/TiO_2$. Pt on $TiO_2$ enhances the formation rate of ·OH compared to bare $TiO_2$, whereas Ag depresses it. Supported Au seems to have no influence on activity of $TiO_2$ for ·OH formation.

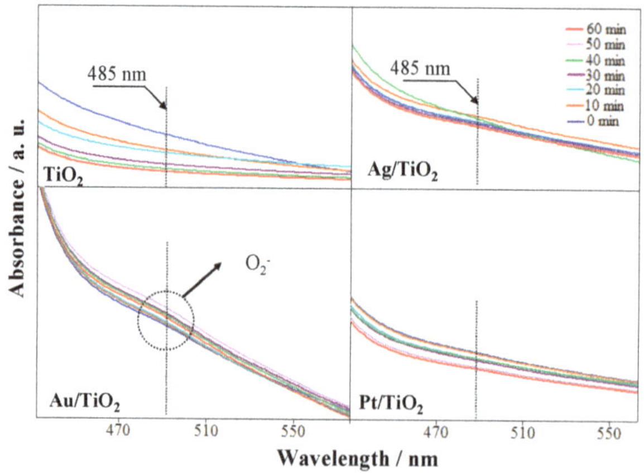


**Figure 9.** Survey of $O_2^-$ production, by monitoring formazan specific absorbance, over the catalysts exposed to visible light ($\lambda > 420$ nm).

In visible region ($\lambda > 420$ nm), both catalyst types (bare and metal-loaded $TiO_2$) show negligible photocatalytic activity because neither ·OH nor $O_2^-$ are produced. The tiny amounts of $O_2^-$ generated on $Au/TiO_2$ are originate from Surface Plasmon Resonance (SPR) shown by Au nanoparticles. Hot electrons on surface of Au particles reduce small amount of $O_2$. The absence of ROS production in visible light is most likely due to the fact that the light absorption edge is at 400 nm (see the UV-VIS spectra in Figure 2), consistent with a band gap of ≈ 3.1 eV. The visible light absorbed by Ag/$TiO_2$ and Au/$TiO_2$ is capable of triggering the formation of tiny amounts of $O_2^-$ by electron donation to adsorbed $O_2$, only in case of Au/$TiO_2$ (see Figure 9).

To get additional experimental evidence on the nonselective Ph degradation route by ·OH radicals, we have designed a new series of experiments, aiming to hinder the formation of umbelliferone from coumarin. The concentration of Ph was chosen to be high enough (2 mM) to consume the majority of ·OH radicals formed in 30 min of exposure to light, thus lowering the probability of coumarin to quench ·OH radicals. In this way, the photoluminescence of umbelliferone was expected to diminish in presence of Ph.

The results of Figure 10 confirm that the ·OH radicals produced by $TiO_2$ are able to react with Ph. When the concentration of Ph is small (0.2 mM), the ·OH radicals react preferentially with coumarine, yielding the photoluminescent umbelliferone. When Ph concentration is raised to 2mM, the formation of umbelliferone is depressed by the competing reaction between ·OH and Ph (see orange trace in Figure 10). In case of Au/$TiO_2$ and Ag/$TiO_2$ catalysts, the small residual PL maxima indicate that tinny amount of ·OH radicals are still able to react with coumarine even in presence of Ph in high concentration. Other studies [27] reported that the addition of alcohols have only a limited influence on umelliferone formation because the alcohols are preferentially adsorbed and oxidized by holes on the surface of the photocatalyst, without significant interference of ·OH radicals.

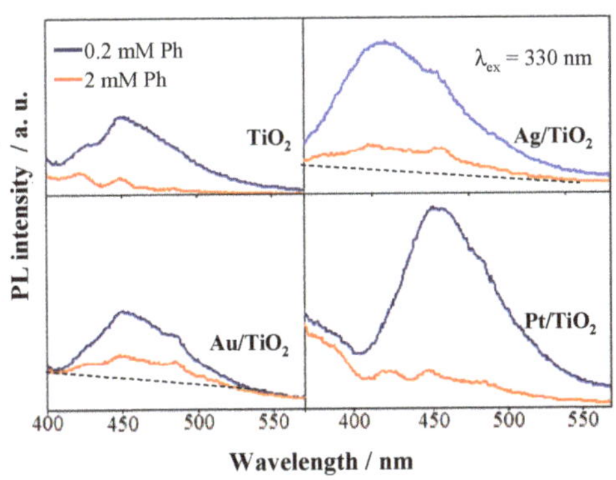


**Figure 10. The** survey of ·OH formation in low (trace 0.2 mM Ph) and high Ph concentration (trace 2 mM Ph) over photocatalyst exposed to simulated solar light for 30 min. Experimental conditions: 0.6 mg photocatalyst was dispersed by ultrasonication in 40 mL solution of coumarin (11 mM)-Ph (0.2 mM) (blue trace 0.2 mM). In second case, the concentration of Ph in 11 mM coumarine solution was increased to 2 mM (orange trace 2 mM).

### *2.5. Photoelectric Properties of Bare and Metal-Modified $TiO_2$*

It is documented that the energies of valence and conduction bands of metal-modified semiconductors are shifted upward with a value depending on the height of the Shottky barrier, forcing the electrons and holes to move in different directions [14,28]. The quick charge carrier recombination, the time scale varying from μs to ns [29], is hindered, allowing the time for a charge transfer to occur at the interface. The space separation of photogenerated charge carriers inherently leads to the appearance of a certain surface photovoltage (SPV), the measurement of which can provide valuable information concerning the transfer dynamic of such carriers [30–32]. Thus, it was previously demonstrated that, for an n-type semiconductor, photoinduced electrons migrate towards the illuminated side of the material, giving rise to a negative SPV signal. Conversely, a positive SPV signal corresponds to a p-type semiconductor, in which case holes are directed from the surface to the bulk [33,34]. However, in both cases the surface photovoltage is wavelength dependent, being affected by the particular features of the semiconducting material, in terms of light absorption and transport of excess carriers [35,36].

The surface photovoltage was measured for each sample at several wavelengths and, as expected, SPV spectra (Figure 11) revealed in all the cases an n-type semiconducting character. Obviously, SPV signals measured under the actual experimental conditions correspond, in fact, to the potential difference between the Fermi level of ITO (indium tin oxide, + 0.35 V vs. NHE [34]) and that of the irradiated sample. Since the conduction band of non-stoichiometric $TiO_2$ is located above its Fermi level with an average value of ca. 0.5 V [37], and by taking into account a value of around −0.33 V for the $O_2/O_2^-$ level [1], it appears that the formation of $O_2^-$ species at the surface of the irradiated samples requires an SPV value higher than ca. −0.18 V. As the results from Figure 11 indicate, this condition is not fulfilled in the case of pristine $TiO_2$, whereas at noble metal-modified samples $O_2^-$ formation is possible, at least in principle, for irradiation wavelengths lower than 380 nm. Nevertheless, the probability for this process increases with SPV signal, in the order $Pt/TiO_2$ > $Au/TiO_2$ > $Ag/TiO_2$, as schematically illustrated in the inset in Figure 11. These findings are in excellent agreement with the activity for $O_2^-$ formation deduced from radical quenching experiments (see Figure 7).

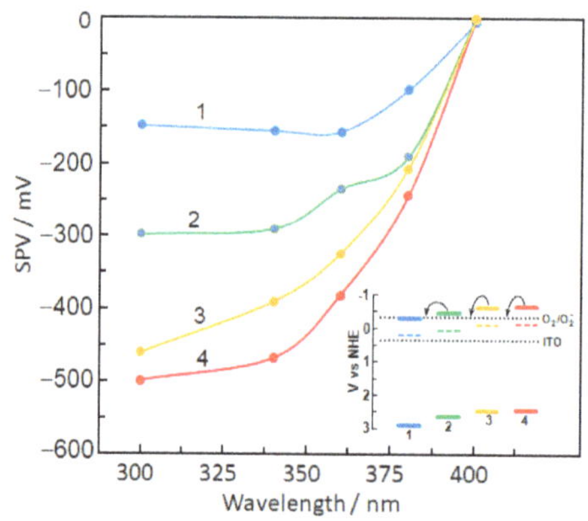


**Figure 11.** Surface photovoltage (SPV) spectra of pristine $TiO_2$ (1), $Ag/TiO_2$ (2), $Au/TiO_2$ (3), and $Pt/TiO_2$ (4). Inset: corresponding energy diagram at 300 nm; the dashed lines indicating the Fermi levels of samples.

To emphasize the effect of noble metal modification of the titanium oxide on the $O_2^-$ generation process, chronoamperometric experiments were performed in dark, at an applied voltage of −1V. Figure 12 shows the time-variation of the oxygen reduction current, estimated as the difference between the current recorded in $O_2$ atmosphere and that observed under Ar conditions. For easier comparison, the currents were expressed in terms of mass activity (oxygen reduction current normalized to the amount of the investigated powder sample). Pristine $TiO_2$ exhibited negligible response (see curve 1 from Figure 12), which clearly demonstrates that the presence of noble metal particles is a prerequisite for $O_2$ reduction. It was interesting to observe that, up to ca. 50 s, the current recorded at $Au/TiO_2$ is higher than that at $Pt/TiO_2$, although during further polarization the decrease in the current tends to become much slower for the latter (compare curves 3 and 4 from Figure 12). To better put into perspective the role of the noble metal nature, inset (a) in Figure 12 illustrates the decay of the oxygen reduction current on a log–log scale. Linear dependences were found in all cases, which could indicate a Langmuir adsorption kinetic control of $O_2$ on the overall reduction process [38]. However, $Pt/TiO_2$ exhibited the slowest current decrease, whereas for $Au/TiO_2$ a change in slope was observed, the decline of the current becoming much steeper after only ca. 10 s, probably as result of a more sluggish adsorption of oxygen reactant species. Consequently, after about 200 s of continuous polarization, oxygen reduction current at $Pt/TiO_2$ is more than twice as high as that observed with $Au/TiO_2$. These results are important because they can provide an explanation for the fact that, compared to the case of Au/$TiO_2$, the total amount of $O_2^-$ produced at Pt/$TiO_2$ is much higher (see inset in Figure 12) than would have been expected for rather small difference in terms of SPV signals between the two materials. Integration of the current responses from Figure 13 over the entire polarization time, yielded oxygen reduction charges of ca. 0.78, ca. 1.10, and ca. 1.46 mC $g^{-1}$ for Ag/$TiO_2$, Au/$TiO_2$, and Pt/$TiO_2$, respectively. As illustrated by the inset (b) in Figure 12, based upon these values, corresponding amounts of $O_2^-$ species of 7.8, 11.4, and 15.2 nmol $g^{-1}$ were estimated as being formed at the investigated active samples. The maximum amount of oxygenated compounds (HQ + BQ + 1,2 DHBz) formed over Ag/$TiO_2$ (4.7 μmoles), Au/$TiO_2$ (5.1 μmoles), and Pt/$TiO_2$ (9.5 μmoles) after 1 h of reaction time (0.5 h in case of Ag/$TiO_2$), follows closely the tendency observed in polarization measurements of $O_2$ reduction (Figure 12). The precise correlation between quantitative polarization and photocatalytic data concerning oxygenated compounds is difficult because the formation and depletion of $O_2^-$ by reaction with organic substrate(s) is a dynamic process compared to $O_2$ adsorption on polarized surface. To build up a reliable kinetic, it is necessary to find out the rate of $O_2^-$ formation in reaction conditions. However, a close relationship between formation of $O_2^-$ species and mild oxidation of Ph is demonstrated by two independent experimental techniques (selective radical trapping and chronoamperometric experiments).

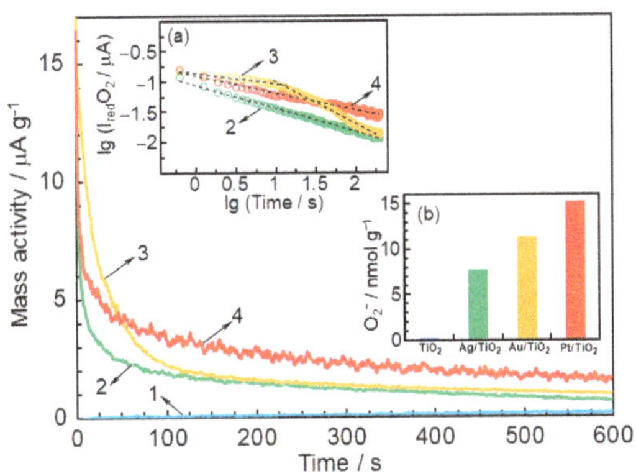


**Figure 12.** Time-variation of the mass activity for oxygen reduction at pristine $TiO_2$ (1), Ag/$TiO_2$ (2), Au/$TiO_2$ (3), and Pt/$TiO_2$ (4), at an applied voltage of −1 V. Insets: (**a**), log–log plots for the oxygen reduction current decay at noble metal-modified $TiO_2$; (**b**), estimated $O_2^-$ amounts formed during 10 min of continuous polarization.

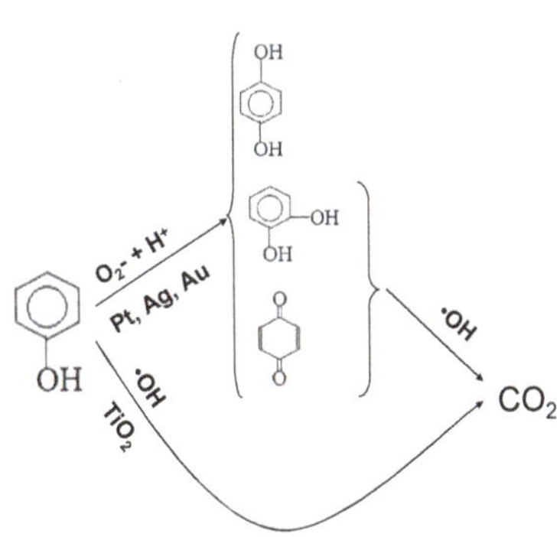


**Figure 13.** Distinct mechanisms of Phenol (Ph) photocatalytic oxidative conversion over bare and noble metal-modified $TiO_2$.

The corroboration of entire experimental evidences collected in this research lead to the reaction scheme presented below.

We assumed, based on experimental facts, that Ph is oxidized non-selectively by ·OH radicals directly to $CO_2$, apparently without producing in our experimental conditions detectable long-lived intermediates. Supported noble metals are responsible for $O_2$ reduction to $O_2^-$, by mediating the transfer of photoelectron from $TiO_2$ to adsorbed $O_2$. The subsequent $O_2^-$ reaction with Ph leads to formation of oxygenated products (HQ, BQ, 1,2 DHBz). The literature focuses mostly on reactivity $O_2^-$ in organic protic and aprotic solvents and less on the reactivity in aqueous solutions [21]. However, it is recognized that, $O_2^-$ disproportionates spontaneously in water, forming $O_2$ and hydroperoxide anion ($HO_2^-$). One possibility is that the reaction follows the superoxide dismutase (SOD) pathway (Equation (1)), proposed to explain the biological function of superoxide ion [39]:

$$2\,O_2^- + 2H^+ \rightarrow O_2 + H_2O_2 \quad (1)$$

$$\text{R-H} + H_2O_2 \rightarrow \text{R-OH} + H_2O \quad (2)$$

Equation (2) suggests that the insertion of oxygen in ortho and para position takes place by the reaction of organic substrate with $H_2O_2$ in vicinity or on supported noble metal(s). Alternatively, the oxidation mechanism may occur through activation of Ph by transfer of photogenerated hole followed by nucleophil attack of $O_2^-$ to generate organic peroxyl radicals which react further with $H^+$ and $e^-$ to form finally the hydroxylated organic compound (see Equation (3)) [40].

$$\text{R-H} + h^+ \rightarrow (\text{R-H})^+ + O_2^- \rightarrow (\text{R-H})\text{-O-O}^\cdot + 2\,H^+ + 2\,e^- \rightarrow \text{R-OH} + H_2O \quad (3)$$

Actually, there is a limited knowledge on the reaction mechanism concerning the interaction in aqueous media between the adsorbed $O_2^-$ and organic substrate in the presence of a catalytic metal. The published literature gives no information on the eventual role played an active metal in the above proposed reaction mechanisms.

## 3. Experimental

### *3.1. Preparation of Catalytic Materials*

The photocatalysts employed in this study are $TiO_2$, $Pt/TiO_2$, $Ag/TiO_2$, and $Au/TiO_2$. High surface area $TiO_2$, formed of nanometric sized particles, were prepared by laser pyrolysis using $TiCl_4$ as precursor material. The detailed experiental setup and procedures are described elswhere [41]. Metal deposition was perfomed by impregnating $TiO_2$ with $H_2PtCl_6$, $AgNO_3$, and $HAuCl_4$ precursors followed by reduction with $NaBH_4$ [42]. For additional information on preparation of photocatalytic materials, see the Supplementary Information. Finally, the catalysts were dried and calcined 100 and 300 °C, respectively. The elemental composition, identification of crystalline phases and measurements of average crystallite size of photocatalytic materials was made by Energy Dispersive X-ray analysis (EDAX) and XRD, respectively. In the case of $Pt/TiO_2$, the dispersion and average Pt particle size was estimated by CO chemisorption measurements, carried out with a ChemBet−3000 Quantachrome Instrument (Odelzhausen, Germany) equipped with a thermal conductivity detector (TCD) apparatus [43].

The values of optical band gap are estimated from the Tauc plots of diffuse reflectance UV-VIS data converted into absorbance by Kubelka–Munk function [44].

The X-ray Photoelectron Spectroscopy (XPS) measurements were performed in an ESCALAB Xi+ (Thermo SCIENTIFIC Surface Analysis, Baltimore, MD, USA) setup equipped with a multichannel hemispherical electron Analyzer (dual X-ray source) working with Al Kα radiation (hν = 1486.2 eV), using C 1s (284.8 eV) as the energy reference. The chemical compositions of surface and oxidation states were estimated from the XPS spectra by calculating the integral of each peak after subtraction of the "S-shaped" Shirley-type background using the appropriate experimental sensitivity factors.

The Photoluminescence (PL) emission spectra were recorded with Cary Eclipse Fluorescence Spectrophotometer (Agilent Technologies, Santa Clara, CA, USA) apparatus [45].

Experimental setup for photocatalytic tests. Photocatalytic experiments were conducted in a batch-type photo reactor depicted in Figure 14. The AM 1.5 (1000 W $m^{-2}$) light beam of 4.5 × 4.5 $cm^2$ was provided by a solar light simulator (Peccell-L01, Yokohama, Japan) equipped with a 150 W xenon short-arc lamp. The double-walled photoreactor was provided with optical degree quartz window. For each test, 110 mL of 50 mg·$L^{-1}$ phenol aqueous solution containing the suspended catalyst powder (0.05 g) were placed into the photoreactor, thermostated at 18 °C with a chiller. Prior tests, the suspension was kept in dark for 30 min, under stirring, to attain equilibration of experimental system. Typically, one experiment consisted of light irradiation of liquid-suspended photocatalysts for 360 min.

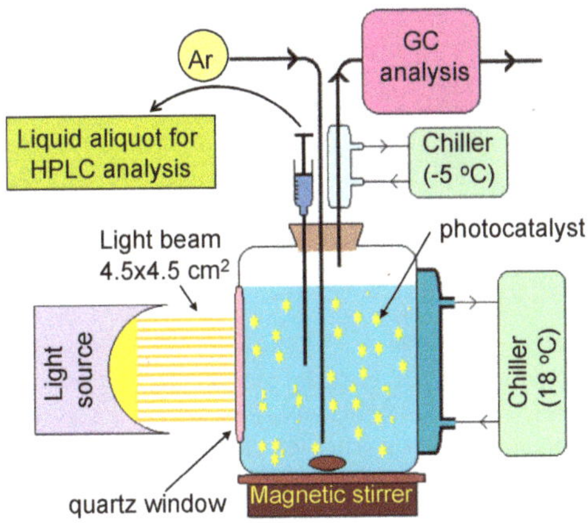


**Figure 14.** Experimental setup used for photocatalytic tests composed of quartz reactor provided with quartz window, AM 1.5 light source, liquid and gas sampling systems, gas and liquid chromatographs, and chillers to control the temperatures of glass condenser (−5 °C) and of reactor (18 °C).

During tests, the Ar carrier gas was purged continuously into the phenol aqueous solution at a flow rate of 10 mL·$min^{-1}$, passed through a refrigerant cooled to −5 °C with a chiller to remove liquid vapors, and then sent to GC for on-line composition analysis at 30 min time interval with a gas chromatograph (Buck Scientific, Norwalk, CT, USA) equipped with TCD detectors. The $H_2$ and $O_2$ were separated and quantified on Molecular Sieve 5Å, whereas $CO_2$ and the eventually formed $C_2H_6$ and $C_2H_4$ on the Hayesep column. Meanwhile, aliquots of 2 mL were extracted every 30 min from the liquid phase, filtered through 0.22 μm Q-Max membrane filter, and then injected for analysis into a liquid chromatograph (Alliance e2659, Waters, Milford, MA, USA). The organic components of liquid phase (phenol (Ph), hydroquinone (HQ), benzoquinone (BQ), and 1,2-dihydroxibenzene (1,2 DHBz)) were separated on HPLC column (C18–3.5 μm Symmetry, Waters), identified, and then quantified using the UV-VIS detector set at 273 nm. The mobile phase of HPLC (isocratic elution program) was a mixture of Milli-Q ultrapure water (18 MΩ) and methanol (80/20 $v/v$). The flow rate of the mobile phase was 1 mL·$min^{-1}$ and the sample injection volume was 2 μL.

### 3.2. Detection of ·OH Radicals

Coumarin was used as selective trap for the ·OH radicals formed under photocatalysts exposure to light [12,27]. The 0.001 g of powder catalysts were first suspended in 40 mL of 11 mM coumarin (Merck) aqueous solution and then exposed either to simulated solar light AM 1.5 or to visible light. A cut off filter (L42, Asahi Spectra, California, USA) was in the case of visible light ($\lambda > 420$ nm). Aliquots of 1.5 mL solution were sampled at 10 min time interval for fluorescence measurements (Cary Eclipse Fluorescence Spectrophotometer, Agilent Technologies, Santa Clara, CA, USA) to monitor umbelliferone, formed by reaction between coumarin and ·OH radicals. Umbelliferone gives a specific fluorescence peak at $\approx$ 450 nm for $\lambda_{exc} = 330$ nm.

### 3.3. Detection of $O_2^-$

In a typical experiment, 0.004g of catalyst was suspended into 4 mL of 3 mM solution of XTT sodium salt (2, 3-bis(2-methoxi-4-nitro-5sulfophenyl)-2H-tetrazolium-5-carboxanilide) (Alfa Aesar). Then, the samples were exposed to simulated solar or visible light for 10, 20, and 30 min to induce the formation of $O_2^-$. XTT reduction by $O_2^-$ produces XTT-formazan, which can be evidenced by a broad absorption peak at $\approx$ 470 nm [12,44]. The measurements were carried out with UV-VIS spectrophotometer (Analytik Jena Specord, 200 Plus, Jena, Germany).

For the surface photovoltage (SPV) measurements, a sandwich-like photovoltaic cell was built, according to a method previously described in the literature [30]. Briefly, a small amount (ca. 12 mg) of pristine or metal-modified titanium oxide was firmly pressed in between two ITO electrodes, to obtain a confined film composed of the investigated powder sample. The upper face of the cell was irradiated (under chopped conditions) with a monochromatic light (the light beam of 300 W Xe lamp of Asahi Spectra MAX-350 (Tokyo, Japan) light source was passed through high transmission bandpass filters with FWHM = 11 nm) and the SPV signal was measured by means of a computer-driven Keithley 2425 source-meter (Cleveland, Ohio, USA). The energy of the monochromated light beam was measured with Newport optical power meter (Model 1830-R, Irvine, CA, USA) equipped a calibrated photodiode detector (Newport, 918D series). For the chronoamperometric experiments, the same cell was used and the measurements were carried out in an air-tight reactor by means of a PAR 273A (Princeton Applied Research Walpole, MA, USA) potentiostat, both under pure $O_2$ and Ar atmospheres.

## 4. Conclusions

This study gives a comprehensive view on the light-initiated photocatalytic oxidation pathways of a model organic substrate with an aromatic ring (Ph) over bare and noble metal-loaded $TiO_2$. The analysis of complex phenomena associated with photocatalytic reaction focuses on particular roles played by oxide support and by noble metals on light absorption, charge separation, formation of ROS (·OH and $O_2^-$), as well as on reaction mechanism of oxidative conversion of Ph.

We have found out that $TiO_2$ support generates only ·OH as ROS when it is exposed to light with $\lambda < 400$ nm. These radicals are responsible for deep oxidation of Ph directly to $CO_2$, apparently without the formation of detectable long-lived intermediates. The formation of ·OH, and consequently the photocatalyst activity, cease in visible light domain.

Deposited noble metals (Ag, Au, Pt) (i) adsorb the visible light (SPR phenomenon), (ii) assist effectively the charge separation, and the (iii) $O_2$ reduction to $O_2^-$. The deposited metal raises the Fermi level of $TiO_2$ allowing the reduction of adsorbed $O_2$ to $O_2^-$. The $O_2^-$ produced on metals oxidizes mildly Ph to oxygenated products (HQ, BQ, 1,2 DHBz). In a parallel process, ·OH radicals produced by $TiO_2$ support mineralize Ph directly to $CO_2$ by fast reaction sequences. At this stage, it is not clear the precise function of metals in the reaction between organic substrate (Ph) and $O_2^-$. With the exception of Au, the hot electrons produced by SPR at $\lambda > 400$ nm are not active to produce measurable amounts of $O_2^-$.

This study demonstrates, by two complementary experimental methods (radical quenching and photo electrochemical measurements), that production of ·OH and $O_2^-$ over the investigated catalysts correlates well with the activity showed for oxidative conversion of Ph. According to our data, the oxidation of Ph by photo charges is intermediated by ROS.

In light of our results, the bare $TiO_2$ suits the best the photocatalytic depollution purposes, where the aim is to mineralize the harmful organic substrate to $CO_2$. When noble metals are deposited on $TiO_2$, intermediate oxygenated compounds are formed by mild oxidation of organic substrate(s) by $O_2^-$, via photo induced electron transfer from metals to $O_2$. Thus, from a depollution point of view, the modification of $TiO_2$ with noble metals is not beneficial. In addition, the metal-modified photocatalyst in powder form dispersed in water can be harmful to the environment. Same assessment can be made for photo water splitting, where the consumption of photo-generated electron by adsorbed $O_2$ hinders $H^+$ reduction. On the other hand, should be the practical aim of valuable oxygenated compounds synthesis by mild selective oxidation of organic compounds, the use of catalytic metals is mandatory.

**Supplementary Materials:** The following are available online at https://www.mdpi.com/article/10.3390/catal11040487/s1, Figure S1: High resolution XPS spectra of $TiO_2$ in O1s and Ti2p binding energy regions, Figure S2: High resolution XPS spectra of Ag/$TiO_2$ in O 1s, Ti 2p and Ag 3d binding energy regions, Figure S3: High resolution XPS spectra of Au/$TiO_2$ in O 1s, Ti 2p and Au 4f binding energy regions, Figure S4: S4 High resolution XPS spectra of Pt/$TiO_2$ in O 1s, Ti 2p and Pt 4f binding energy regions, Figure S5: Comparative XRD difraction patterns of simple and metal-modified $TiO_2$. •-anatase, +-rutile, Table S1: Elemental composition obtained from EDAX analysis of simple and metal -modified $TiO_2$, Table S2: XPS survey of elemental composition of simple and noble metal-modified $TiO_2$, Table S3: Chemical state of titanium in the investigated materials, Table S4: Crystalline phase composition and average crystallite size of simple and metal-modified $TiO_2$.

**Author Contributions:** Conceptualization, I.B., Investigation, A.S., C.A., F.P., M.R., A.V., T.S., M.S., C.F., C.N.M., V.S.T., N.S., M.Z. and I.B., Methodology, A.S., C.A., F.P., M.R., A.V., T.S., M.S., C.F., C.N.M., V.S.T., N.S., M.Z. and I.B., Resources, A.S., C.A., F.P., M.R., A.V., T.S., M.S., C.F., C.N.M., V.S.T., N.S., M.Z. and I.B., Supervision, N.S. and I.B., Visualization, I.B., Writing—original draft, N.S. and I.B., Writing—review & editing, N.S. and I.B. All authors have read and agreed to the published version of the manuscript.

**Funding:** This research received no external funding.

**Acknowledgments:** This work was supported by Grants 46 PCCDI/2018 MALASENT.

**Conflicts of Interest:** The authors declare no conflict of interest.

## References

1. Li, X.; Xie, J.; Jiang, C.; Yu, J.; Zhang, P. Review on design and evaluation of environmental photocatalysts. *Front. Environ. Sci. Eng.* **2018**, *12*, 14. [CrossRef]
2. Paz, Y. Application of $TiO_2$ photocatalysis for air treatment: Patents' overview. *Appl. Catal. B* **2010**, *99*, 448–460. [CrossRef]
3. Shaham-Waldmann, N.; Paz, Y. Away from $TiO_2$: A critical minireview on the developing of new photocatalysts for degradation of contaminants in water. *Mat. Sci. Semicond. Process.* **2016**, *42*, 72–80. [CrossRef]
4. Di Paola, A.; García-López, E.; Marcì, G.; Palmisano, L. A survey of photocatalytic materials for environmental remediation. *J. Hazard. Mater.* **2012**, *211*, 3–29. [CrossRef]
5. Chen, L.; Tang, J.; Song, L.N.; Chen, P.; He, J.; Au, C.T.; Yin, S.F. Heterogeneous photocatalysis for selective oxidation of alcohols and hydrocarbons. *Appl. Catal. B* **2019**, *242*, 379–388. [CrossRef]
6. Meng, S.; Ye, X.; Zhang, J.; Fu, X.; Chen, S.; Chen, S. Effective use of photogenerated electrons and holes in a system: Photocatalytic selective oxidation of aromatic alcohols to aldehydes and hydrogen production. *J. Catal.* **2018**, *367*, 159–170. [CrossRef]
7. Kou, J.; Lu, C.; Wang, J.; Chen, Y.; Xu, Z.; Varma, R.S. Selectivity Enhancement in Heterogeneous Photocatalytic Transformations. *Chem. Rev.* **2017**, *117*, 1445–1514. [CrossRef] [PubMed]
8. Hou, W.; Cronin, S.B. A Review of Surface Plasmon Resonance-Enhanced Photocatalysis. *Adv. Funct. Mater.* **2013**, *23*, 1612–1619. [CrossRef]
9. Carp, O.; Huisman, C.L.; Reller, A. Photoinduced reactivity of titanium dioxide. *Progress Solid State Chem.* **2004**, *32*, 33–177. [CrossRef]
10. Fujishima, A.; Rao, T.N.; Tryk, D.A. Titanium Dioxide Photocatalysis. *J. Photochem. Photobiol. C Photochem. Rev.* **2000**, *1*, 1–21. [CrossRef]
11. Linsebigler, A.L.; Lu, G.; Yates, J.T., Jr. Photocatalysis on $TiO_2$ Surfaces: Principles, Mechanisms and Selected Results. *Chem. Rev.* **1995**, *95*, 735–758. [CrossRef]

12. Anastasescu, C.; Negrila, C.; Angelescu, D.G.; Atkinson, I.; Anastasescu, M.; Spataru, N.; Zaharescu, M.; Balint, I. Particularities of photocatalysis and formation of reactive oxygen species on insulators and semiconductors: Cases of $SiO_2$, $TiO_2$ and their composite $SiO_2$-$TiO_2$. *Catal. Sci. Technol.* **2018**, *8*, 5657–5668. [CrossRef]
13. Fujishima, A.; Zhang, X.; Tryk, D.A. Heterogeneous photocatalysis: From water photolysis to applications in environmental cleanup. *Int. J. Hydrog. Energy* **2007**, *32*, 2664–2672. [CrossRef]
14. Bumajdad, A.; Madkour, M. Understanding the superior photocatalytic activity of noble metals modified titania under UV and visible light irradiation. *Phys. Chem. Chem. Phys.* **2014**, *16*, 7146–7158. [CrossRef] [PubMed]
15. Yang, D.; Lee, S.W. Photocatalytic activity of Ag, Au-Deposited $TiO_2$ nanoparticles prepared by sonochemical reduction method. *Surface Rev. Lett.* **2010**, *17*, 21–26. [CrossRef]
16. Tian, Y.; Tatsuma, T. Mechanisms and Applications of Plasmon-Induced Charge Separation at $TiO_2$ Films Loaded with Gold Nanoparticles. *J. Am. Chem. Soc.* **2005**, *127*, 7632–7637. [CrossRef]
17. Tanaka, A.; Sakaguchi, S.; Hashimoto, K.; Kominami, H. Preparation of Au/$TiO_2$ with Metal Cocatalysts Exhibiting Strong Surface Plasmon Resonance Effective for Photoinduced Hydrogen Formation under Irradiation of Visible Light. *ACS Catal.* **2013**, *3*, 79–85. [CrossRef]
18. Zhai, W.; Xue, S.; Zhu, A.; Luo, Y.; Tian, Y. Plasmon-Driven Selective Oxidation of Aromatic Alcohols to Aldehydes in Water with Recyclable Pt/$TiO_2$ Nanocomposites. *Chem. Cat. Chem.* **2011**, *3*, 127–130. [CrossRef]
19. Osterloh, F.E. Nanoscale Effects in Water Splitting Photocatalysis. *Top. Curr. Chem.* **2016**, *371*, 105–142. [CrossRef]
20. Yoffe, A.D. Semiconductor quantum dots and related systems: Electronic, optical, luminescence and related properties of low dimensional systems. *Adv. Phys.* **2001**, *50*, 1–208. [CrossRef]
21. Hayyan, M.; Hashim, M.A.; AlNashef, I.M. Superoxide Ion: Generation and Chemical Implications. *Chem. Rev.* **2016**, *116*, 3029–3085. [CrossRef]
22. Suh, M.; Bagus, P.S.; Pak, S.; Rosynek, M.P.; Lunsford, J.H. Reactions of Hydroxyl Radicals on Titania, Silica, Alumina and Gold Surfaces. *J. Phys. Chem. B* **2000**, *104*, 2736–2742. [CrossRef]
23. Goto, H.; Hanada, Y.; Ohno, T.; Matsumura, M. Quantitative Analysis of Superoxide Ion and Hydrogen Peroxide Produced from Molecular Oxygen on Photoirradiated $TiO_2$ Particles. *J. Catal.* **2004**, *225*, 223–229. [CrossRef]
24. Zhu, S.; Liang, S.; Gu, Q.; Xie, L.; Wang, J.; Ding, Z.; Liu, P. Effect of Au supported $TiO_2$ with dominant exposed facets on the visible-light photocatalytic activity. *Appl. Catal. B* **2012**, *119*, 146–155. [CrossRef]
25. Watanabe, K.; Menzel, D.; Nilius, N.; Freund, H.J. Photochemistry on Metal Nanoparticles. *Chem. Rev.* **2006**, *106*, 4301–4320. [CrossRef]
26. Ueno, K.; Misawa, H. Surface plasmon-enhanced photochemical reactions. *J. Photochem. Photobiol. C Photochem. Rev.* **2013**, *15*, 31–52. [CrossRef]
27. Nosaka, Y.; Nosaka, A.Y. Generation and Detection of Reactive Oxygen Species in Photocatalysis. *Chem. Rev.* **2017**, *117*, 11302–11336. [CrossRef]
28. Tung, R.T. Recent Advances in Schottky Barrier Concepts. *Mat. Sci. Eng. R* **2001**, *35*, 1–138. [CrossRef]
29. Henderson, M.A. A surface science perspective on $TiO_2$ photocatalysis. *Surface Sci. Rep.* **2011**, *66*, 185–297. [CrossRef]
30. Fan, H.; Jiang, T.; Li, H.; Wang, D.; Wang, L.; Zhai, J.; He, D.; Wang, P.; Xie, T. Effect of $BiVO_4$ Crystalline Phases on the Photoinduced Carriers Behavior and Photocatalytic Activity. *J. Phys. Chem. C* **2012**, *116*, 2425–2430. [CrossRef]
31. Zhang, Y.; Xie, T.; Jiang, T.; Wei, X.; Pang, S.; Wang, X.; Wang, D. Surface photovoltage characterization of a ZnO nanowire array/CdS quantum dot heterogeneous film and its application for photovoltaic devices. *Nanotechnology* **2009**, *20*, 155707. [CrossRef] [PubMed]
32. Wu, P.; Wang, J.; Zhao, J.; Guo, L.; Osterloh, F.E. Structure defects in g-$C_3N_4$ limit visible light driven hydrogen evolution and photovoltage. *J. Mat. Chem. A* **2014**, *2*, 20338–20344. [CrossRef]
33. Duzhko, V.V.; Timoshenko, V.; Koch, F.; Dittrich, T. Photovoltage in nanocrystalline porous $TiO_2$. *Phys. Rev. B* **2001**, *64*, 075204. [CrossRef]
34. Osterloh, F.E.; Holmes, M.A.; Zhao, J.; Chang, L.; Kawula, S.; Roehling, J.D.; Moule, A.J. P3HT:PCBM Bulk-Heterojunctions: Observing Interfacial and Charge Transfer States with Surface Photovoltage Spectroscopy. *J. Phys. Chem. C* **2014**, *118*, 14723–14731. [CrossRef]
35. Kronik, L.; Shapira, Y. Surface photovoltage spectroscopy of semiconductor structures: At the crossroads of physics, chemistry and electrical engineering. *Surf. Interface Anal.* **2001**, *31*, 954–965. [CrossRef]
36. Frame, F.A.; Townsend, T.K.; Chamousis, R.L.; Sabio, E.M.; Dittrich, T.; Browning, N.D.; Osterloh, F.E. Photocatalytic Water Oxidation with Nonsensitizied $IrO_2$ Nanocrystals under Visible and UV Light. *J. Am. Chem. Soc.* **2011**, *133*, 7264–7267. [CrossRef]
37. Shi, F.; Baker, L.R.; Hervier, A.; Somorjai, G.A.; Komvopoulos, K. Tuning the electronic structure of titanium oxide support to enhance the electrochemical activity of platinum nanoparticles. *Nano Lett.* **2013**, *13*, 4469–4474. [CrossRef]
38. Mirceski, V.; Guziejewski, D.; Bozem, M.; Bogeski, I. Characterizing electrode reactions by multisampling the current in square-wave voltammetry. *Electrochim. Acta* **2016**, *213*, 520–528. [CrossRef]
39. Sawyer, D.T.; Valentine, J.S. How super is superoxide? *Acc. Chem. Res.* **1981**, *14*, 393–400. [CrossRef]
40. TBui, D.; Kimura, A.; Ikeda, S.; Matsumura, M. Determination of Oxygen Sources for Oxidation of Benzene on $TiO_2$ Photocatalysts in Aqueous Solutions Containing Molecular Oxygen. *J. Am. Chem. Soc.* **2010**, *132*, 8453–8458. [CrossRef]

41. Ilie, A.G.; Scarisoreanu, M.; Dutu, E.; Dumitrache, F.; Banici, A.M.; Fleaca, C.T.; Vasile, E.; Mihailescu, I. Study of phase development and thermal stability in as synthesized $TiO_2$ nanoparticles by laser pyrolysis: Ethylene uptake and oxygen enrichment. *Appl. Surf. Sci.* **2018**, *427*, 798–806. [CrossRef]
42. Gupta, N.; Bansal, P.; Pal, B. Metal ion-$TiO_2$ nanocomposites for the selective photooxidation of benzene to phenol and cycloalkanol to cycloalkanone. *J. Experim. Nanosci.* **2015**, *10*, 148–160. [CrossRef]
43. State, R.; Scurtu, M.; Miyazaki, A.; Papa, F.; Atkinson, I.; Munteanu, C.; Balint, I. Influence of metal-support interaction on nitrate hydrogenation over Rh and Rh-Cu nanoparticles dispersed on $Al_2O_3$ and $TiO_2$ supports. *Arab. J. Chem.* **2017**, *10*, 975–984. [CrossRef]
44. Preda, S.; Anastasescu, C.; Balint, I.; Umek, P.; Sluban, M.; Negrila, C.C.; Angelescu, D.G.; Bratan, V.; Rusu, A.; Zaharescu, M. Charge separation and ROS generation on tubular sodium titanates exposed to simulated solar light. *Appl. Surf. Sci.* **2019**, *470*, 1053–1063. [CrossRef]
45. Anastasescu, C.; Zaharescu, M.; Angelescu, D.; Munteanu, C.; Bratan, V.; Spataru, T.; Negrila, C.; Spataru, N.; Balint, I. Defect-related light absorption, photoluminescence and photocatalytic activity of $SiO_2$ with tubular morphology. *Solar Energ. Mater. Solar Cells* **2017**, *159*, 325–335. [CrossRef]

catalysts

MDPI

Article

# Novel Heterostructures of Noble Plasmonic Metals/Ga-Substituted Hydrotalcite for Solar Light Driven Photocatalysis toward Water Purification

**Eugenia Corina Ignat [1], Doina Lutic [2,*], Gabriel Ababei [3] and Gabriela Carja [1,*]**

1. Department of Chemical Engineering, Faculty of Chemical Engineering and Environmental Protection "Cristofor Simionescu", Technical University of Iasi, 71 D. Mangeron, 700050 Iasi, Romania
2. Faculty of Chemistry, Al. I. Cuza University of Iasi, 11 Bd. Carol I, 700501 Iasi, Romania
3. National Institute of Research and Development for Technical Physics, 47 Mangeron Boulevard, 700050 Iasi, Romania

* Correspondence: doilub@uaic.ro (D.L.); carja@uaic.ro (G.C.); Tel./Fax: +40-740-23-6565 (D.L.); +40-232-20-1231 (G.C.)

**Abstract:** Heterostructures formed by close conjunctions of plasmonic metal nanoparticles and non-plasmonic (2D) lamellar nanostructures are receiving extensive interest as solar-light-driven photocatalysts for environmental pollutant remediation. Herein, the conjunction of plasmonic Au or Ag and Ga-substituted hydrotalcite are obtained by exploiting the manifestation of the structural "memory effect" of Ga-substituted hydrotalcite in the aqueous solutions of $Au(CH_3COO)_3$ and $Ag_2SO_4$, respectively. The 2D layered matrix of MgGaAl plays a dual function; it is involved in the synthesis of the plasmonic metal nanoparticles, and further, is acting as a support. The compressive investigations using X-ray diffraction (XRD), UV-diffuse reflectance spectroscopy (UVDR), infrared spectroscopy (FT-IR), transmission electron microscopy (TEM/HRTEM), high-angle annular dark-field imaging/scanning transmittance electron microscopy (HAADF/STEM) and X-ray photoelectron spectroscopy (XPS) describe structural, composition and nano/micromorphology characteristics of the novel heterostructures, while UVDR analysis afforded to study the features of their plasmonic responses. Results reveal that the catalysts are formed by close conjunction of small nanoparticles of Au or Ag (with a mean size less than 20 nm) that are formed on the larger particles of MgGaAl and own plasmonic features within the visible range. The catalysts performances were tested towards photocatalytic degradation of p-dichlorobenzene and 4-nitrophenol under solar light irradiation. Results revealed that the degradation of the pollutants is entangled to the plasmonic response of the heterostructured catalysts that is the key functionality in promoting photocatalysis and degrading the pollutants, under solar light irradiation. MgGaAl showed a very low photocatalytic activity when irradiated by UV or solar light. Notably, the heterostructured catalysts proceeded in good to excellent yield to remove the tested pollutants, under solar light irradiation. The sustainability of the novel catalysts was assessed through the kinetic analysis of the degradation processes of the tested pollutants and their mixture.

**Keywords:** layered double hydroxides; plasmonic metals; solar light photocatalysis

**Citation:** Ignat, E.C.; Lutic, D.; Ababei, G.; Carja, G. Novel Heterostructures of Noble Plasmonic Metals/Ga-Substituted Hydrotalcite for Solar Light Driven Photocatalysis toward Water Purification. *Catalysts* **2022**, *12*, 1351. https://doi.org/10.3390/catal12111351

Academic Editors: Ioan Balint and Monica Pavel

Received: 9 October 2022
Accepted: 27 October 2022
Published: 2 November 2022



## 1. Introduction

Deterioration of the environment and a shortage of sustainable energy supply have become major societal issues that are threating the development of human society and the preservation of our planet [1]. Whereas water shortage was for many decades associated to certain regions from Africa and central Asia, nowadays becomes a worrying problem worldwide, being entangled to the climate changes from the last decades [2]. Nowadays a large variety of organic pollutants have been found in effluents of sewage treatment plants, rivers, surface and ground waters [3]. Among them, man-made consumables containing phenols, dyes, nitrobenzene or halobenzene compounds are the leading sources of water pollution [4]. The quest to provide clean water has led to a tremendous boost in the scientific efforts to develop novel performant technologies for environmental remediation [5]. Toward this, photocatalysis has received much attention because, among the traditional physical techniques, it provides a powerful tool for the removal of organic contaminants by completely degrading them [6]. Inside this, solar-light-driven photocatalysis is an effective and very promising way to meet both energy demands and water pollution issues [7]. It utilizes photogenerated carriers (electrons and holes) to initiate redox reactions and realize solar-to-chemical energy conversion.

The demonstrations of solar-light-driven chemical transformations on plasmonic nanostructures have led to the emergence of a new field in heterogeneous catalysis known as plasmonic catalysis [8]. Plasmonic metals are light-harvesting nanostructures that interact with visible light through the excitation of localized surface plasmon resonance (LSPR) [9]. A question that has emerged recently is whether it is possible to take advantage of the functionality of the plasmonic behavior in multicomponent catalysts. These are formed by close conjunctions of a plasmonic metal, which amplifies and concentrates the photons' energy within the material and, a non-plasmonic component that is able to play the role of support to stabilize the nanometal and, further, to extract the plasmon energy in the form of electronic excitations to perform a targeted catalytic function [10]. On such a plasmonic/non-plasmonic interface the light energy harvested by the plasmonic metal can modulate specific interactions with the support that are entangled to the rearrangement of electrons, transfer of photogenerated carriers and their prolonged lifetime and the extended light-response range within heterostructured components [11].

LDH are 2D layered matrices with a brucite-like structure that are conventionally described by the general formula $[M^{II}_{1-x}M^{III}_{x}(OH)_2]^{x+} \cdot A^{n-}_{x/n} \cdot mH_2O$, where the divalent $M^{II}$ and trivalent $M^{III}$ cations might be defined as $Mg^{2+}$, $Zn^{2+}$, $Cu^{2+}$, $Al^{3+}$, $Ga^{3+}$, etc. and, the $A^{n-}$ can be almost any organic or inorganic anion [12]. The possibility to incorporate specific cations in the 2D layers of LDH delivers a generous palette of semiconductive materials useful for photocatalytic devices [13–15]. Further, LDH are relatively simple and cheap to prepare and own the ability to reconstruct its layered structure when the calcined LDH are introduced in the aqueous solutions containing anions [16,17]. By virtue of their unique 2D-layered structure, tuned optical absorption, their hydroxylated surfaces and ease of preparation, LDHs have emerged as very promising candidates for obtaining versatile and robust catalysts for many actual and potential applications and several reviews on this topic are available [18–25]. Importantly, the close conjunction of the LDH matrix and plasmonic nanoparticles affords to obtain multicomponent catalysts in which the LDH unit plays multifarious roles including immobilization and stabilization of nanoparticles [26], and further provides a unique interface-space confined in a 2D matrix for controlling nanoparticle spatial distribution [27]. Additionally, serving as a support, LDH will afford to stabilize low nuclearity nanospecies on its surface by minimizing atom diffusion, controlling the nanoparticle morphology and tuning active nanometal electronic structure [28].

4-Nitrophenol (4-NPh) is one of the highly toxic organic pollutants found among the substances bearing nitro groups, which are common components of industrial effluents. It has been detected in urban and agricultural waste and is recognized as a priority hazardous pollutant by the Environmental Protection Agency (EPA) due to its poisonous and volatile nature. [29]. Furthermore, para-Dichlorobenzene (p-DCB) is the main component of moth balls, disinfection fumigants and toilet deodorization cakes. It is considered a low toxicity compound, but it causes skin, eyes and gastrointestinal tract irritation, causing nausea, vomiting and diarrhea. Neurotoxic effects (retardation, dysarthria, ataxia, cognitive decline, memory disorders) were reported in cases of ingestion [30].

Herein, by exploiting the "structural reconstruction" of the LDH in the aqueous solutions of $Au(CH_3COO)_2$ and $Ag_2SO_4$, respectively, we successfully constructed novel plasmonic heterostructures defined by the close conjunction of nanoparticles of Au or Ag with gallium partially substituted hydrotalcite-like 2D matrix, denoted as Au_MgGaAl and Ag_MgGaAl, respectively. Next, we present their structural, morphological and plasmonic characteristics and applications in plasmon-induced photocatalysis toward degradation of both 4-NPh and p-DCB from water. Results point out the enhanced catalytic performances of the synthesized plasmonic heterostructures in comparison to their calcined forms and the basic LDH, and further reveal the advantage of plasmonic metals in catalyst composition. A discussion of the kinetic models that govern the studied plasmonic catalysis is also included.

## 2. Results and Discussion

### *2.1. Synthesis Procedures and Structure Characterization*

In our method, MgGaAl was obtained by coprecipitation at constant pH and 65 °C, while Ag_MgGaAl and Au_MgGaAl were obtained, at room temperature, after the reconstruction of the calcined MgGaAl in the aqueous solutions of $Au(CH_3COO)_3$ and $Ag_2SO_4$, respectively [31]. In fact, the synthesis procedures exploited the LDH capability to manifest its structural memory in the aqueous solution containing $CH_3COO^-$ and $SO_4^{2-}$ [31]. The purities and crystalline phases of the as-prepared samples were analyzed by X-ray diffraction (XRD). Figure 1a shows the XRD diffractograms of MgGaAl as "as synthesized" and after the reconstruction processes. The recovery of the LDH structure by reconstruction is shown by the XRD analysis, revealing patterns that could be perfectly indexed to the LDHs phase (ICDD file No. 22-700), with a series of sharp and symmetric basal reflections of the ($00\ell$, $\ell$ = 3, 6, 9) planes and broad, less intense, reflections for the nonbasal ($01\ell$, $\ell$ = 2,5,8) planes [32]. Particularly, MgGaAl shows a well crystalized LDH structure though, other phases such as gallium oxy-hydroxide GaOOH are easily identified, as indicated as their characteristic reflections at 2θ = 34.5, 39 and 46.88 (JCPDS file no. 36-1451) denoted in Figure 1a as (*). Importantly, after the reconstruction, the structural features of Ag_MgGaAl and Au_MgGaAl are defined as a single crystalline LDH-like phase, pointing out that the reconstruction procedure, at ambient temperature, promoted the reconstruction of the LDH.

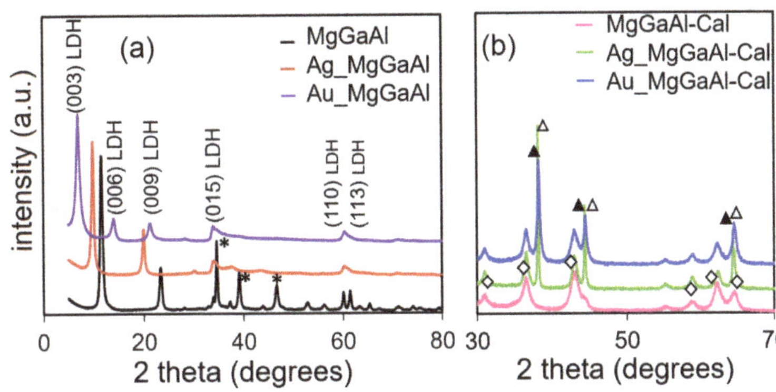


**Figure 1.** XRD patterns of (**a**). MgGaAl, Ag_MgGaAl and Au_MgGaAl; (**b**) after calcination at 870 °C. (*) GaOOH; Δ Au and ▲ Ag; ◊ $MgGa_2O_4$.

The most intense peak corresponds to (003) reflection and is associated with the distance between two consecutive brucite-like layers in the LDH structure. On the contrary, for Ag_MgGaAl and Au_MgGaAl the position of (003) reflection shifted to lower 2θ degrees. The (003) peak is related to the interlayer distance between the brucite-like layers that is established by the size of the anions of the interlayers and the 2θ values of (003), which are 11.52°, 9.92° and 6.94° for MgGaAl, Ag_MgGaAl and Au_MgGaAl, respectively. This shows that the reconstruction process altered the LDH interlayer space. Therefore, the replacement of carbonate anions of MgGaAl with the acetate and sulfate anions, after the reconstruction, promoted the increase of the interlayer spaces from 7.675 nm for MgGaAl to 8.909 nm for Ag_MgGaAl and 12.727 nm for Au_MgGaAl, as shown in Table 1. The position of the diffraction maximum is seldom varying with the nature of the cation, since the distance in the layer is depending on fitting the cations in the octahedral cage defined by six hydroxyl groups [19,33]. The local lack of order or deformations issued in the brucite-like sheet are due to discrepancies in the ordered arrangement of the octahedral units and promoted the overlap of (110) and (113) peaks for Ag_MgGaAl and Au_MgGaAl. As consequence, the characteristics of the diffraction patterns of Ag_MgGaAl and Au_MgGaAl demonstrate the reconstruction of the LDH structure, but further points out that this procedure may affect the ordering inside the layers leading to the formation of the structural defects. The intensity of the peaks due to (015) and (018) reflections further indicate the distortion of the layered structure. The "*a*" and "c" structural parameters [17,31] were calculated as: a = 2 × d (110) and c = 3 × d (006), where d (110) and d (006) are given by the Bragg relation and presented in Table 1. The increases of parameter "c" indicate the different nature of anions in Ag_MgGaAl and Au_MgGaAl, and agrees well to the previously reported values for the LDH containing $CH_3COO^-$ and $SO_4^{2-}$ in the interlayers, as previously reported [33,34].

**Table 1.** The XRD structural characteristics of the MgGaAl based catalysts.

| Sample | d (*003*) nm | d (*110*) nm | *a* nm | *c* nm |
|---|---|---|---|---|
| MgGaAl | 7.675 | 1.538 | 3.076 | 23.026 |
| Ag_MgGaAl | 8.909 | 1.536 | 3.072 | 26.728 |
| Au_MgGaAl | 12.727 | 1.533 | 3.066 | 38.180 |

Characteristic reflections of gold or silver phases are not observed in the XRD patterns of Ag_MgGaAl and Au_MgGaAl catalysts. This can be due to the small sizes and/or the low content of Au and Ag nanoparticles. Hence, to promote the growing of the small sizes nanoparticles and to study them by XRD, we further calcinated the heterostructured samples at 870 °C. After the calcination, the XRD patterns of the heterostructures (see Figure 1b) show the formation of $MgGa_2O_4$ (denoted with ◊ in Figure 1b) though further reveal the specific reflections of the face-centered cubic (fcc) of Ag (denoted with (▲) in Figure 1b), namely (111), (200) and (220) (JCPDS data no. 04-0783) and the diffraction lines of the (111) and (200) planes of (fcc) of Au (JCPDS Card No. 65-2870), denoted with (Δ) in Figure 1b.

Next, we used FT-IR analysis to get information about the nature of the anions of the LDH structure. The FT-IR spectra of the catalysts (see Figure 2) resemble those of the LDH phases [33,35]. Typical of all spectra are the strong broad absorbance band between 3600 and 3200 $cm^{-1}$ associated with the stretching mode of the hydroxyl groups, both from the brucite-like layers and the interlayer water molecules, as well as, the water molecules physisorbed on the external surface of the crystallites [36]. For MgGaAl, the ν3 mode of interlayer carbonate is responsible for the intense band at 1384 $cm^{-1}$. This is a degenerated mode for the D3h symmetry of the original carbonate anion. However, it can be noticed that the band at 1384 $cm^{-1}$ shows a clearly identified shoulder at 1491 $cm^{-1}$, which can be considered a result of the splitting of the ν3 mode [37]. The replacement of carbonate anions by acetate anions in Au_MgGaAl and by sulphate anions in Ag_MgGaAl is clearly disclosed by FT-IR analysis. For Au_MgGaAl, the bands at 1575 and 1410 $cm^{-1}$ indicate the presence of acetate anions, after the reconstruction; for Ag_MgGaAl, the peak around 1110 $cm^{-1}$ reveal the presence of the $SO_4^{2-}$ in the interlayers. Weaker bands below 1000 $cm^{-1}$ correspond mainly to vibration of lattice bonds of Me–OH while the characteristic peak at 668 $cm^{-1}$ is due to the vibration of the Me−O−Me bonds associated with the cations of the LDH layers [36].

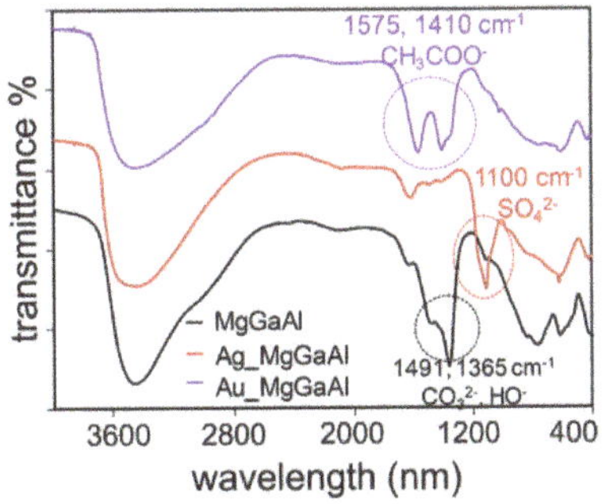


**Figure 2.** FT-IR spectra of MgGaAl, Ag_MgGaAl and Au_MgGaAl.

## *2.2. Morphology and Composition Analyses*

The morphology and crystalline features of the as-prepared catalysts were investigated by HRTEM and HAADF STEM images, as shown in Figure 3a–k. The results indicate that Ag_MgGaAl and Au_MgGaAl were composed by small nanoparticles, with a relatively narrow size, that are randomly distributed on the larger nanoparticles of the LDH, as shown by TEM images in Figure 3a,c. For both Ag_MgGaAl and Au_MgGaAl, almost spherical dark spots defined by an average size less than 20 nm are clearly seen on the larger nanoparticles of MgGaAl (av. size ~130 nm).

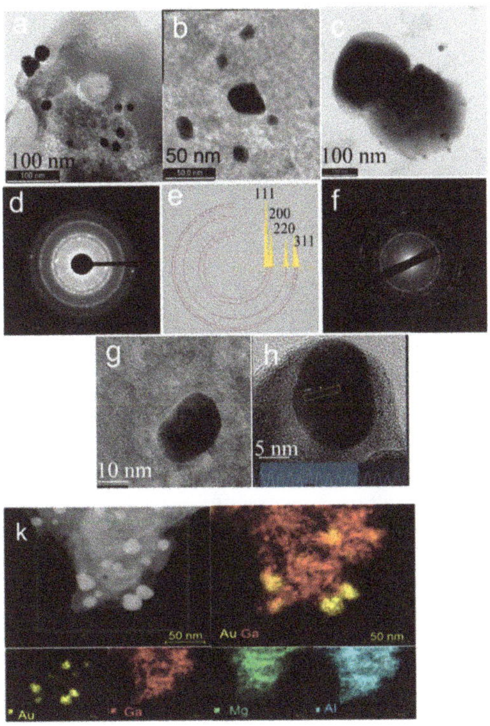


**Figure 3.** (**a**) TEM image of Au_MgGaAl; (**b**) Higher magnification of TEM image of Au_MgGaAl; (**c**) low magnification TEM image of Ag_MgGaAl; (**d**,**e**) SAED pattern of Au_MgGaAl; (**f**) SAED pattern of Ag_MgGaAl; (**g**) HRTEM image of Ag_MgGaAl; (**h**) HRTEM image of a single nanoparticle on Au_MgGaAl with identification of the crystal plane of gold; (**k**) HAADF STEM image of Au_MgGaAl and the corresponding STEM-EDS element maps.

Notably, HRTEM and SAED images in Figure 3b,d–k clearly identified Au and Ag nanoparticles that are well crystalized with lattice fringes defined by an interplanar spacing of 0.24 and 0.27 nm, respectively. This corresponds to the face-centered cubic (fcc) structure of Au (111) planes and Ag (111), being consistent with the SAED patterns [38]. The typical selected area electron diffraction (SAED) patterns display bright circular rings, featuring a complex pattern in which the diffraction rings of gold (see Figure 3d,e) and silver (Figure 3f), where (1 1 1), (2 0 0), (2 2 0) and (3 1 1) planes of Au and Ag, respectively are combined with the set of the diffraction pattern derived from the LDH [34]. These results demonstrate that in the heterostructured catalysts nanoparticles of Au and Ag, with an average size lower than 20 nm are in close conjunction [37] with the MgGaAl matrix. In addition, high-angle annular dark-field scanning transmission electron microscopy (HAADF-STEM) was employed to analyze the element distribution in Ag_MgGaAl and Au_MgGaAl heterostructures. The elemental mapping by energy-dispersive spectroscopy (EDS) was performed under STEM and we present the EDS map of Au_MgGaAl in Figure 3k. Results confirm that Mg, Ga, and Al are uniformly distributed over the entire catalysts, while Au is dispersed on the catalysts surface. In addition, the EDS analysis (see Supplementary Figure S1) indicates that Au and Ag content of the catalysts are 3.8 wt% for Au in Au_MgGaAl and 3.5 wt% for Ag in Ag_MgGaAl, which is close to the calculated content for the synthesis.

### 2.3. Optical Characteristics and Plasmonic Response

The formation of the plasmonic metals-LDH heterostructures was further tracked by analyzing the optical behavior of Ag_MgGaAl and Au_MgGaAl by UVDR analysis and the results are shown in Figure 4. The absorption edge of MgGaAl lies in the low UV region while the heterostructured catalysts reveal enhanced absorption in the visible range due to the LSPR response of plasmonic silver and gold. For Au_MgGaAl, the peak with a maximum at 550 nm is attributed to the SPR band of well dispersed nanoparticles of Au. It originates from the intraband excitation of electrons in the outer orbital (6sp) of the Au species and confirms the production of energized electrons [38,39]. For Ag_MgGaAl, the broad peak at 360–600 nm is assigned to the plasmonic response of silver, covering all the visible range. Moreover, the absorption bands of the calcined catalysts displayed higher intensity in comparison to the as-synthesized ones. This can be a consequence of the increase of the nanoparticle size during the calcination process [40–42].

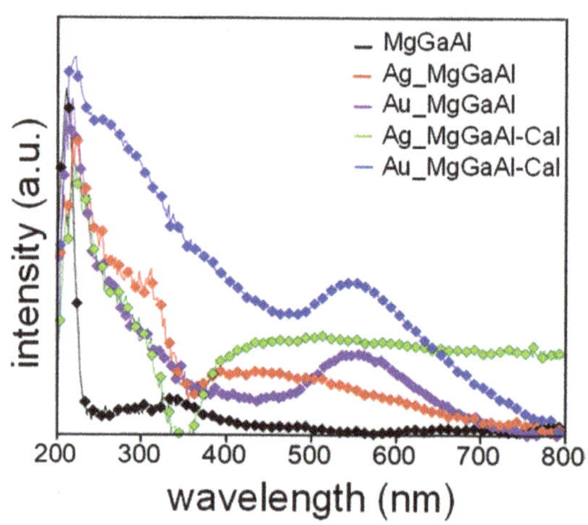


**Figure 4.** UVDR spectra of the studied photocatalysts.

## 2.4. XPS Analysis

The surface state of gold and silver in the heterostructured catalysts was explored by XPS analysis and the corresponding spectra are shown in Figure 5.

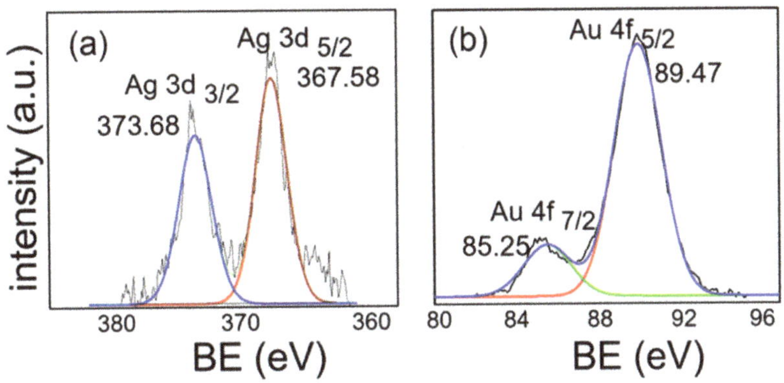


**Figure 5.** High-resolution XPS spectra of (**a**) Ag3d in Ag_MgGaAl; (**b**) Au4f in Au_MgGaAl.

The peaks at 373.68 and 367.58 eV, associated to Ag 3d 3/2 and Ag 3d 5/2 transitions, respectively (see Figure 5a), indicate that on the surface of Ag_MgGaAl silver is present as $Ag^0$. This is supported by the fact that the formation of silver nanoparticles was achieved under light that promoted the reduction of $Ag^+$ to $Ag^0$ [43]. The state of gold nanoparticles is described by the peaks at 85.25 and 89.47 eV in Figure 5b, which come from Au4f 7/2 and Au 4f 5/2 transitions, respectively, revealing the presence of a mixed states of $Au^{\delta+}$ ($\delta$ = 1, 1, 3) on the surface of Au_MgGaAl [44–46]. In both heterostructures, the signals generated by gallium (Supplementary Figure S2) formed peaks at 1145.2 eV and 1118 eV, revealing the presence of Ga 2p 1/2 and Ga 2p 3/2 as $Ga^{3+}$ cations in the LDH layer.

## 2.5. Photocatalytic Performance in 2-PNh and 2-DCB Photodegradation in Aqueous Solutions

The photocatalytic activity of the plasmonic catalysts was evaluated in the photodegradation of 4-NPh and p-DCB. The characteristic absorptions of 4-NPh at 390 nm and p-DCB at 220 nm were used to monitor the photocatalytic degradation process [47]. Among the examined catalysts, Ag_MgGaAl exhibited the best performance, completely decomposing the tested pollutants. As shown in Figure 6a,b for Ag_MgGaAl, the main adsorption peaks of the pollutants decreased gradually as solar irradiation proceeded, such that the peak almost vanished after 90 min for p-DCB while for 4-NPh the corresponding peak vanished after 360 min.

Figure 7 presents the effective mineralization of 4-NPh (Figure 7i) and 2-DCB (Figure 7ii) evaluated by TOC measurements for the catalysts irradiated by solar and UV light. Firstly, it reveals that the studied catalysts have not been active when irradiated by UV light. This clearly reveals the role of plasmonic characteristics in harvesting the light energy to promote photocatalysis. Thus, under solar light the presence of silver in Ag_MgGaAl highly increased the TOC removal up to 87% as compared to Au_MgGaAl, which mineralized only 71% of 4-NPh in the same experimental conditions. Further, the degradation of 2-DCB Ag_MgGaAl showed the strongest photocatalytic activity under solar irradiation while, almost no photodegradation was achieved under UV irradiation. Thus, the photocatalytic activity of the tested catalysts followed the orders Ag_MgGaAl > Au_MgGaAl > Ag_MgGaAlcal > Au_MgGaAlcal with degradation efficiencies of 2-DCB calculated to be 91%, 76%, 61% and 47% respectively, after 90 min of irradiation with solar light. MgGaAl was not active for the degradation of 2-DCB, while the degradation efficiency for 4-NPh reached only 11% on the LDH precursor. These results reveal the essential roles of the plasmonic responses on promoting the photons harvesting from the visible range of solar light and promoting pollutant degradation [48], but further disclose that the silver-based photocatalyst led to the best performance. This can be due to the much wider plasmonic peak of nanosilver that points out that Ag_MgGaAl is able to harvest light over a more extended visible light-responsive range in comparison to Au_MgGaAl [49]. The excellent activity of the Ag_MgGaAl sample was a good reason to test the performance of this photocatalyst using a mixture of 4-NPh and p-DCB (1/1 molar ratio).

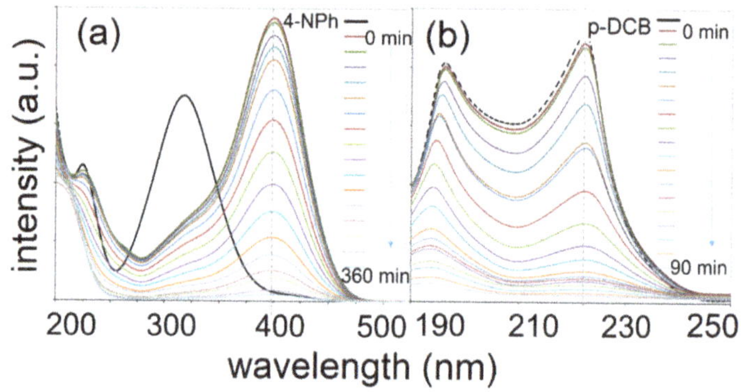


**Figure 6.** UV−vis absorption spectra of (**a**) 4-NPh; and (**b**) p-DCB solutions in the presence of Ag_MgGaAl and solar irradiation.

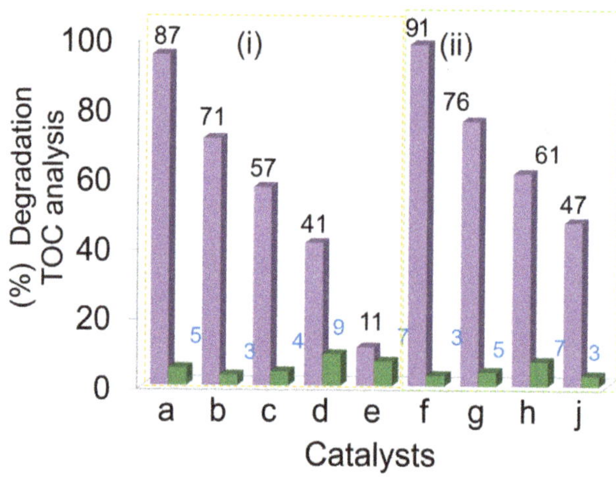


**Figure 7.** Photocatalytic performance evaluated by TOC measurements for (**i**) 4-NPh removal, after 360 min of irradiation, shown as: (a) Ag_MgGaAl; (b) Au_MgGaAl; (c) Ag_MgGaAcal; (d) Au_MgGaAlcal; (e) MgGaAl; (**ii**) p-DCB removal after 100 min of irradiation for: (f) Ag_MgGaAl; (g) Au_MgGaAl; (h) Ag_MgGaAlcal (j) Au_MgGaAl, by ( ) solar and ( ) UV light.

For 4-NPh (see Figure 8a) a delay of about 120 min was noticed for 4-NPh when found in the mixture of the pollutants. Thus, after 360 min the mixture exposed to the photocatalyst reveal a degradation yield of 75% of 4-NPh after 360 min of irradiation by solar light. This reveals a competition between the pollutants for the adsorption sites or for the HO· radicals involved in the photocatalytic process. On the other hand, p-DCB shows a different degradation profile (Figure 8b), with a fast degradation fitting a Langmuir type allure. After this, the degradation was quite slow, reaching almost 50% after 360 min of irradiation. The different behavior further points out that in their aqueous mixture p-DCB is highly preferred to 4-NPh in the adsorption process in the beginning of irradiation by solar light.

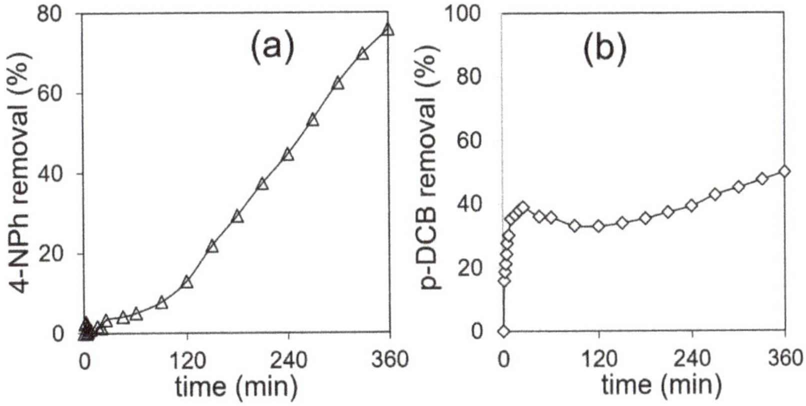


**Figure 8.** Pollutant % removal of: (**a**) 4-NPh; and (**b**) p-DCB from an aqueous solution containing both pollutants on Ag_MgGaAl.

The reusability of the catalysts was investigated via cycle experiments using a Ag_MgGaAl hybrid photocatalyst. For each test, the reusability of Ag_MgGaAl was evaluated by successive addition of the pollutants such that, before each cycle, the concentration of 4-NPh and p-DCB was equal to 0.125 mmol/L for each. As exhibited in Figure 9, there was just a slight decrease after testing three times, indicating that Ag_MgGaAl can be considered to be a stable and recyclable photocatalyst for the degradation of 4-NPh and p-DCB, respectively.

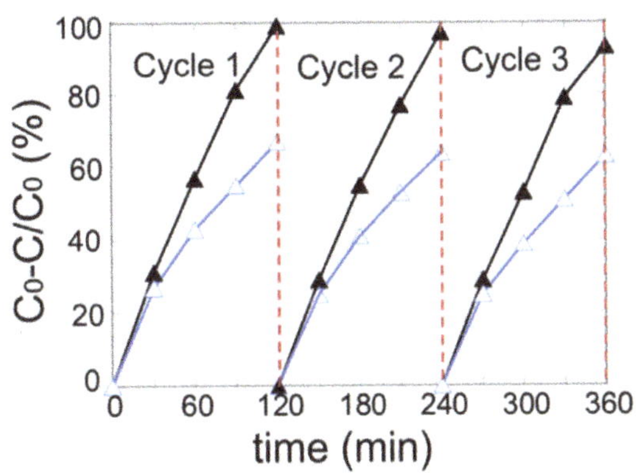


**Figure 9.** Recycling of Ag_MgGaAl photocatalyst for 4-NPh (Δ) and p-DCB (▲) degradation.

Information about the reactive species formed during the degradation of 4-NPh was obtained by using quenchers, such as isopropanol, that acts as efficient radical scavengers for •OH and benzoquinone, which is acting as an $O_2\bullet^-$ radical scavenger [37,48]. The results are given in Supplementary Figure S3 and show that in the presence of isopropanol the TOC removal (%) of 4-NPh decreased from 87% to 51%, pointing out that •OH radicals act as major reactive species in the photocatalytic system. Further, when benzoquinone was added to trap $O_2\bullet^-$, an obvious inhibition of degradation process was observed, from 87% to 59%, revealing that the addition of benzoquinone decreased the amount of involved $O_2\bullet^-$ during the degradation process. Based on these results, the possible photocatalytic mechanism for the degradation of the studied pollutants is illustrated in Scheme 1 where the pollutants are denoted as Ar-X. It describes that, under solar irradiation, nanoparticles of silver have given rise to photogenerated electron–hole pairs and, some of the photogenerated electrons at the CB of the plasmonic silver could be transferred to the CB of the LDH, due to the more negative CB position of the LDH than that of nanosilver [50]. Thus, the dissolved oxygen formed $\bullet O_2^-$ radicals, and then further •OH by combining with $H_2O$. These are preliminary results regarding the mechanism of the studied photocatalytic degradation and will be explored further in future work.

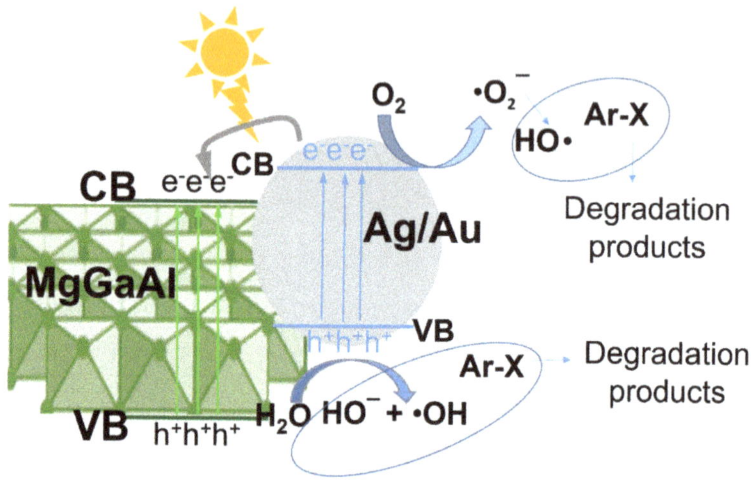


**Scheme 1.** Possible photocatalytic mechanism for the pollutant's degradation.

Next, we used the experimental data fitting to the pseudo-first, pseudo-second, Weber intraparticle diffusion and Elovich kinetic models, which were used to further investigate the kinetics of the photodegradation of 4-NPh, p-DCB and their mixture, under solar light. Each data set was fitted to the above-described models and the accuracy of the proposed models and the corresponding correlation coefficients are described in Figure 10 and Table 2. The kinetic analysis reveals that the pseudo-first order model fits relatively well when just one pollutant is present in the aqueous solution. As shown by the conversion data presented in Figure 8, the reaction rate increased almost six times for p-DCB degradation in comparison to that of 4-NPh degradation. Therefore, when the pollutants are mixed together, the pseudo-first order model is almost acceptable to fit the data concerning 4-NPh degradation ($R^2$ = 0.8245), while for p-DCB degradation, the pseudo-second order model fits the decomposition of p-DCB in the mixture with 4-NPh. However, to point out the above proposed kinetic models we present in Figure 10a,b the linear fit for all the systems describing the pseudo-first and pseudo-second order kinetics. Furthermore, the Weber intraparticle diffusion model [51] was chosen to highlight the influence of the intraparticle diffusion on the reaction rate, considering that the adsorption on the photocatalyst surface of the transformed species is a key step of the process, and results are shown in Figure 10c. If data fitted to a linear plot defined as qt versus $t^{0.5}$ with the plot passing through the origin, the intraparticle diffusion is the only rate-controlling step [45]. In the case of our reactions, the plots contain two linear portions for the conversion of p-DCB, alone or mixed with 4-NPh. This reveals that there is more than one rate-controlling step; thus, the adsorption of p-DCB occurs by weaker binding on the photocatalyst surface, followed by the oxidation through the HO radicals, while the ionic structure of the 4-NPh could favor binding forces through electrostatic attraction with the catalysts. Moreover, the deactivating effect the nitro group exerted on the aromatic ring might contribute to a certain inertia of this molecule in the oxidation step. We further reveal that the Elovich model did not fit for any of the studied catalysts because there are no linear dependencies between ln t and qt, as shown by results in Figure 10d hence, the $\alpha$ and $\beta$ parameters were not calculated for this model.

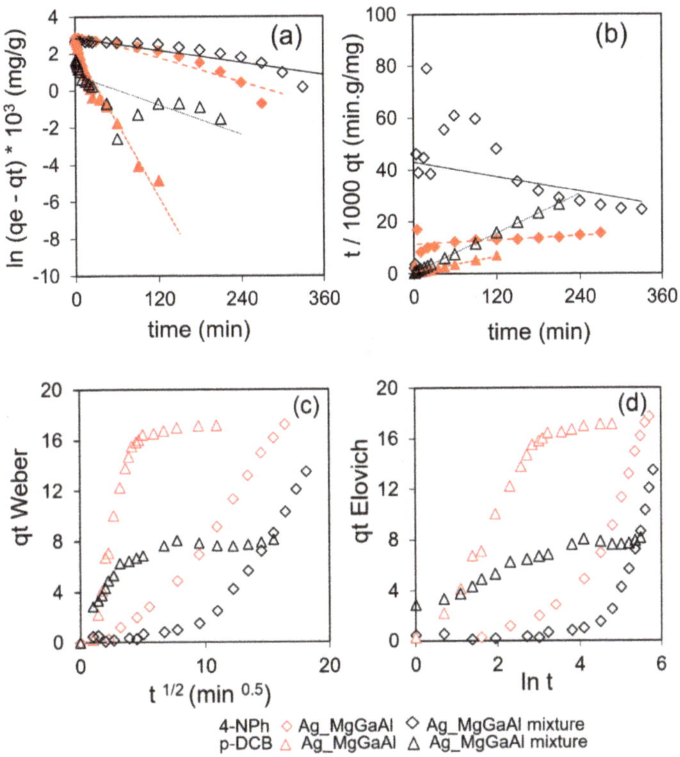


**Figure 10.** Kinetic analysis as fitted to: (**a**) pseudo-first order; (**b**) pseudo-second order; (**c**) Weber intraparticle diffusion; (**d**) Elovich kinetic models.

**Table 2.** Kinetic parameters and the correlation coefficients $R^2$.

| Reaction System | | Ag_MgGaAl_4-NPh | Ag_MgGaAl_4-NPh (Mix) | Ag_MgGaAl_p-DCB | Ag_MgGaAl_p-DCB (Mix) |
|---|---|---|---|---|---|
| Pseudo-first order | $k_1$ | 0.011 | 0.0056 | 0.0665 | 0.0133 |
| | $R^2$ | 0.9019 | 0.8245 | 0.9435 | 0.5288 |
| Pseudo-second order | $k_2$ | 0.0014 | 0.0046 | 0.0044 | 0.0112 |
| | $R^2$ | 0.292 | 0.0773 | 0.8294 | 0.9994 |
| Weber intraparticle diffusion | $k_{id}$ | 1.1821 | - | 4.412 | 1.6169 |
| | C | −1.407 | - | −2.422 | 1.1056 |
| | $R^2$ | 0.9862 | - | 0.9478 * | 0.9904 * |

* calculated for the first linear range.

## 3. Materials and Methods

### 3.1. Fabrication and Characterization of the Catalysts

Ga partially substituted hydrotalcite defined by a $M^{2+}/M^{3+}$ ratio of 2 (where $M^{2+}$ = $Mg^{2+}$ and $M^{3+}$ = $Al^{3+}$ and $Ga^{3+}$), are obtained by a standard co-precipitation method at a constant pH [13,15] and denoted as MgGaAl. More precisely, the aqueous solutions of $Mg(NO_3)_2{\cdot}6H_2O$, $Al(NO_3)_3{\cdot}9H_2O$ and $GaSO_4{\cdot}7H_2O$ were added to a solution of $Na_2CO_3/NaOH$ at a constant pH of 9.5 and the precipitate was aged for 24 h at 45 °C. Further, after the calcination at 550 °C for 8 h, the evolved mixed oxides were denoted as MgGaAlcal. Heterostructures of the LDH with plasmonic Au and Ag was achieved in aqueous medium during the structural reconstruction of the LDHs, in the aqueous solutions of $Au(CH_3COO)_3$ and $Ag_2SO_4$, respectively, at room temperature [33,40] and Au and Ag mass content were calculated to reach ~4.0% mass loading.

The heterostructured catalysts were denoted as Au_MgGaAl and Ag_MgGaAl, respectively. After calcination at 550 °C for 8 h the derived catalysts were denoted as Au_MgGaAlcal and Ag_MgGaAlcal.

X-ray diffractions were performed on a PANalytical X'Pert ((Malvern PANalytical Ltd., Almelo, The Netherlands) PROMPD diffractometer equipped with a filtered Cu Kα radiation; measurements were done in the 2θ mode using a bracket sample holder with a scanning speed of 0.04°/4 s in continuous mode. Diffuse Reflectance Infrared Fourier Transform spectra (DRIFT) were measured on a Nicolet 6700 FT-IR spectrometer (Thermo, Waltham, MA, USA), equipped with a mid-IR source (400–4000 $cm^{-1}$), a KBr-beam splitter and about 200 scans were taken with a 4 $cm^{-1}$ resolution. TEM/HRTEM images and results were obtained on Ultra-High resolution TEM microscope, UHR-TEM-Libra 200 MC/Carl Zeiss GmbH (Carl Zeiss Microscopy GmbH, Oberkochen, Germany). X-ray photoelectron spectroscopy (XPS, PerkinElmer Inc., Waltham, MA, USA) spectra were recorded using a Perkin-Elmer Model 5500-MT spectrometer equipped with Mg Kα radiation (1253.6 eV), operating at 15 kV and 20 mA; the binding energies (BE) were corrected by referencing the C1s peak to 284.8 eV. The UV-vis analysis profiles were acquired with a Jasco 500 UV-vis spectrophotometer within the wavelength range of 200–800 nm.

### 3.2. Photocatalytic Experiments

The photocatalytic activity of the LDH, the heterostructured catalysts and the calcined derivates were tested for the photodegradation of 4-NPh and 2 DCP in aqueous solutions. For this, appropriate amounts of catalyst powder were dispersed in the pollutant solution with the initial concentration of 0.025 g/L (of single or mol/mol mixture of pollutants) and a solid/liquid ration at 1 g/L. The solutions were stirred in the dark in order to realize the adsorption–desorption equilibrium between the catalytic surface and the pollutants, until no modification of the UV-vis absorption spectra was observed. Afterwards, the catalyst/pollutants mixtures were irradiated using an UNNASOL US 800 solar simulator, using a Xenon Lamp as a light source reaching 870 $W/m^2$, equipped with UV or Vis block filters. The photocatalytic activity of all catalysts was tested under solar and UV irradiation. The reactions were carried out in a 200 mL reactor with water recirculation in order to avoid the evaporation and to keep the system at a constant volume and temperature. The pollutants' photodegradation profile was monitored by measuring the UV-Vis absorption spectra and by total organic carbon (TOC). TOC of 4-NPh and 2-DCB aqueous solutions were analyzed with a Shimadzu TOC-VCPH (Shimadzu, Kyoto, Japan) analyzer equipped with a NDIR detector, to evaluate the mineralization degree of the organic material. The TOC values were obtained by subtracting IC (inorganic carbon) values from TC (total carbon) values. Removal yield (%) was calculated as: 100 × $(Ao-At)/Ao$, where $Ao$ and $At$ were the absorbance values measured on solutions taken before the illumination (counted at time zero value) and at time $t$ from irradiation start, respectively.

The r kinetics analysis was studied by fitting the experimental data on four models: pseudo-first-order model, pseudo-second-order model, Weber intraparticle diffusion model and Elovich model. The corresponding equations and significance of the calculated parameters are given in Table 3.

**Table 3.** Equations of the kinetic models applied in this study.

| Model | Equation | Parameters Significance and Defining Relations |
|---|---|---|
| Pseudo-first order | $\ln(q_e - q_t) = \ln q_e - k_1 t$ | $q_e = V(C_o - C_e)/m$; $q_t = V(C_o - C_t)/m$<br>$q_e$ and $q_t$—measured species at equilibrium and time t; V—volume of solution; m—adsorbent weight; $k_1$, $k_2$—rate constant for pseudo-first and pseudo-second order; $k_{id}$—rate constant for intraparticle diffusion; α—initial adsorption rate; β—desorption constant |
| Pseudo-second order | $t/q_t = 1/(k_2 q_e^2) + t/q_e$ | |
| Weber diffusion | $q_t = k_{id}\, t^{0.5} + C$ | |
| Elovich | $q_t = 1/\beta\, (\ln \alpha\beta) + 1/\beta \ln t$ | |

## 4. Conclusions

We have successfully achieved to obtain novel plasmonic heterostructures by the close conjunction of plasmonic metals, specifically Au and Ag nanoparticles and gallium partially substituted hydrotalcite layered matrix. The catalysts were obtained by a facile synthesis procedure, via calcination-reconstruction of MgGaAl in the solutions of gold acetate and silver sulphate, respectively. The analysis of photocatalytic degradation of 4-NPh and 2-DCB from aqueous solutions, under irradiation by solar light, revealed that the plasmonic behavior of the newly designed catalysts plays a key role in establishing high photocatalytic performances for the removal of the studied pollutants from the aqueous mediums.

**Supplementary Materials:** The following supporting information can be downloaded at: https://www.mdpi.com/article/10.3390/catal12111351/s1, Figure S1: EDS analysis for (a) Ag_MgGaAl and (b) Au_MgGaAl catalysts; Figure S2: High resolution XPS spectra of Ga2p in both Ag_MgGaAl and Au_MgGaAl catalysts. Figure S3: Effect of different scavengers on the degradation of 4-NP in the presence of Ag-MgGaAl.

**Author Contributions:** Conceptualization, G.C.; Methodology, E.C.I.; Validation, E.C.I., G.A.; Formal analysis, D.L., G.C.; Investigation, G.A., E.C.I.; Data curation, D.L.; Writing—original draft preparation, D.L.; Writing—review and editing, G.C., D.L.; Visualization, G.C.; Supervision, G.C.; Funding acquisition, E.C.I. All authors have read and agreed to the published version of the manuscript.

**Funding:** This research received no external funding.

**Data Availability Statement:** Not applicable.

**Acknowledgments:** The authors are grateful for the financial supports from the Grant-in-Aid for Scientific Research from Romanian National Authority for Scientific Research, CNCS-UEFISCDI (PN-II-ID-1751).

**Conflicts of Interest:** The authors declare no conflict of interest.

## References


1. Available online: https://www.dw.com/en/water-scarcity-eu-countries-forced-to-restrict-drinking-water-access/a-62363819 (accessed on 27 July 2022).
2. Available online: https://www.statista.com/chart/26140/water-stress-projections-global/ (accessed on 21 July 2022).
3. Lin, L.; Jiang, W.; Chen, L.; Xu, P.; Wang, H. Treatment of produced water with photocatalysis: Recent advances, affecting factors and future research prospects. *Catalysts* **2020**, *10*, 924. [CrossRef]
4. Zimmerman, J.B.; Anastas, P.T.; Erythropel, H.C.; Leitner, W. Designing for a green chemistry future. *Science* **2020**, *367*, 397–400. [CrossRef] [PubMed]
5. Murgolo, S.; De Ceglie, C.; Di Iaconi, C.; Mascolo, G. Novel $TiO_2$-based catalysts employed in photocatalysis and photoelectrocatalysis for effective degradation of pharmaceuticals (PhACs) in water: A short review. *Curr. Opin. Green Sustain. Chem.* **2020**, *30*, 100473. [CrossRef]
6. Dong, H.; Zeng, G.; Tang, L.; Fan, C.; Zhang, C.; He, X.; He, Y. An overview on limitations of $TiO_2$-based particles for photocatalytic degradation of organic pollutants and the corresponding countermeasures. *Water Res.* **2015**, *79*, 128–146. [CrossRef] [PubMed]
7. Marcu, I.C.; Pavel, O.D. Layered Double Hydroxide-Based Catalytic Materials for Sustainable Processes. *Catalysts* **2022**, *12*, 816. [CrossRef]
8. Liu, C.; Mao, S.; Shi, M.; Wang, F.; Xia, M.; Chen, Q.; Ju, X. Peroxymonosulfate activation through 2D/2D Z-scheme CoAl-LDH/BiOBr photocatalyst under visible light for ciprofloxacin degradation. *J. Hazard. Mater.* **2021**, *420*, 126613. [CrossRef]
9. Brongersma, M.L.; Halas, N.J.; Nordlander, P. Plasmon-induced hot carrier science and technology. *Nat. Nanotechnol.* **2015**, *10*, 25–34. [CrossRef]
10. Liu, C.; Mao, S.; Shi, M.; Hong, X.; Wang, D.; Wang, F.; Xia, M.; Chen, Q. Enhanced photocatalytic degradation performance of BiVO4/BiOBr through combining fermi level alteration and oxygen defect engineering. *Chem. Eng. J.* **2022**, *449*, 137757. [CrossRef]
11. Chen, Y.; Wang, Y.; Li, W.; Yang, Q.; Hou, Q.; Wei, L.; Liu, L.; Huang, F.; Ju, M. Enhancement of photocatalytic performance with the use of noble-metal-decorated $TiO_2$ nanocrystals as highly active catalysts for aerobic oxidation under visible-light irradiation. *Appl. Catal. B Environ.* **2017**, *210*, 352–367. [CrossRef]
12. Zhang, G.; Zhang, X.; Meng, Y.; Pan, G.; Ni, Z.; Xia, S. Layered double hydroxides-based photocatalysts and visible-light driven photodegradation of organic pollutants: A review. *Chem. Eng. J.* **2020**, *392*, 123684. [CrossRef]
13. Li, C.; Wei, M.; Evans, D.G.; Duan, X. Layered double hydroxide-Based nanomaterials as highly efficient catalysts and adsorbents. *Small* **2014**, *10*, 4469–4486. [CrossRef]

14. Puscasu, C.-M.; Seftel, E.M.; Mertens, M.; Cool, P.; Carja, G. ZnTiLDH and the Derived Mixed Oxides as Mesoporous Nanoarchitectonics with Photocatalytic Capabilities. *J. Inorg. Organomet. Polym. Mater.* **2014**, *25*, 259–266. [CrossRef]
15. Mohapatra, L.; Parida, K. A review on the recent progress, challenges and perspective of layered double hydroxides as promising photocatalysts. *J. Mater. Chem. A* **2016**, *4*, 10744–10766. [CrossRef]
16. Du, H.; Fan, J.; Miao, C.; Gao, M.; Liu, Y.; Li, D.; Feng, J. Recent advances in constructing interfacial active catalysts based on layered double hydroxides and their catalytic mechanisms. *Trans. Tianjin Univ.* **2021**, *27*, 24–41. [CrossRef]
17. Gu, Z.; Atherton, J.J.; Xu, Z.P. Hierarchical layered double hydroxide nanocomposites: Structure, synthesis and applications. *Chem. Comm.* **2015**, *51*, 3024–3036. [CrossRef]
18. Wu, M.J.; Wu, J.Z.; Zhang, J.; Chen, H.; Zhou, J.Z.; Qian, G.R.; Xu, Z.P.; Du, Z.; Rao, Q.L. A review on fabricating heterostructures from layered double hydroxides for enhanced photocatalytic activities. *Catal. Sci. Technol.* **2018**, *8*, 1207–1228. [CrossRef]
19. Ng, S.F.; Lau, M.Y.L.; Ong, W.J. Engineering layered double hydroxide–based photocatalysts toward artificial photosynthesis: State-of-the-Art progress and prospects. *Solar RRL* **2021**, *5*, 2000535. [CrossRef]
20. Prasad, C.; Tang, H.; Liu, Q.Q.; Zulfiqar, S.; Shah, S.; Bahadur, I. An overview of semiconductors/layered double hydroxides composites: Properties, synthesis, photocatalytic and photoelectrochemical applications. *J. Mol. Liq.* **2019**, *289*, 111114. [CrossRef]
21. Li, Z.; Chen, M.; Hu, H.; Zhang, Q.; Tao, D. Mechanochemical synthesis of novel Pt modified ZnAl-LDH for effective ciprofloxacin photodegradation. *J. Solid State Chem.* **2020**, *290*, 121594. [CrossRef]
22. Li, Z.; Zhang, Q.; Liu, X.; Wu, L.; Hu, H.; Zhao, Y. One-step mechanochemical synthesis of plasmonic Ag/Zn–Al LDH with excellent photocatalytic activity. *J. Mater. Sci.* **2018**, *53*, 12795–12806. [CrossRef]
23. Espinal, R.; Taboada, E.; Molins, E.; Chimentao, R.J.; Medina, F.; Llorca, J. Ethanol steam reforming over hydrotalcite-derived Co catalysts doped with Pt and Rh. *Top. Catal.* **2013**, *56*, 1660–1671. [CrossRef]
24. Liu, D.; Barbar, A.; Najam, T.; Javed, M.S.; Shen, J.; Tsiakaras, P.; Cai, X. Single noble metal atoms doped 2D materials for catalysis. *Appl. Catal. B Environ.* **2021**, *297*, 120389. [CrossRef]
25. Katsumata, K.I.; Sakai, K.; Ikeda, K.; Carja, G.; Matsushita, N.; Okada, K. Preparation and photocatalytic reduction of $CO_2$ on noble metal (Pt, Pd, Au) loaded Zn–Cr layered double hydroxides. *Mat. Lett.* **2013**, *107*, 138–140. [CrossRef]
26. Zhang, Z.; Li, P.; Zhang, X.; Hu, C.; Li, Y.; Yu, B.; Zeng, N.; Lv, C.; Song, J.; Li, M. Recent advances in layered-double-hydroxides based noble metal nanoparticles efficient electrocatalysts. *Nanomaterials* **2021**, *11*, 2644. [CrossRef] [PubMed]
27. Tsyganok, A.I.; Inaba, M.; Tsunoda, T.; Uchida, K.; Suzuki, K.; Takehira, K.; Hayakawa, T. Rational design of Mg–Al mixed oxide-supported bimetallic catalysts for dry reforming of methane. *Appl. Catal. A Gen.* **2005**, *292*, 328–343. [CrossRef]
28. Lestari, P.R.; Takei, T.; Yanagida, S.; Kumada, N. Facile and controllable synthesis of Zn-Al layered double hydroxide/silver hybrid by exfoliation process and its plasmonic photocatalytic activity of phenol degradation. *Mater. Chem. Phys.* **2020**, *250*, 122988. [CrossRef]
29. Bates, N. Mothball poisoning. *Emerg. Nurse* **2002**, *10*, 24–28. [CrossRef]
30. Weidman, E.K.; Tsiouris, A.J.; Heier, L.A. Toxic encephalopathy due to paradichlorobenzene toxicity: A case report and review of imaging characteristics. *Clin. Imaging* **2015**, *39*, 1095–1098. [CrossRef]
31. Evans, D.G.; Slade, R.C. Structural aspects of layered double hydroxides. In *Layered Double Hydroxides*; Xue, D., Evans, G.D., Eds.; Springer: Berlin, Germany, 2006; pp. 1–87.
32. Braterman, P.S.; Xu, Z.P.; Yarberry, F. Layered Double Hydroxides (LDHs). In *Handbook of Layered Materials*; Auerbach, S.M., Carrado, K.A., Dutta, P.K., Marcel, D., Eds.; Marcel Dekker, Inc.: New York, NY, USA, 2004; Chapter 8; pp. 363–457.
33. Gilea, D.; Radu, T.; Muresanu, M.; Carja, G. Plasmonic photocatalysts based on silver nanoparticles-Layered double hydroxides for efficient removal of toxic compounds using solar light. *Appl. Surf. Sci.* **2018**, *444*, 407–413. [CrossRef]
34. Mikami, G.; Grosu, F.; Kawamura, S.; Yoshida, Y.; Carja, G.; Izumi, Y. Harnessing self-Supported Au nanoparticles on layered double hydroxides comprising Zn and Al for enhanced phenol decomposition under solar light. *Appl. Catal. B Environ.* **2016**, *199*, 260–271. [CrossRef]
35. Govindappa, M.; Farheen, H.; Chandrappa, C.P.; Rai, R.V.; Raghavendra, V.B. Mycosynthesis of silver nanoparticles using extract of endophytic fungi, Penicillium species of Glycosmis mauritiana, and its antioxidant, antimicrobial, anti-inflammatory and tyrokinase inhibitory activity. *Adv. Natur. Sci. Nanosci. Nanotechnol.* **2016**, *7*, 035014. [CrossRef]
36. Gomes, J.F.; Garcia, A.C.; Gasparotto, L.H.S.; De Souza, N.E.; Ferreira, E.B.; Pires, C.; Tremiliosi-Filho, G. Influence of silver on the glycerol electro-oxidation over AuAg/C catalysts in alkaline medium: A cyclic voltammetry and in situ FTIR spectroscopy study. *Electrochim. Acta* **2014**, *144*, 361–368. [CrossRef]
37. Liu, C.; Mao, S.; Wang, H.; Wu, Y.; Wang, F.; Xia, M.; Chen, Q. Peroxymonosulfate-assisted for facilitating photocatalytic degradation performance of 2D/2D WO3/BiOBr S-scheme heterojunction. *Chem. Eng. J.* **2022**, *430*, 132806. [CrossRef]
38. Furube, A.; Du, L.; Hara, K.; Katoh, R.; Tachiya, M. Ultrafast Plasmon-Induced Electron Transfer from Gold Nanodots into $TiO_2$ Nanoparticles. *J. Am. Chem. Soc.* **2007**, *129*, 14852–14853. [CrossRef]
39. Kominami, H.; Tanaka, A.; Hashimoto, K. Mineralization of organic acids in aqueous suspensions of gold nanoparticles supported on cerium (iv) oxide powder under visible light irradiation. *Chem. Commun.* **2010**, *46*, 1287–1289. [CrossRef]
40. Carja, G.; Birsanu, M.; Okada, K.; Garcia, H. Composite plasmonic gold/layered double hydroxides and derived mixed oxides as novel photocatalysts for hydrogen generation under solar irradiation. *J. Mater. Chem. A* **2013**, *1*, 9092–9098. [CrossRef]
41. Lin, C.H.; Chu, H.L.; Hwang, W.S.; Wang, M.C.; Ko, H.H. Synthesis and optical properties of Mg-Al layered double hydroxides precursor powders. *AIP Adv.* **2017**, *7*, 125005. [CrossRef]

42. Al-Sarraj, A.; Saoud, K.M.; Elmel, A.; Mansour, S.; Haik, Y. Optoelectronic properties of highly porous silver oxide thin film. *SN Appl. Sci.* **2021**, *3*, 15. [CrossRef]
43. Wu, Y.; Lin, Y.; Xu, J. Synthesis of Ag–Ho, Ag–Sm, Ag–Zn, Ag–Cu, Ag–Cs, Ag–Zr, Ag–Er, Ag–Y and Ag–Co metal organic nanoparticles for UV-Vis-NIR wide-range bio-tissue imaging. *Photochem. Photobiol. Sci.* **2019**, *18*, 1081–1091. [CrossRef]
44. Klyushin, A.Y.; Rocha, T.C.; Hävecker, M.; Knop-Gericke, A.; Schlögl, R. A near ambient pressure XPS study of Au oxidation. *Phys. Chem. Chem. Phys.* **2014**, *16*, 7881–7886. [CrossRef]
45. Pramanik, G.; Humpolickova, J.; Valenta, J.; Kundu, P.; Bals, S.; Bour, P.; Dracinsky, M.; Cigler, P. Gold nanoclusters with bright near-Infrared photoluminescence. *Nanoscale* **2018**, *10*, 3792–3798. [CrossRef] [PubMed]
46. Peters, S.; Peredkov, S.; Neeb, M.; Eberhardt, W.; Al-Hada, M. Size-dependent XPS spectra of small supported Au-Clusters. *Surf. Sci.* **2013**, *608*, 129–134. [CrossRef]
47. Chen, H.; Yang, M.; Tao, S.; Chen, G. Oxygen vacancy enhanced catalytic activity of reduced $Co_3O_4$ towards p-nitrophenol reduction. *Appl. Catal. B Environ.* **2017**, *209*, 648–656. [CrossRef]
48. Chen, Y.; Li, W.; Wang, J.; Yang, Q.; Hou, Q.; Ju, M. Gold nanoparticles-modified $TiO_2$/SBA-15 nanocomposites as active plasmonic photocatalysts for the selective oxidation of aromatic alcohols. *RSC Adv.* **2016**, *6*, 70352–70363. [CrossRef]
49. Gu, S.; Lu, Y.; Kaiser, J.; Albrecht, M.; Ballauff, M. Kinetic analysis of the reduction of 4-nitrophenol catalyzed by Au/Pd nanoalloys immobilized in spherical polyelectrolyte brushes. *Phys. Chem. Chem. Phys.* **2015**, *17*, 28137–28143. [CrossRef]
50. Seftel, E.M.; Puscasu, M.; Mertens, M.; Cool, P.; Carja, G. Photo-responsive behavior of γ-Fe2O3 NPs embedded into ZnAlFe-LDH matrices and their catalytic efficiency in wastewater remediation. *Catal. Today* **2015**, *252*, 7–13. [CrossRef]
51. Cai, J.; Chen, C.R.; Zeng, H.Y.; Yi, M.Y.; Xiao, G.F.; Zhu, R.L.; Cao, X.J.; Shen, S.-G.; Peng, J.W. Fabrication of Ag2O/Ag decorated ZnAl-Layered double hydroxide with enhanced visible light photocatalytic activity for tetracycline degradation. *Ecotoxicol. Environ. Saf.* **2019**, *172*, 423–431.

*catalysts* MDPI

*Article*

# Using Gd-Enhanced β-$NaYF_4$:Yb,Er Fluorescent Nanorods Coupled to Reduced $TiO_2$ for the NIR-Triggered Photocatalytic Inactivation of *Escherichia coli*

**Huang Zhou and Fengjiao He ***

State Key Laboratory of Chemo/Biosensing and Chemometrics, College of Chemistry and Chemical Engineering, Hunan University, Changsha 410082, China; huangzhou527@hnu.edu.cn
* Correspondence: fengjiaohe@hnu.edu.cn

**Abstract:** β-$NaYF_4$:Yb,Er,Gd fluorescent nanorods were successfully coupled to a reduced $TiO_2$ (UCNPs@R-$TiO_2$) nanocomposite and applied to visible-light catalytic sterilization under 980 nm near-infrared (NIR) light illumination. The UCNPs (β-$NaYF_4$:Yb,Er,Gd) absorb the NIR light and emit red and green light. The visible light can be absorbed by the R-$TiO_2$ (Eg = 2.8 eV) for the photocatalytic reaction. About 98.1% of *Escherichia coli* were effectively killed upon 12 min of NIR light irradiation at a minimum inhibitory concentration (MIC) of 40 μg/mL UCNPs@R-$TiO_2$ nanocomposite. The bactericidal properties were further evaluated by matrix-assisted laser desorption/ionization time-of-flight mass spectrometry (MALDI-TOF MS) analysis. We found that the high bactericidal activity was due to the synergistic effect between the UCNPs and R-$TiO_2$. Moreover, the UCNPs show excellent upconversion luminance properties, and the introduction of visible-light-absorbed R-$TiO_2$ nanoparticles (2.8 eV) was conducive to the efficient separation and utilization of photogenerated electron-hole pairs.

**Keywords:** β-$NaYF_4$:Yb,Er,Gd nanorods; reduced $TiO_2$ nanoparticles; near-infrared light; visible light; *Escherichia coli*

**Citation:** Zhou, H.; He, F. Using Gd-Enhanced β-$NaYF_4$:Yb,Er Fluorescent Nanorods Coupled to Reduced $TiO_2$ for the NIR-Triggered Photocatalytic Inactivation of *Escherichia coli*. *Catalysts* **2021**, *11*, 184. https://doi.org/10.3390/catal11020184

Academic Editors: Ioan Balint and Monica Pavel

Received: 29 December 2020
Accepted: 26 January 2021
Published: 31 January 2021



## 1. Introduction

Bacterial infections can cause great harm to public health and have attracted considerable research attention from scientists. To date, many traditional bactericidal agents manifest a wide range of potential applications in antibacterial disinfection [1,2]. However, ordinary bactericidal agents have been shown to induce risks to the environment and encourage antibiotic resistance [3]. Contemporarily, the photocatalytic sterilization created by utilizing some reactive species produced under ultraviolet or visible light illumination has become increasingly valuable [2,4–6]. In short, photocatalytic nanocomposites harvest optical energy to produce positive and negative charge carriers that are involved in photoredox reactions [7–9]. It has been reported that $TiO_2$ is one of the most representative and dominant photocatalysts on account of its non-toxicity and chemical stability. It, however, has a wide-band energy gap of 3.2 eV, which endows it with a high light absorption capacity within the ultraviolet (UV) light region to enhance photocatalytic efficiency with hardly any absorption within the visible region [10]; UV light and visible light account for only 5% and 45% of sunlight, respectively [10,11]. To improve the efficiency of sunlight, it would be interesting to develop $TiO_2$-based materials that can absorb visible light. The ion doping or morphology-engineering of $TiO_2$ nanoparticles can narrow their energy bandgap, resulting in the efficient absorption of visible light. For example, two-dimensional reduced $TiO_2$ nanosheets with an energy bandgap of 2.86 eV have been used to efficiently inactivate bacteria under visible light irradiation [11]. Hydrothermally synthesized $TiO_2$ nanosheets doped with N, C, and/or S also exhibit visible light absorption capacities [12–16]. However, ultraviolet and visible light can injure healthy tissue and display a short light penetration depth in the human body, both of which hinder their further use in in vivo antibacterial applications [17].

Recently, near-infrared (NIR) light-induced upconversion particles have been introduced to photocatalyst sterilization systems. The upconversion particles can absorb low NIR photons, convert them to higher energy photons, and emit ultraviolet or visible light [18,19]. Coupling $TiO_2$ with upconversion particles can remarkably extend the utilization of the whole solar spectrum [20]. It has been suggested that β-$NaYF_4$ is the most ideal upconversion matrix for the doping of different lanthanide ions, on account of its high refractive index and transparency [21]. It may act as an intermedium for transferring NIR light energy to UV-Vis light that can be absorbed by the $TiO_2$ nanoparticles to produce oxidative holes ($h^+$) and reductive electrons ($e^-$). Effective $e^-$-$h^+$ pairs are able to react with $O_2$, OH, and $H_2O$ in a mixed solution to generate various reactive species that are helpful for sterilization [22,23]. For instance, β-$NaYF_4$:Yb,Er nanomaterial has been widely used in biological imaging analysis and photocatalytic applications. However, its luminous intensity is not satisfactory [24,25].

In this study, Gd-enhanced β-$NaYF_4$:Yb,Er,Gd fluorescent nanorods with a high fluorescence intensity were coupled to reduced $TiO_2$ nanoparticles with excellent visible light absorption abilities (UCNPs@R-$TiO_2$ nanocomposite) using electrostatic assembly (Figure 1). As expected, the created UCNPs@R-$TiO_2$ nanocomposite exhibits an effective photocatalytic sterilization performance against *Escherichia coli* (*E. coli*) under 980 nm NIR light illumination. The in vitro cellular cytotoxicity and antibacterial performance of the obtained UCNPs@R-$TiO_2$ were also evaluated by MTT (3-(4,5)-dimethylthiahiazo (-z-y1)-3,5-di-phenytetrazoliumromide) assay and MALDI-TOF MS analysis, respectively.

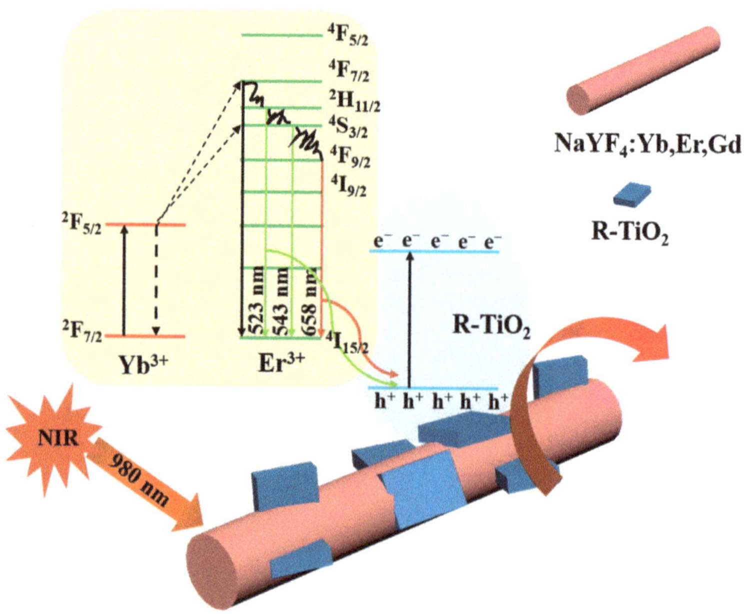


**Figure 1.** Schematic illustration of UCNPs@R-$TiO_2$ nanocomposite for photocatalytic sterilization under 980 nm NIR light irradiation.

## 2. Results and Discussion

### *2.1. Characterization of the UCNPs@R-$TiO_2$ Nanocomposites*

The transmission electron microscopy (TEM) image shown in Figure 2A was recorded on a Tecnai G2 F20 microscope (USA) and was used to observe the crystal morphology and size of the β-$NaYF_4$:Yb,Er,Gd fluorescent nanorods (UNCPs). The UNCPs were homodispersed and rod-shaped with a length of ~500 nm (Figure 2(Ba)) and a diameter of ~50 nm (Figure 2(Bb)). Moreover, the fast Fourier transform pattern indicated a (100) zone axis (Figure 2A). We found that the reduced $TiO_2$ (R-$TiO_2$) nanoparticles exhibited a square shape and were uniformly scattered (Figure 2C). The HRTEM (insert in Figure 2C) showed that R-$TiO_2$ belongs to pure anatase [11]. Additionally, the TEM image in Figure 2(Da) shows that the R-$TiO_2$ nanoparticles were successfully assembled on the UNCPs. Furthermore, we confirmed the crystal structure of the UCNPs@R-$TiO_2$ nanocomposite and found that the average lattice spacings that can be measured are 0.521 nm and 0.352 nm (Figure 2(Db)), matching well with a (100) facet and (101) facet lattice distance for the β-$NaYF_4$ and anatase $TiO_2$, respectively [11,21]. From energy-dispersive X-ray spectroscopy (EDS) measurements (shown in Figure S1), the elemental composition of the UNCPs@R-$TiO_2$ was obtained, and, as shown in the table of Figure 2(Dc), Na, Ti, Cu, Yb, F, Er, Gd, Y, and O could be detected; Cu originated from the Cu grid used for TEM measurements [26]. Taken together, these results clearly illustrate that the R-$TiO_2$ nanoparticles were successfully assembled on the UCNPs. The FT-IR spectra of the UCNPs@R-$TiO_2$ nanocomposite was obtained on a Nexus 670 spectrophotometer and shown in Figure S2. A strong and broad absorption band at 464 $cm^{-1}$ was assigned to Ti–O and O–Ti–O flexion vibration originating from the $TiO_2$ crystals [27,28]. The wide band of around 3435 $cm^{-1}$ was attributed to the H–O stretching, which helped to enhance photocatalytic activity [27]. The band around 1557 $cm^{-1}$ was attributed to the carbonyl group (−C=O−) vibration [21]. The FT-IR spectrum analysis indicates that the R-$TiO_2$ nanoparticles were successfully assembled onto the surface of the as-synthesized UCNPs through electrostatic attraction.

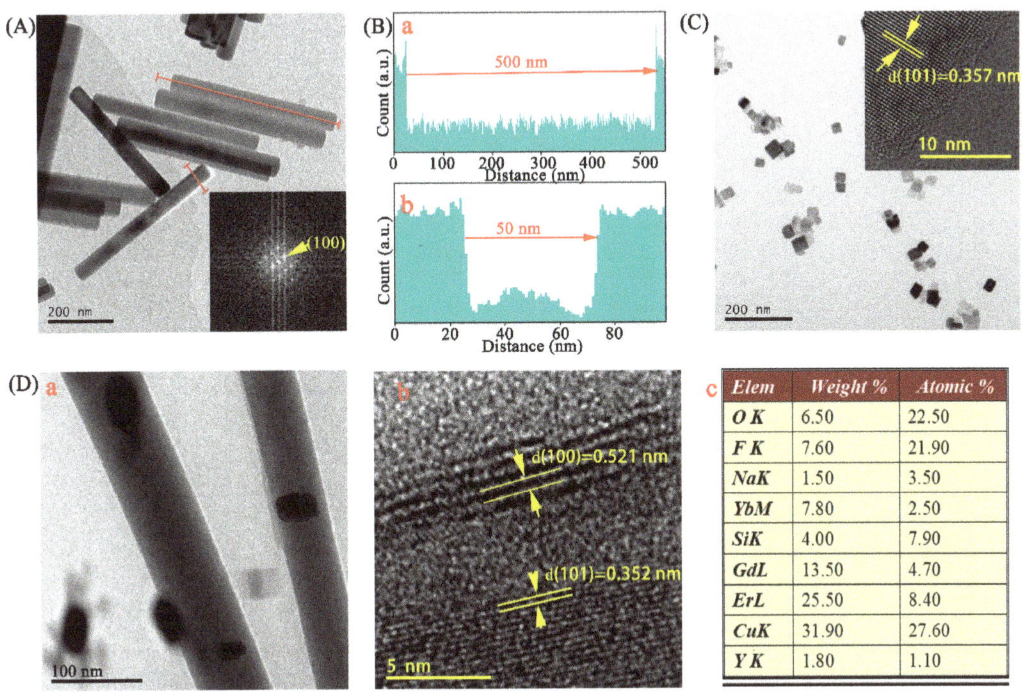


| Elem | Weight % | Atomic % |
|---|---|---|
| O K | 6.50 | 22.50 |
| F K | 7.60 | 21.90 |
| NaK | 1.50 | 3.50 |
| YbM | 7.80 | 2.50 |
| SiK | 4.00 | 7.90 |
| GdL | 13.50 | 4.70 |
| ErL | 25.50 | 8.40 |
| CuK | 31.90 | 27.60 |
| Y K | 1.80 | 1.10 |

**Figure 2.** (**A**) TEM image of UCNPs, with insert showing the fast Fourier transform pattern. (**B**) Length (**a**) and diameter (**b**) analyses of (**A**). (**C**) TEM and HRTEM (high resolution TEM) images (inset) of R-$TiO_2$. (**D**) TEM image (**a**), HRTEM image (**b**), and elemental compositions analysis (**c**) of UCNPs@ R-$TiO_2$.

The XRD patterns were recorded on an X-ray diffractometer (D8 Advance, Brucker, Germany) with the Cu K radiation (λ = 0.155 nm) operating at 40 kV and 80 mA; see Figure 3A and Figure S3. Figure S3 shows that the UCNPs and R-$TiO_2$ nanoparticles were pure hexagonal- (Joint Committee on Powder Diffraction Standards JCPDS 00-016-0334) and anatase-phase (JCPDS 01-021-1272), respectively. The sharp diffraction peaks indicate that the UCNPs and R-$TiO_2$ nanoparticles were highly crystallized hexagonal-and anatase-structured. The XRD pattern of the UCNPs@R-$TiO_2$ nanocomposite is shown in Figure 3A. And the XRD pattern analysis confirmed that the UCNPs@R-$TiO_2$ nanocomposite had a high degree of crystallization. The XRD pattern further showed that the UCNP nanorods were in a pure hexagonal phase (JCPDS 00-016-0334), which has been previously shown to have a higher luminous efficiency than the cubic phase $NaYF_4$ [29]. The distinct peaks at 25.3° of the prepared R-$TiO_2$ nanoparticles are likely ascribed to the (101) facet of the anatase $TiO_2$ when compared to the JCPDS 01-021-1272 database, indicating that the R-$TiO_2$ nanoparticles were present on the surface of the UCNP fluorescent nanorods by a form of substitutional doping. Compared to the peak locations of pure hexagonal and anatase phase, we observed that all of the diffraction peaks of the UCNPs@R-$TiO_2$ nanocomposite shifted to lower diffraction angles due to an expansion in unit-cell volume as a result of the partial substitution of $Ti^{4+}$ (65 pm) by the larger $Y^{3+}$ (104 pm) in the lattice [26].

As shown in Figure S4, the upconversion luminescence (UCL) intensity of the Gd-doped β-$NaYF_4$:Yb,Er (UCNPs) fluorescent nanorods recorded on a Hitachi F-7000 spectrometer was higher than that of the β-$NaYF_4$:Yb,Er fluorescent nanorods, which was attributed to the Gd dopant [21,30]. The UCL spectra of the UCNP nanorods and the UCNPs@R-$TiO_2$ nanocomposite were analyzed and shown in Figure 3(Ba,b). We found that, under 980 nm irradiation, the UCNP nanorods emit intense UCL emissions at 523, 542, and 658 nm, which were assigned to the $^2H_{11/2}$–$^4I_{15/2}$, $^4S_{3/2}$–$^4I_{15/2}$, and $^4F_{9/2}$–$^4I_{15/2}$ transitions of $Er^{3+}$ (Figure 1), respectively [21]. By contrast, the UCNPs@R-$TiO_2$ exhibited a drastic reduction to the UCL intensity because of the energy transfer from UCNPs to R-$TiO_2$ [20]. The absorption spectrum in the UV-Vis range (Figure 3C) was recorded on a UV-visible Cary 300 spectrophotometer and indicated that the R-$TiO_2$ (Figure 3(Ca)) possesses a higher absorption in the visible light region than pure anatase $TiO_2$ (Figure 3(Cb)), which was caused by the oxygen vacancies and lower bandgap of the R-$TiO_2$ (2.8 eV) than that of $TiO_2$ (3.2 eV) (inset in Figure 3C and Equation (S1)) [10]. The oxygen vacancies and low bandgap of the R-$TiO_2$, which arose from the dopant of $Ti^{3+}$ under an argon atmosphere, are helpful to enhance the absorption of visible light [10,11]. In addition, we observed that the emission peaks of UCNPs can match the enhanced visible light absorption of the R-$TiO_2$ nanoparticles (see the dotted boxes in Figure 3C), based on the UV-Vis absorption spectrum data (Figure 3C). We also found that the zeta potentials of the UCNP nanorods (Figure 3(Da)) and R-$TiO_2$ nanoparticles (Figure 3(Db)) were negative and positive, respectively, indicating that R-$TiO_2$ can be coupled to the surface of UCNP nanorods in a solution by electrostatic attraction. As a result, the UCNPs@R-$TiO_2$ composites possessed a positive zeta potential (Figure 3(Dc)) that helps its binding to the negatively charged surface of *E.coli* bacteria [31].

### 2.2. Antibacterial Performance

The plate-counting bacteria colonies of *E. coil* were used to evaluate the bacteriostatic ability of UCNPs@R-$TiO_2$ composites under 980 nm light irradiation (1 W). Under the different preparation conditions (Table S1), the highest antibacterial efficiency (98.1%) was achieved for the UCNPs@R-$TiO_2$ (180 °C, 20 h) composites. The low reaction temperature (180 °C) was unfavorable to the crystal growth of R-$TiO_2$, and the high reaction temperature (300 °C) caused the agglomeration of R-$TiO_2$, which was not good for photocatalytic reactions [11,31]. Simultaneously, the high crystallinity of R-$TiO_2$ was achieved at the optimal reaction time (20 h) [11]. The highest antibacterial efficiency (98.7%) was also achieved on the UCNPs@R-$TiO_2$ (40%) composites among the different mass ratios of R-$TiO_2$ and UCNPs (Figure S5). Consequently, the optimal UCNPs@R-$TiO_2$ (30%, 180 °C, 20 h) composites were used for further research. As shown in Figure 4A, the bacterial photoinactivation effect was suitably correlated with the dosage of nanomaterials and the UCNPs@R-$TiO_2$ nanocomposite, resulting in the highest bactericidal effect (97.3%) at the concentration of 50 μg/mL. Interestingly, we further found that both the UCNPs (Figure 4(Ba)) and R-$TiO_2$ (Figure 4(Bb)) were capable of killing *E. coil* colonies on the agar plate under 980 nm laser irradiation for 12 min when compared to the saline control (Figure 4B). Strikingly, the UCNPs@R-$TiO_2$ nanocomposite (Figure 4(Bc) and Figure S6) treatment eliminated about 98.1% of *E. coil* colonies on the plate, reflecting its enhanced bactericidal activity. Simultaneously, the bactericidal performance of the optimized UCNPs@R-$TiO_2$ nanocomposite was compared to the previously reported works (Table S2). We also found that the UCNPs@R-$TiO_2$ nanocomposite had the best bactericidal effect among these antibacterial agents.

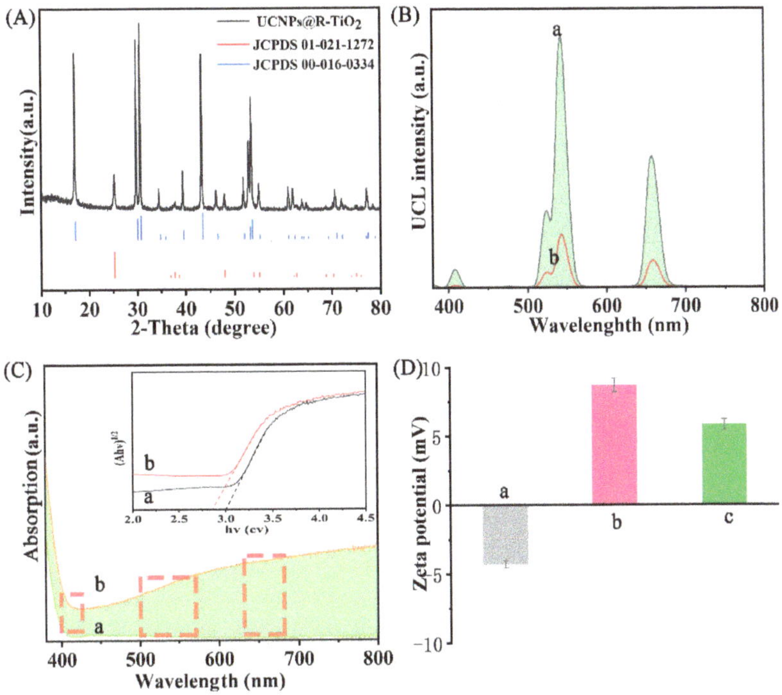


**Figure 3.** (**A**) X-ray powder diffraction (XRD) patterns of UCNPs@R-$TiO_2$ and the standard hexagonal phase (JCPDS 00-016-0334) and anatase phase (JCPDS 01-021-1272). (**B**) The upconversion luminescence (UCL) spectrum of the UCNPs (a) and UCNPs@R-$TiO_2$ (b). (**C**) The UV-Vis absorption spectrum of $TiO_2$ (a) and R-$TiO_2$ (b), with insert showing the corresponding bandgap determined by Tauc plot. (**D**) The zeta potential of UCNPs (a), R-$TiO_2$ (b), and UCNPs@R-$TiO_2$ (c).

The sterilization of these three materials under the same condition was also investigated via matrix-assisted laser desorption/ionization time-of-flight mass spectrometry (MALDI-TOF MS) analysis. As shown in Figure 5a, there were distinct characteristic peaks ($m/z$ = 4331, 5060, 6220, 7250, 9190, 9519) of *E. coli* K12 [32–34] present in our analysis, indicating the massive survival of *E. coli* K12. However, the number of peaks was dramatically decreased upon adding either the UCNP nanorods (Figure 5b) or R-$TiO_2$ nanoparticles (Figure 5c), indicating that some of the bacteria were killed. Importantly, there were no characteristic peaks detected between the 4000 to 14,000 Da region after the treatment of the UCNPs@R-$TiO_2$ composites, indicating that the *E. coli* K12 was nearly entirely killed.

All these data demonstrate that the UCNPs@R-$TiO_2$ nanocomposites possessed a highly effective bactericidal ability under the 980 nm NIR illumination.

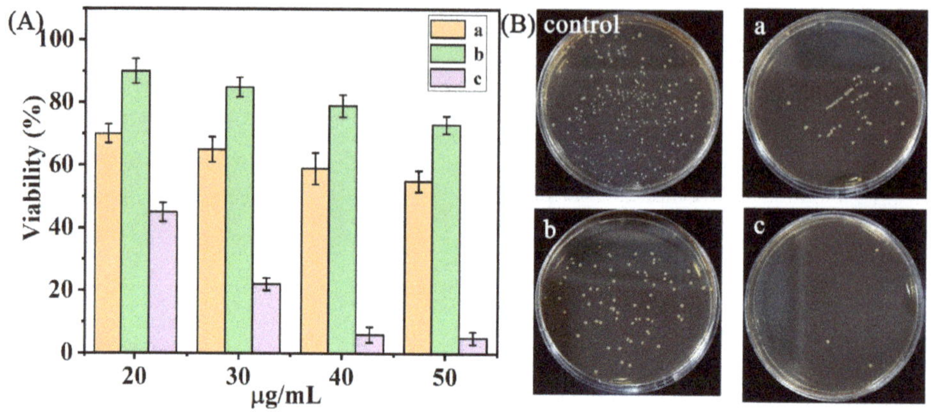


**Figure 4.** (**A**) *E. coli* viability under all different sample concentrations and (**B**) photographs of agar plates of *E. coli* incubated with 40 μg/mL of UCNPs (**a**), R-$TiO_2$ (**b**), and UCNPs@R-$TiO_2$ nanocomposite (**c**) using a 980 nm laser (1 W, 12 min).

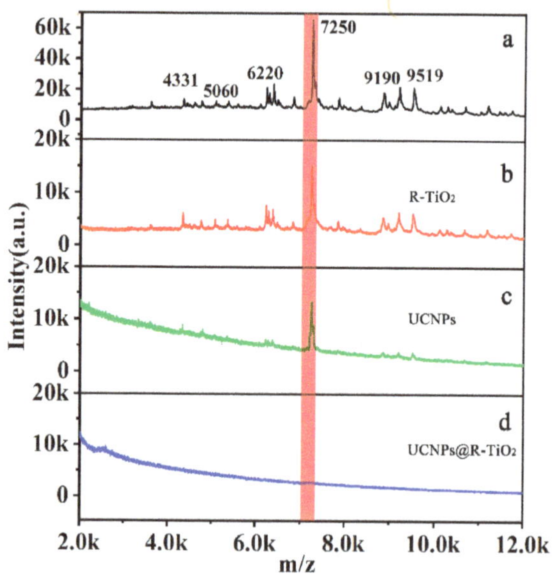


**Figure 5.** MALDI-TOF MS analysis of *E. coli* without (**a**) and with R-$TiO_2$ (**b**), UCNPs (**c**), and UCNPs@R-$TiO_2$ composites (**d**) under 980 nm NIR light irradiation for 20 min (40 μg/mL). The red band is the characteristic peak of *E. coli* at $m/z = 7250$.

### 2.3. *Cytotoxicity Assessment*

We evaluated the potential cytotoxicity to cells (HEK 293) of the as-prepared nanocomposites via MTT assay. There was about 80%, 87%, and 98% viability of HEK293 cells upon the treatment of the UCNPs, R-$TiO_2$, and UCNPs@R-$TiO_2$ materials at a concentration of 12 μg/mL, indicating low cytotoxicity to mammalian cells (Figure 6). It was noted that the cell viability was largely decreased to around 68% when the UCNP concentration was up to 50 μg/mL. By contrast, R-$TiO_2$ showed almost no toxicity to HEK 293 cells, because the HEK 293 cells still manifested a high survival rate (more than 91%) even under a high concentration (100 μg/mL) condition. Notably, the cell viability of the UCNPs@R-$TiO_2$ treatment was improved, possibly due to the low toxicity of R-$TiO_2$. About 80% of cells survived under a high concentration (40 μg/mL) of UCNPs@R-$TiO_2$ incubation, demonstrating its potential for bactericide material application [35].

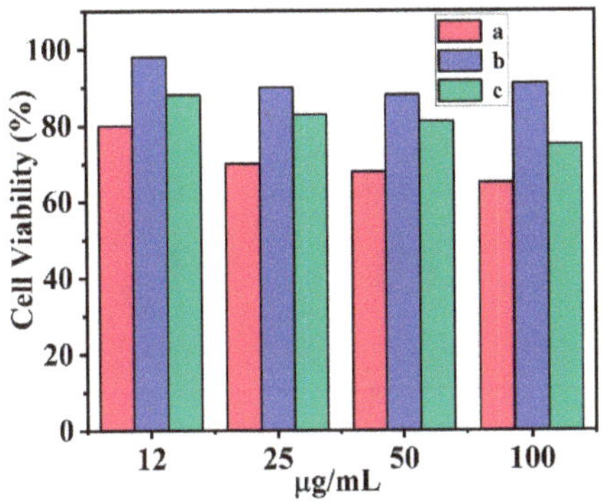


**Figure 6.** Cytotoxicity of UCNPs (**a**), R-$TiO_2$ (**b**), and UCNPs@R-$TiO_2$ nanocomposite (**c**) in HEK 293 cells.

*2.4. Antibacterial Mechanism*

Figure 7 depicts the possible 980 nm NIR light-driven antibacterial mechanism of the UCNPs@R-$TiO_2$ nanocomposites. After the 980 nm NIR light irradiation, the $^2F_{7/2}$ state electrons of $Yb^{3+}$ would be promoted into the $^2F_{5/2}$ excited state band of $Yb^{3+}$, then the $Yb^{3+}$ $^2F_{5/2}$ state would be relaxed by energy transfer to a neighboring $Er^{3+}$ ion. The energy promotes the valence band ($^4I_{15/2}$) electrons of $Er^{3+}$ into the excited state band ($^2H_{11/2}$, $^4S_{3/2}$, or $^4F_{9/2}$). The electrons in the states H, S, or F in $Er^{3+}$ are unstable and would be relaxed to the $^4I_{15/2}$ state by releasing energy, emitting at 523 nm, 542 nm, and 658 nm, respectively [21,35,36]. The visible light energy would be further absorbed by the valence band electrons of neighboring R-$TiO_2$ nanoparticles. The high energy valence electrons would jump into the stable conduction band, producing electron-hole pairs.

The visible light (red and green light) emitted by UCNPs is absorbed by the R-$TiO_2$ to produce strongly reductive electrons ($e^-$) and oxidative holes ($h^+$). The valid $h^+/e^-$ pairs afterwards could react with $H_2O$ and $O_2$ in an aqueous solution (Equations (1)–(7)) to produce reactive species. As it is known, the generated hydroxyl radical (·OH) species can be used as a strong oxidizer for the non-selective killing of bacteria [37]. The amount of ·OH is detected by the fluorescent intensity of 2-hydroxyterephthalic acid (λ = 420 nm) which is a product of the reaction of ·OH with terephthalic acid [38]. As shown in Figure 7, the UCNPs@R-$TiO_2$ nanocomposite has the highest fluorescent intensity among these materials, indicating the amount of OH that was generated. These results show that the visible light emitted by UCNPs can be effectively absorbed by R-$TiO_2$ nanoparticles. The UCNPs@R-$TiO_2$ composites possess excellent photocatalytic performance and the specific process of the NIR photocatalytic sterilization of *E. coli* is summarized using the following reactions:

$$\text{UNCPs} + \text{NIR light} \rightarrow \text{Visible light}, \tag{1}$$

$$\text{R-TiO}_2 + \text{Visible light} \rightarrow h^+ + e^- \tag{2}$$

$$O_2 + e^- \rightarrow \cdot O_2^{\,-}, \tag{3}$$

$$2\cdot O_2^{\,-} + 2H^+ \rightarrow H_2O_2 + O_2, \tag{4}$$

$$\cdot O_2^{-} + H_2O_2 \rightarrow \cdot OH + OH^{-} + O_2, \quad (5)$$

$$H_2O + h^{+} \rightarrow \cdot OH + H^{+}, \quad (6)$$

$$\cdot OH + E.\ coli \rightarrow \text{Inactivated } E.\ coli. \quad (7)$$

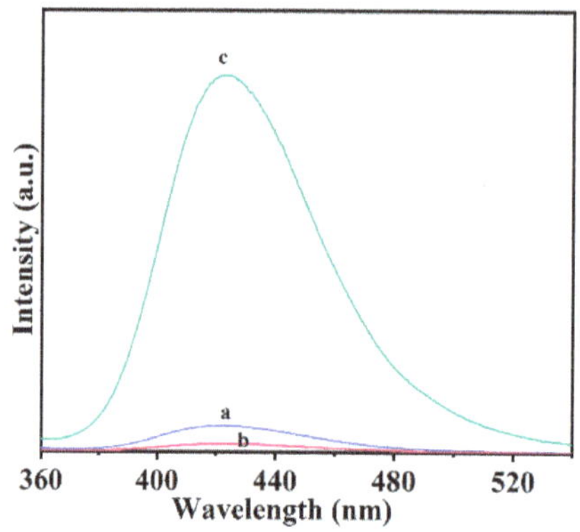


**Figure 7.** Photoluminescence spectra of UCNPs (**a**), R-$TiO_2$ (**b**), and UCNPs@R-$TiO_2$ nanocomposite (**c**), respectively, measured under 980 nm NIR illumination for 20 min.

## 3. Experimental Designs

### *3.1. Reagents and Materials*

All of the chemical reagents—at analytical grade, unless otherwise noted—were used without further purification. Ytterbium(III) chloride hexahydrate ($YbCl_3{\cdot}6H_2O$), Yttrium(III) chloride hexahydrate ($YCl_3{\cdot}6H_2O$), Gadolinium(III) chloride ($GdCl_3{\cdot}6H_2O$), and Erbium(III) chloride ($ErCl_3$) were purchased from Alfa Aesar (Shanghai, China). Titanium isopropoxide, sodium hydroxide (NaOH, 96%), oleic acid (OA), sodium citrate, hydrofluoric acid, acetonitrile, ethanol, and chloroform of analytical grade were purchased from Sinopharm Chemical Reagent Co., Ltd. from Shanghai, China. $NH_4F$ and Thiazolyl Blue Tetrazolium Bromide (MTT) were obtained from Aladdin (Hang Kong, China) and Bomei Biotechnology (Hefei, China), respectively. Ultrapure water (18.2 MΩ·cm, Mili-Q, Millipore, Burlington, MA, USA) was used throughout the experiment.

### *3.2. Preparation of β-$NaYF_4$:Yb,Er,Gd (60, 18, 2, 20 mol%) Fluorescent Nanorods*

The preparation of β-$NaYF_4$:Yb,Er,Gd (60, 18, 2, 20 mol%) highly fluorescent nanorods (UCNPs) is summarized as follows [21,39]: Firstly, a mixed solution with 10 mL of ethanol, 3 mL of deionized water, and 0.6 g of NaOH was obtained. A total of 10 mL of oleic acid was then dropped in the above mixture under vigorous stirring. The stirring was continued for 20 min; the obtained mixture was named A solution. Moreover, 145.6 mg of $YCl_3{\cdot}6H_2O$, 59.47 mg of $GdCl_3{\cdot}6H_2O$, 4.38 mg of $ErCl_3$, and 55.8 mg of $YbCl_3{\cdot}6H_2O$ were dissolved in 4 mL of deionized water; this mixture was named B solution. Then, B solution was slowly dropped into the A solution under vigorous stirring. After 10 min, 2.0 mL of $NH_4F$ (2 M) solution was added to the above mixture dropwise. Finally, a milky colloidal solution was produced. The obtained solution was encapsulated in a 50 mL autoclave flask and heated to 180 °C at a heating rate of 3 °C/min, then it was kept at 200 °C for 2 h. The prepared UCNPs were collected and washed with ethanol and ultrapure water, respectively.

### *3.3. Synthesis of UCNPs@R-$TiO_2$ Nanocomposite*

UCNPs@R-$TiO_2$ were prepared as follows [11,20,40]. Firstly, the R-$TiO_2$ nanoparticles ($TiO_{2-x}$) were prepared as follows: 4 μL of titanium isopropoxide was dropped into a solution containing 1.2 mL of HF and 20 mL of isopropanol. After the obtained mixture was stirred for 10 min under an argon atmosphere, 5 μL of $TiCl_3$ solution was added to it [11]. Then, 20 mg of UCNPs were also added to the obtained mixture under an ultrasonic bath and the mixture solution continued to sonicate for 30 min under an argon atmosphere. The obtained mixture was then encapsulated into a 50 mL autoclave flask, heated to 180 °C at a heating rate of 3 °C/min, and kept at 180 °C for 20 h under an argon atmosphere. The produced UCNPs@R-$TiO_2$ (40%) nanocomposite was collected and washed, first with ethanol and then with ultrapure water. It was then dried at 60 °C for 10 h in a vacuum environment. The R-$TiO_2$ nanoparticles and UCNPs@R-$TiO_2$ nanocomposites with different mass ratios were prepared according to the above method by changing the amount of UCNPs.

### 3.4. Vitro Cell Viability Assay

A standard MTT assay was used to assess the cytotoxicity of the UCNPs, R-$TiO_2$, and UCNPs@R-$TiO_2$. Human Embryonic Kidney 293 cells (HEK293) were selected for the assay. The temperature of the whole incubation process was controlled at 37 °C. The HEK293 cells present in a 96-well plate (with 10,000 cells per well) were first cultured for 12 h. After that, the different calculated concentrations of UCNPs, R-$TiO_2$, and UCNPs@R-$TiO_2$ nanomaterials were added to the above plate and the mixture was further incubated for 24 h. Afterwards, 100 μL of MTT solution was added to each well. After incubation for 2 h, the sediment was retained. Then, 100 μL of DMSO (dimethyl sulfoxide) was added into each sample mentioned above and the mixture was shaken for 20 min. The absorbance at 595 nm detected by a microplate reader was used to calculate the cell viability rate.

### 3.5. Bacteria (E. coli K12) Culture and Preparation

*E. coli* K12 were inoculated into the Luria-Bertani broth and shaken (300 rpm) constantly in an incubator shaker at 37 °C. After overnight incubation, the *E. coli* K12 suspensions (8 mL) were centrifuged (8000 rpm) for 2 min and the sediments were retained. The obtained precipitates were resuspended by adding 6 mL of sterile normal saline. By measuring the optical density (OD) value at 600 nm, the bacterium liquid concentration was adjusted to a proper level for after use. Different colonies were distributed on LB plates and incubated at 37 °C overnight. The relevant colony-forming units (CFU) were calculated to obtain the number of bacteria per milliliter.

### 3.6. Antibacterial Properties

The Gram-negative bacterium *E. coli* K-12 was used to investigate the in vitro antibacterial abilities of the UCNPs, R-$TiO_2$, and UCNPs@R-$TiO_2$. An amount of 900 μL of *E. coli* suspension (~$10^6$ CFU/mL) and 100 μL of these nanomaterials at different concentrations were mixed. The resulting concentrations were 20, 30, 40, and 50 μg/mL, respectively. After incubation at 37 °C for 2 h, the obtained bacterial suspension was diluted by a factor of $10^3$. The resulting bacterial samples were irradiated for 20 min by a 980 nm NIR light (1 W). Afterwards, 100 μL of the above suspension was spread on the Luria-Bertani medium and hatched at 37 °C for 16 h. In the end, the antibacterial abilities were assessed on the LB agar plates using the colony counting method. Simultaneously, instead of the added nanocomposites, an isotonic saline solution was added into the *E. coli* suspension as a blank control.

### 3.7. MALDI-TOF MS Analysis

The characterization changes of *E. coli* K12 bacterial strains were analyzed using a MALDI-TOF MS [11,41–43]. Firstly, 300 μL of ultrapure water and 900 μL of ethanol were mixed, and then 20 mg of the *E. coli* sample was added into the mixture under mild shaking. The sediment was retained after centrifuging (13,000 rpm) for 3 min. Subsequently, 50 μL of $CH_3CN$ (acetonitrile) and 50 μL of 70% HCOOH (formic acid) were added. The mixture was centrifuged at 13,000 rpm for 2 min again. The above mixture (0.5 μL) and a DHB matrix solution (2,5-dihydroxybenzoic acid solution, 0.5 μL) were dropped on a plate and allowed to dry. The experiments of the MALDI-TOF MS were performed on an UltrafleXtreme TOF/TOF operating system equipped with a 355 nm $N_2$ laser. The operating conditions were as follows: positive ion mode, mass range (5–20 kDa), and acceleration voltage (20 kV).

## 4. Conclusions

The UCNPs@R-$TiO_2$ nanocomposite, an effective antibacterial material, was prepared with the electrostatic assembly strategy. The reduced $TiO_2$ nanoparticles with a bandgap of 2.8 eV and the ability to absorb visible light were successfully assembled onto the surface of Gd-enhanced β-$NaYF_4$:Yb,Er fluorescent nanorods. The UCNPs@R-$TiO_2$ composite with an MIC of 40 μg/mL can kill more than 98.1% of *E. coli* within 12 min under 980 nm NIR light irradiation (1 W). The good antibacterial properties are mostly attributed to the efficient light energy transfer from UCNPs to R-$TiO_2$ nanoparticles. The low-toxicicity UCNPs@R-$TiO_2$ nanocomposite shows great potential for creating efficient NIR-responsive photocatalysts.

**Supplementary Materials:** The following are available online at https://www.mdpi.com/2073-4344/11/2/184/s1: Figures S1–S6: EDX, FT-IR, XRD, UCL and Antibacterial efficiency analysis of the prepared nanomaterials, Table S1: Antibacterial efficiency of UCNPs@R-$TiO_2$ nanocomposite under different preparation conditions, Table S2: Comparison of the performance of UCNPs@R-$TiO_2$ photocatalytic sterilization system with that of some antibacterial agents.

**Author Contributions:** H.Z.: the acquisition and analysis of data for the work; Drafting the work; Final approval of the version to be published; Agreement to be accountable for all aspects of the work. F.H.: the conception or design of the work; revising the work; Final approval of the version to be published; Agreement to be accountable for all aspects of the work. All authors have read and agreed to the published version of the manuscript.

**Funding:** This research funded by National Natural Science Foundation of China, grant number 21275042.

**Institutional Review Board Statement:** The study was conducted and approved by the Ethics Committee of University of South China (SYXK(湘)2020-0002 and January, 2020).

**Informed Consent Statement:** Informed consent was obtained from all subjects involved in the study.

**Data Availability Statement:** Data sharing not applicable.

**Conflicts of Interest:** There are no conflict of interest to declare.

## References

1. Qian, Y.; Qi, F.; Chen, Q.; Zhang, Q.; Qiao, Z.; Zhang, S.; Wei, T.; Yu, Q.; Yu, S.; Mao, Z.; et al. Surface Modified with a Host Defense Peptide-Mimicking β-Peptide Polymer Kills Bacteria on Contact with High Efficacy. *ACS Appl. Mater. Interfaces* **2018**, *10*, 15395–15400. [CrossRef] [PubMed]
2. Cruz, J.; Flórez, J.; Torres, R.; Urquiza, M.; Gutiérrez, J.A.; Guzmán, F.; Ortiz, C.C. Antimicrobial activity of a new synthetic peptide loaded in polylactic acid or poly (lac-tic-co-glycolic) acid nanoparticles against Pseudomonas aeruginosa, Escherichia coli O157:H7 and methicillin resistant Staphylococcus aureus (MRSA). *Nanotechnology* **2017**, *28*, 135102–135110. [CrossRef] [PubMed]
3. Chait, R.; Craney, A.; Kishony, R. Antibiotic interactions that select against resistance. *Nat. Cell Biol.* **2007**, *446*, 668–671. [CrossRef]
4. Yin, Q.; Tan, L.; Lang, Q.; Ke, X.; Bai, L.; Guo, K.; Qiao, R.; Bai, S. Plasmonic molybdenum oxide nanosheets supported silver nanocubes for enhanced near-infrared antibacterial activity: Synergism of photothermal effect, silver release and photocatalytic reactions. *Appl. Catal. B Environ.* **2018**, *224*, 671–680. [CrossRef]
5. Ao, Z.; Sun, H.; Li, G.; Zhao, H.; Wong, P.K. Differences in photoelectrocatalytic inactivation processes between *E. coli* and its isogenic single gene knockoff mutants: Destruction of membrane framework or associated proteins? *Appl. Catal. B Environ.* **2016**, *188*, 360–366. [CrossRef]
6. Song, J.; Yu, J.; Sun, G.; Si, Y.; Ding, B. Visible-light-driven, hierarchically heterostructured, and flexible silver/bismuth oxyiodide/titania nanofibrous membranes for highly efficient water disinfection. *J. Colloid Interface Sci.* **2019**, *555*, 636–646. [CrossRef]
7. Liu, Y.; Luo, X.; Zhou, C.; Du, S.; Zhen, D.; Chen, B.; Li, J.; Wu, Q.; Iru, Y.; Chen, D. A modulated electronic state strategy designed to integrate active HER and OER components as hy-brid heterostructures for efficient overall water splitting. *Appl. Catal. B-Environ.* **2019**, *260*, 118197–118210. [CrossRef]

8. Zhen, D.; Shi, S.; Gao, C.; Kang, Q.; Xiao, X.; Grimes, C.A.; Cai, Q. Bi, Fe and Ti ternary co-doped $ZrO_2$ nanocomposites as a mass spectrometry matrix for the deter-mination of bisphenol A and tetrabromobisphenol A in tea. *Microchim. Acta* **2020**, *187*, 582–593. [CrossRef]
9. Nie, Y.C.; Yu, F.; Wang, L.C.; Xing, Q.J.; Liu, X.; Pei, Y.; Zou, J.P.; Dai, W.L.; Li, Y.; Suib, S.L. Photocatalytic degradation of organic pollutants coupled with simultaneous photocatalytic H2 evolution over graphene quantum dots/Mn-N-$TiO_2$/g-$C_3N_4$ composite catalysts: Performance and mechanism. *Appl. Catal. B-Environ.* **2018**, *227*, 312–321. [CrossRef]
10. Sorcar, S.; Hwang, Y.; Grimes, C.A.; In, S.-I. Highly enhanced and stable activity of defect-induced titania nanoparticles for solar light-driven $CO_2$ reduction into CH. *Mater. Today* **2017**, *20*, 507–515. [CrossRef]
11. Zhen, D.; Liu, Y.A.; Grimes, C.; Cai, Q. Reduced titania nanosheets as an effective visible-light germicide. *Nanotechnology* **2019**, *30*, 405602. [CrossRef] [PubMed]
12. Liu, G.; Yang, H.G.; Wang, X.; Cheng, L.; Pan, J.; Lu, G.Q.; Cheng, H.M. Visible light responsive nitrogen doped anatase $TiO_2$ sheets with dominant {001} facets de-rived from TiN. *J. Am. Chem. Soc.* **2009**, *131*, 12868–12869. [CrossRef]
13. Xiang, Q.; Yu, J.; Wang, W.; Jaroniec, M. Nitrogen selfdoped nanosized $TiO_2$ sheets with exposed {001} facets for enhanced visible-light photocatalytic activity. *Chem. Commun.* **2011**, *47*, 6906–6908. [CrossRef] [PubMed]
14. Yu, J.; Dai, G.; Xiang, Q.; Jaroniec, M. Fabrication and enhanced visible-light photocatalytic activity of carbon selfdoped $TiO_2$ sheets with exposed {001} facets. *J. Mater. Chem.* **2011**, *21*, 1049–1057. [CrossRef]
15. Zhang, J.; Wu, Y.; Xing, M.; Leghari, S.A.K.; Sajjad, S. Development of modified N doped $TiO_2$ photocatalyst with metals, nonmetals and metal oxides. *Energy Environ. Sci.* **2010**, *3*, 715–726. [CrossRef]
16. Miyauchi, M.; Takashio, M.; Tobimatsu, H. Photocatalytic Activity of $SrTiO_3$ Codoped with Nitrogen and Lanthanum under Visible Light Illumination. *Langmuir* **2004**, *20*, 232–236. [CrossRef] [PubMed]
17. Chang, J.; Ning, Y.; Wu, S.; Niu, W.; Zhang, S. Effectively Utilizing NIR Light Using Direct Electron Injection from Up-Conversion Nanoparticles to the $TiO_2$ Photoanode in Dye-Sensitized Solar Cells. *Adv. Funct. Mater.* **2013**, *23*, 5910–5915. [CrossRef]
18. Wang, F.; Deng, R.; Wang, J.; Wang, Q.; Han, Y.; Zhu, H.; Chen, X.; Liu, X. Tuning upconversion through energy migration in core–shell nanoparticles. *Nat. Mater.* **2011**, *10*, 968–973. [CrossRef]
19. Sun, M.; Dong, H.; Dougherty, A.W.; Lu, Q.; Peng, D.; Wong, W.-T.; Huang, B.; Sun, L.-D.; Yan, C. Nanophotonic energy storage in upconversion nanoparticles. *Nano Energy* **2019**, *56*, 473–481. [CrossRef]
20. Tang, Y.; Di, W.; Zhai, X.; Yang, R.; Qin, W. NIR-Responsive Photocatalytic Activity and Mechanism of $NaYF_4$:Yb,Tm@$TiO_2$ Core–Shell Nanoparticles. *ACS Catal.* **2013**, *3*, 405–412. [CrossRef]
21. Huang, X.; Wang, L.; Zhang, X.; Yin, X.; Bin, N.; Zhong, F.; Liu, Y.; Cai, Q. Dye-assembled nanocomposites for rapid upconversion luminescence sensing of $Cu^{2+}$. *Sens. Actuators B Chem.* **2017**, *248*, 1–8. [CrossRef]
22. Zhen, D.; Gao, C.; Yang, D.; Zhu, X.; Grimes, C.A.; Liu, Y.; Cai, Q. Blue $Ti^{3+}$ self-doped $TiO_2$ nanosheets with rich {001} facets for photocatalytic performance. *New J. Chem.* **2019**, *43*, 5759–5765. [CrossRef]
23. Sun, J.; Song, L.; Fan, Y.; Tian, L.; Luan, S.; Niu, S.; Ren, L.; Ming, W.; Zhao, J. Synergistic Photodynamic and Photothermal Antibacterial Nanocomposite Membrane Triggered by Single NIR Light Source. *ACS Appl. Mater. Interfaces* **2019**, *11*, 26581–26589. [CrossRef]
24. Chen, G.; Qiu, H.; Prasad, P.; Chen, X. Upconversion Nanoparticles: Design, Nanochemistry, and Applications in Theranostics. *Chem. Rev.* **2014**, *114*, 5161–5214. [CrossRef] [PubMed]
25. González-Béjar, M.; Liras, M.; Francés-Soriano, L.; Voliani, V.; Herranz-Pérez, V.; Duran-Moreno, M.; Garciaverdugo, J.M.; Alarcon, E.I.; Scaiano, J.C.; Pérez-Prieto, J. NIR excitation of upconversion nanohybrids containing a surface grafted Bodipy induces oxygen-mediated cancer cell death. *J. Mater. Chem. B* **2014**, *2*, 4554–4563. [CrossRef] [PubMed]
26. Zhou, Q.; Li, C.; Chen, P.; Cai, Q. Preparation of $Bi_{0.15}Fe_{0.15}TiO_2$ Nanocomposites for the Highly Selective Enrichment of Phospho-peptides. *Anal. Chem.* **2018**, *90*, 12414–12421.
27. Li, J.L.; Zhen, D.S.; Sui, G.; Zhang, C.; Deng, Q.; Jia, L. Nanocomposite of Cu–$TiO_2$–$SiO_2$ with high photoactive performance for degradation of rhoda-mine B dye in aqueous wastewater. *J. Nanosci. Nanotech.* **2012**, *12*, 6265–6270. [CrossRef]
28. Hu, Z.; Zhan, Y.; She, J. The role of Nd on the microstructural evolution and compressive behavior of Ti–Si alloys. *Mater. Sci. Eng. A* **2013**, *560*, 583–588. [CrossRef]
29. Krämer, K.W.; Biner, D.; Frei, G.; Güdel, H.U.; Hehlen, M.P.; Lüthi, S.R. Hexagonal sodium yttrium fluoride based green and blue emitting upconversion phos-phors. *Chem. Mater.* **2004**, *16*, 1244–1251.
30. Watkins, Z.; Taylor, J.; D'Souza, S.; Britton, J.; Nyokong, T. Fluorescence Behaviour and Singlet Oxygen Production of Aluminium Phthalocyanine in the Presence of Upconversion Nanoparticles. *J. Fluoresc.* **2015**, *25*, 1417–1429. [CrossRef]
31. Yang, C.; Xie, H.; Li, Q.-C.; Sun, E.-J.; Su, B.-L. Adherence and interaction of cationic quantum dots on bacterial surfaces. *J. Colloid Interface Sci.* **2015**, *450*, 388–395. [CrossRef] [PubMed]
32. Ryzhov, V.; Fenselau, C. Characterization of the protein subset desorbed by MALDI from whole bacterial cells. *Anal. Chem.* **2001**, *73*, 746–750. [CrossRef] [PubMed]
33. Arnold, R.J.; Karty, J.A.; Ellington, A.D.; Reilly, J.P. Monitoring the growth of a bacteria culture by MALDI-MS of whole cells. *Anal. Chem.* **1999**, *71*, 1990–1996. [CrossRef] [PubMed]
34. Gedda, G.; Wu, H.F. Fabrication of surface modified ZnO nanorod array for MALDI-MS analysis of bacteria in a nanoliter droplet: A multiple function biochip. *Sens. Actuat. B-Chem.* **2019**, *288*, 667–677. [CrossRef]

35. Li, S.; Cui, S.; Yin, D.; Zhu, Q.; Ma, Y.; Qian, Z.; Gu, Y. Dual antibacterial activities of a chitosan-modified upconversion photodynamic therapy system against drug-resistant bacteria in deep tissue. *Nanoscale* **2017**, *9*, 3912–3924. [CrossRef]
36. Luo, Z.; Zhang, L.; Zeng, R.; Tang, D. Near-infrared light-excited core–core–shell UCNP@Au@CdS upconversion nanospheres for ultrasensitive photoelectrochemical enzyme immunoassay. *Anal. Chem.* **2018**, *90*, 9568–9575. [CrossRef]
37. Yin, X.; Sheng, P.; Zhong, F.; Nguyen, V.; Cai, Q.; Grimes, C. CdS/$ZnIn_2S_4$/$TiO_2$ 3D-heterostructures and their photoelectrochemical properties. *New J. Chem.* **2016**, *40*, 6675–6685. [CrossRef]
38. Sheng, P.; Li, W.; Cai, J.; Wang, X.; Tong, X.; Cai, Q.; Grimes, C.A. A novel method for the preparation of a photocorrosion stable core/shell CdTe/CdS quantum dot $TiO_2$ nanotube array photoelectrode demonstrating an AM 1.5G photoconversion efficiency of 6.12%. *J. Mater. Chem. A* **2013**, *1*, 7806–7815. [CrossRef]
39. Wang, F.; Han, Y.; Lim, C.S.; Lu, Y.; Wang, J.; Xu, J.; Chen, H.; Zhang, C.; Hong, M.; Liu, X. Simultaneous phase and size control of upconversion nanocrystals through lanthanide doping. *Nat. Cell Biol.* **2010**, *463*, 1061–1065. [CrossRef]
40. Su, W.; Zheng, M.; Li, L.; Wang, K.; Qiao, R.; Zhong, Y.; Hu, Y.; Li, Z. Directly coat $TiO_2$ on hydrophobic $NaYF_4$:Yb,Tm nanoplates and regulate their photocatalytic ac-tivities with the core size. *J. Mater. Chem. A* **2014**, *2*, 13486–13491. [CrossRef]
41. Jones, J.J.; Stump, M.J.; Fleming, R.C.; Lay, J.O.; Wilkins, C.L. Investigation of MALDI-TOF and FT-MS techniques for analysis of Escherichia coli whole cells. *Anal. Chem.* **2003**, *75*, 1340–1347. [CrossRef] [PubMed]
42. Russell, S.C.; Edwards, N.; Fenselau, C. Detection of plasmid insertion in Escherichia coli by MALDI-TOF mass spectrometry. *Anal. Chem.* **2007**, *79*, 5399–5406. [CrossRef] [PubMed]
43. Sutherland, J.B.; Heinze, T.M.; Holder, C.L.; Voorhees, K.J.; Lay, J.O., Jr. Identification of bacterial proteins observed in MALDI TOF mass spectra from whole cells. *Anal. Chem.* **1999**, *71*, 3226–3230.

MDPI
St. Alban-Anlage 66
4052 Basel
Switzerland
Tel. +41 61 683 77 34
Fax +41 61 302 89 18
www.mdpi.com

*Catalysts* Editorial Office
E-mail: catalysts@mdpi.com
www.mdpi.com/journal/catalysts

www.ingramcontent.com/pod-product-compliance
Lightning Source LLC
Chambersburg PA
CBHW040837250926
51850CB00007BA/1337
* 9 7 8 3 0 3 6 5 8 3 8 9 1 *